THE COMING OF THE RAILWAY

THE COMING OF THE RAILWAY

A NEW GLOBAL HISTORY 1750–1850

DAVID GWYN

YALE UNIVERSITY PRESS
NEW HAVEN AND LONDON

Copyright © 2023 David Gwyn

All rights reserved. This book may not be reproduced in whole or in part, in any form (beyond that copying permitted by Sections 107 and 108 of the U.S. Copyright Law and except by reviewers for the public press) without written permission from the publishers.

All reasonable efforts have been made to provide accurate sources for all images that appear in this book. Any discrepancies or omissions will be rectified in future editions.

For information about this and other Yale University Press publications, please contact:
U.S. Office: sales.press@yale.edu   yalebooks.com
Europe Office: sales@yaleup.co.uk   yalebooks.co.uk

Set in Adobe Caslon Pro by IDSUK (DataConnection) Ltd
Printed in Great Britain by TJ Books, Padstow, Cornwall

Library of Congress Control Number: 2023931528

ISBN 978-0-300-26789-1

A catalogue record for this book is available from the British Library.

10 9 8 7 6 5 4 3 2 1

CONTENTS

List of illustrations and maps   vii
Acknowledgements   xiii

Introduction   1

1   Trade, transport and coal 1767–1815   8
2   'Rails best adapted to the road': cast-iron rails and their alternatives in Britain 1767–1832   30
3   Canal feeders, quarry railways and construction sites   49
4   'Art has supplied the place of horses': traction 1767–1815   80
5   War and peace 1814–1834   108
6   'Geometrical precision': wrought-iron rails 1808–1834   126
7   'Most suitable for hilly countries': rope and chain haulage 1815–1834   139
8   'That truly astonishing machine': locomotives 1815–1834   159
9   Coal carriers 1815–1834   192
10   Internal communications 1815–1834   213
11   The first main lines 1824–1834   233
12   Coming of age: the public railway 1830–1834   263
13   'The new avenues of iron road' 1834–1850   285
14   'You can't hinder the railroad'   314

CONTENTS

A note on sources and terminology	*322*
Notes	*326*
Bibliography	*364*
Index	*388*

ILLUSTRATIONS AND MAPS

Plates

1. *Chaldron Wagons opposite Ryton, Northumberland* by Samuel Hieronymus Grimm, eighteenth century. The Picture Art Collection / Alamy Stock Photo.
2. A horse-drawn waggon and waggonway. Licence granted by Paul Jarman, Assistant Director at the Beamish Museum. © Paul Jarman.
3. The uncovered remains of a Haytor tramway 'junction' on Dartmoor, 2016. Christopher Nicholson / Alamy Stock Photo.
4. *View of a Stone Bridge across the Valley and River at Risca in Monmouthshire* by Edward Pugh, 1810. Bridgeman Images.
5. *View of Construction Work at St Katherine's Dock, Stepney, London* by J. Phelps, 1828. Heritage Image Partnership Ltd / Alamy Stock Photo.
6. Grand Surrey Iron Railway, *c.* 1803. Science and Society Picture Library / Getty Images.
7. Portrait of William Reynolds by William Armfield Hobday, eighteenth century. Licence granted by Ironbridge Gorge Museum Trust.

8. The original Edge Hill Station opened in 1830 by the Liverpool and Manchester Railway in Liverpool, *c.* 1830. The Picture Art Collection / Alamy Stock Photo.
9. *View from the Inclined Plane, near Philadelphia* by J. C. Wild, *c.* 1840. Library of Congress Control Number 2021670290, Philadelphia on Stone, World Digital Library.
10. *Near Atherstone* by Edmund John Niemann. © Bradford Museums & Galleries / Bridgeman Images.
11–16. Diagrams by Geoffrey Birse, 2022. © Geoffrey Birse.
17. Set of four woodcuts of the Saint-Étienne to Lyon railway, *c.* 1830.
18. *Opening of the Garnkirk and Glasgow Railway* by David Hill, 1831. Lordprice Collection / Alamy Stock Photo.
19. *Opening of the Liverpool and Manchester Railway* by Isaac Shaw, *c.* 1830. Yale Center for British Art, Paul Mellon Collection (B1981.25.2698).
20. *The March of Intellect* by Robert Seymour, 1828. Heritage Image Partnership Ltd / Alamy Stock Photo.
21. *Travelling on the Liverpool & Manchester Railway: A Train of the First Class Carriages, with the Mail and A Train of the Second Class for Outside Passengers* by S. G. Hughes after Isaac Shaw, 1831. Yale Center for British Art, Paul Mellon Collection (B1993.30.129).
22. *View of the Intersection Bridge on the line of St. Helen's & Runcorn Gap Railway, Crossing the Liverpool and Manchester Railway near the Foot of the Sutton Inclined Plane* by S. G. Hughes, 1832. Yale Center for British Art, Paul Mellon Collection (B1978.43.340).
23. Pferdeeisenbahn in Linz. The History Collection / Alamy Stock Photo.
24. *Ellicott's Mills* by Edward Weber, *c.* 1836. Map reproduction courtesy of the Norman B. Leventhal Map & Education Center at the Boston Public Library (06_01_016719).

25. *Scene of the Baltimore & Ohio Railroad and the Chesapeake & Ohio Canal at Harpers Ferry, Virginia* by George Harvey, *c.* 1837–40. Floyd D. and Anne C. Gottwald Fund, Virginia Museum of Fine Arts (2011.73). Image released via Creative Commons CC-BY-NC.
26. *The First Railroad Trail on the Mohawk and Hudson Road* by Henry Edward Lamson, *c.* 1892.
27. 'The Momentous Question', cartoon published by *Punch*, 1845. World History Archive / Alamy Stock Photo.
28. *The Louth-London Royal Mail Travelling by Train from Peterborough East, Northamptonshire* by James Pollard, 1845. Yale Center for British Art, Paul Mellon Collection (B2001.2.140).
29. Steam pumping station of the St Petersburg–Moscow Railway, 1846.
30. *Ingenio Acana* by Eduardo Havana Laplante, 1857. © British Library Board. All Rights Reserved / Bridgeman Images.
31. *Rain, Steam, and Speed – The Great Western Railway* by J. M. W. Turner, 1844. Eraza Collection / Alamy Stock Photo.
32. *River in the Catskills* by Thomas Cole, 1843. Peter Barritt / Alamy Stock Photo.

In text

1. Reinhold Rücker Angerstein's sketch of a coal waggon way, from *R.R. Angerstein's Illustrated Travel Diary, 1753–1755* (Science Museum, 2001). By permission of Jernkontoret. — 13
2. 'View of Newcastle upon Tyne', line engraving by James Fittler, 1783. © Science Museum / Science & Society Picture Library. — 17

3. Drops at Wallsend, from Thomas H. Hair, *A Series of Views* 24
   of the Collieries in the Counties of Northumberland and
   Durham (Madden & Company, 1844).
4. The Jubilee Pit, Coxlodge Colliery, from Thomas H. Hair, 27
   A Series of Views of the Collieries in the Counties of
   Northumberland and Durham (Madden & Company, 1844).
5a&b. Cast-iron plate rail and Losh rail. Reading Room 41
   2020 / Alamy Stock Photo.
6. Plan of the Fonderie Royale at Le Creusot, 1782. 76
   Bibliothèque nationale de France.
7. Little Eaton gangway, early 1900s. KGPA Ltd / 78
   Alamy Stock Photo.
8. Bottom of plane, Peak Forest Railway, Chapel-en-le-Frith. 86
   Picture the Past.
9. Inclined plane powered by waterwheel used on a canal, 87
   1796. The Print Collector / Alamy Stock Photo.
10. Signed ticket for Richard Trevithick's *Catch Me Who Can*, 97
    1808. Universal Images Group North America LLC /
    Alamy Stock Photo.
11. *Salamanca*, 1812. GL Archive / Alamy Stock Photo. 101
12. Railroad scene, Little Falls (Valley of the Mohawk), 123
    c. 1838. Library of Congress (LC-USZ62-51439).
13. Permanent way of the Liverpool and Manchester Railway, 131
    1830, from George Findlay, *The Working and Management*
    of an English Railway (Whittaker & Co., 1889).
14. Boat ascending plane, Morris Canal, Boonton, NJ, *c.* 1900. 143
    Library of Congress (LC-DIG-det-4a07224).
15. *Globe* featured on Timothy Hackworth's business card, from 174
    William Weaver Tomlinson, *The North Eastern Railway:*
    Its Rise and Development (A. Reid and Company, 1915).
16. The Mohawk and Hudson's *DeWitt Clinton*, 1832. 182
    Library of Congress (LC-DIG-pga-0817).

ILLUSTRATIONS AND MAPS

17. Hetton Colliery railway, Hetton-le-Hole, County Durham, *c*. 1822. Science & Society Picture Library.   195
18. The Australian Agricultural Company's coal works, Newcastle, NSW, attributed to J.C. White, *c*. 1833. Mitchell Library, State Library of New South Wales.   211
19. Waggons at the top of Sheep Pasture incline, on the Cromford and High Peak Railway, Derbyshire, photograph by Eric Bruton, 1952. © National Railway Museum / Science & Society Picture Library.   223
20. 'Viaduct on Baltimore & Washington Railroad', engraving from *The United States Illustrated*, *c*. 1858. Library of Congress (HAER MD,14-ELK,1-17).   242
21. Frontispiece to Francis Bacon's *Novum Organum*, 1620. Volgi Archive / Alamy Stock Photo.   256
22. Matthias S. Weaver, 'Rowing on the Schuylkill River', 1842. Philadelphia on Stone, World Digital Library.   274
23. 'Morgan's Newly Invented Rail Road Carriage', from the *American (Boston) Traveller*, 1829.   279
24. William L. Breton, 'Railroad Depot at Philadelphia', 1832. Philadelphia Museum of Art, Gift of the McNeil Americana Collection, 2012.   282
25. Ding Gongchen, 'Yanpao tushuo jiyao', 1841. Xiamen University Library.   296
26. Compressed air locomotive designed by Arthur Parsey. World History Archive / Alamy Stock Photo.   300
27. Atmospheric propulsion. Reading Room 2020 / Alamy Stock Photo.   301
28. Enslaved men repairing a single-track railroad after the Battle of Stone's River, Murfreesboro, TN, 1863. Library of Congress (LC-DIG-cwpb-02135).   307
29. *Acheron* emerging from a tunnel near Bristol, frontispiece to John Cooke Bourne, *The History and Description of the*   315

ILLUSTRATIONS AND MAPS

Great Western Railway, 1846. © Science and Society Picture Library / Bridgeman Images.

30. John Cooke Bourne, 'Building Retaining Wall and Church near Park Street, Camden Town, Sept. 17, 1836', 1838. Yale Center for British Art, Paul Mellon Collection.   317

Maps

1. The north-east of England coalfield and its railways.   28
2. Railways of the Welsh valleys and the Border Country.   53
3. The Cromford and High Peak Railway and the Budweis–Linz–Gmunden horse railway.   219
4. The first main lines.   234
5. Railways in the USA in 1850.   287
6. The European railway network in 1850.   293

ACKNOWLEDGEMENTS

I must first of all acknowledge an enormous debt to Dr Michael Lewis, whose *Early Wooden Railways*, published in 1970, underlies chapter 2, but whose papers and subsequent book-length studies have done more than anyone else to advance our knowledge and understanding of railways from the Classical period to the early years of the nineteenth century. I briefly met Professor Fred Gamst of the University of Massachusetts at the Durham (UK) Early Railways conference, and corresponded with him subsequently; chapters 11 to 13 depend greatly on his edition of von Gerstner.

For mechanical matters and for locomotive traction in particular I am indebted to the work of, and discussion with, Dr Michael Bailey. Otherwise, I am also extremely grateful to Graham Boyes, Anthony Coulls, Mike Chrimes, Helen Gomersall, Andy Guy, Dr John van Laun, Michael Messenger, Jim Rees, Dr Barrie Trinder and all the other contributors to the conference series. Thanks are also due to Anthony Leslie Dawson for his research and publications, and for advancing the cause on social media.

For encouragement to undertake this study in the first place, together with comments on drafts and a certain amount of goading, I am indebted to Sir Neil Cossons and to Julia Elton. Laura

ACKNOWLEDGEMENTS

McCullagh has read chapters as a non-specialist and has pointed out to me many instances where I blithely and wrongly assumed general knowledge of some technical term or other. Dr Marian Gwyn has been, as ever, an unfailing source of encouragement and advice.

Introduction

This is a history of the early iron railway: an account of how a means of transport running on iron rails and using mechanical traction came into being in the second half of the eighteenth century and the first half of the nineteenth, how it evolved from the earlier forms of railway which were operating in Britain, how it then went through a particularly rapid period of transformation between 1786 and 1830, and how over the next 20 years it became a widely accepted means of moving people and goods over much of Europe and North America.

This is not an inquiry into the origins of the railway. Rather, I set out to explore a crucial period of its history, to see how it met the needs of the Industrial Revolution, that profound change in human circumstances which involved not only new methods of manufacture but also more effective means of distribution. The fundamental concept of the railway, a prepared way that both guides and supports wheeled vehicles, dates from prehistory. By the eighteenth century, railways were already a well-established technology.[1] Practically all were mineral-handling systems. They could be found as far east as Kazakhstan, and as far west as France and Ireland. They were constructed almost entirely of wood, and relied on the power of humans, horses or oxen for propulsion. Most operated in darkness underground, though some might extend from the mouth of a mine level for a short distance to a dump or a crushing plant. Only in some British coalfields were there any railways longer than a few hundred yards. In the north-east of England it was agreed that one of these wooden 'waggonways' could be cost-effective up to 10 miles in length, from the pit-head to staithes on the Tyne or Wear rivers.[2] An increase in the demand for coal made it clear that this was a technology with great potential for adaptation and development which could benefit many other industries and economies.

As a result, railways came to carry not only minerals but also foodstuffs and goods such as cotton, a staple of the emerging global financial system, as well as offering passenger services. Most notable in this respect was the Liverpool and Manchester, opened in 1830, which was the first to connect two great trading cities.

Railways also came to be built in places where they had never existed before, and on a vastly greater scale. In 1832 the first railway connecting two great continental rivers was finally completed, between Budweis and Linz in the Austrian Empire. The following year the first railway over 100 miles in length was put into service, the Charleston and Hamburg in South Carolina.

These changes were made possible by a culture of experimentation and of constant adaptation, and by a growing understanding of new materials, specifically the progressive substitution of iron for wooden elements. The inclined plane was patented in Britain in 1750. Significant use of iron began in 1767, when cast-iron straps on wooden rails are first recorded. In 1786–7 the first all-iron rails were introduced. Within a few years, there was the additional option of using mechanical traction powered by fossil fuel in the form of fixed high-pressure steam engines which wound haulage ropes, or even self-propelling 'locomotive engines'. These became a possibility in the context of the coal and metallurgical industries of England and Wales, but soon spread to other countries, revolutionised in 1829–30 by Robert Stephenson's

INTRODUCTION

Rocket/*Northumbrian*/'Planet'/'Samson' design. It soon became clear that a locomotive and its train running along iron rails could accomplish what the carrier's cart, the stagecoach and the canal barge could not.

Historians have often referred to the years from 1825 to 1830 as the dawn of a 'railway age', but the evolution of railways over the previous 80 years or so remains much less known, despite many excellent specialist histories. There is little collective understanding of their own very considerable economic impact. A French visitor to Newcastle in 1765 could see that the phenomenal growth of the coalfield was due to its wooden waggonways to the Tyne. A short hand-propelled system, running on cast-iron rails a few yards from a canal wharf to an adjacent limekiln, or moving manure in a farmyard, also reduced transport costs significantly. These systems are as much part of the story as the steam networks which evolved from them, but history books have little to say about the changes which led from one to the other, and how the achievements of celebrated engineers such as the Stephensons were based on experience gained earlier.

Technical changes take place as components within an engineered system appear, mature, become obsolete and are replaced. In the course of the nineteenth century, wooden vessels moved by wind gave way to iron steamships; rural watermills were superseded by factories where cereal crops were crushed by mechanically driven rollers. Similarly, the railway underwent an evolutionary change common in the Industrial Revolution, by progressively replacing organic elements activated by natural forces with mechanical technologies made of iron and powered by fossil fuels. Technical changes also relied on improved capacity and on a better understanding of physical phenomena, whether established by rational enquiry or by trial and error. In this respect, the early iron railway could be compared with the electric telegraph, which developed within almost exactly the same time frame.[3] It differs from powered flight, where the theoretical principles were understood well over 50 years before the internal combustion engine made it a possibility.[4]

As these examples remind us, technical changes are also social processes. They depend not only on human knowledge and skill but also on changes in demand for particular goods and services and on the profits that they generate – as well as on a broader cultural context, such as political support and popular acceptance. The period from 1750 to 1850 saw not only remarkable economic development and technical experimentation, but also great political upheaval and social reordering, as pointed out by contemporaries.[5] To understand how the 'iron road' evolved, not only do we have to examine *how* a particular technical transformation took place, but also *why* particular social and economic forces made it necessary and possible.

Precisely because of the transformational nature of the technology at this time, it is important to avoid some historical pitfalls. We learn little from a list of 'firsts', which often prove to be erroneous, and which all too easily feed into a narrative of heroic engineer-inventors and of national exceptionalism.[6] Without a doubt, this hundred-year phase in the evolution of the railway owed a great deal to individual talent, British ingenuity and British capital. But these factors must be contextualised. Great Britain and Ireland, and the successor United Kingdom established in 1800–1, underwent remarkable changes in this period in which military and naval victories both consolidated a sea-based empire and encouraged early industrialisation, underpinned by an effective banking system. These in themselves put the early iron railway in perspective, as the consequence not of innate individual or national genius but of particular historical circumstances within the broader rise and fall of states and jurisdictions. Until 1815, its development was largely a British story. Thereafter, its use spread, but, while other countries certainly considered very carefully what had been accomplished in Britain, they did not as a rule adopt it without significant change.

Another related challenge is to avoid a 'Whig view of history', to use Professor Butterfield's memorable phrase, whereby the past is made to tell a story of essentially non-reversible progress, whether in

INTRODUCTION

the public sphere or in technology.[7] The waggonway which preceded the Stephenson railway in the north-east of England was not in any sense a crude system which only required iron rails and steam locomotives to make it effective. It has rightly been described as 'a complex series of lines, continually fine tuned to take advantage of new pit sinkings and more advantageous wayleaves ... in the terms of the time, a mature and sensitive rail system to which the Coalfield, without the distraction of any competing canals, was utterly committed'.[8] It was a system entirely appropriate to its circumstances. Wooden rails remained for many years a perfectly rational choice in some contexts. Furthermore, though we distinguish for convenience's sake between a 'wooden railway' and an 'iron railway', wood remained a vitally important structural material throughout this period and well beyond, particularly in the USA and Russia, where it was increasingly used rather than coal as a locomotive fuel. Even now, wooden sleepers are widely used on railway systems across the world; trains still cross rivers on timber bridges, or stop at stations with wooden shelters. An 'iron road' is one where iron components are common or where mechanical traction is employed – not one from which organic elements are excluded.

Avoiding a 'Whig view of history' also means taking seriously those who, at a time of rapid experimentation, were cautious about new technologies, before labelling these people as dull-witted or obscurantist. Some opposed railways in principle. Governing classes, all too well aware of the direction in which Parisian *sans-culottes* had taken the French Revolution, feared the mobile, landless working class which they knew the railways would help create.[9] Conservatives such as Metternich and Tsar Nicholas I also believed that allowing foreign investment in countries with minimal capital of their own would weaken hard-won social and political cohesion.[10] Others expressed technical or environmental fears which now seem laughable, but in the 1820s there was no body of learned opinion to challenge the assertion that steam locomotives would affect the milk yield of cows in nearby fields, or the concern that the human frame could not cope with speed.

Then there was the challenge of identifying the technologies worth investment: what would work in the long term and what was already appropriate to particular circumstances. Bankers are more cautious than engineers, and need to know whether a proffered innovation has substance, how much further development it is going to require before producing a return, how long this is likely to take – or whether a completely different but better idea is in prospect. By the same token, apparently crude or dead-end technologies remained rational options in some circumstances, or they might re-emerge in different form. The survival of the wooden rail is a case in point. North American railways used them extensively with, and sometimes even without, a thin iron strap on the running surface. Even Brunel's London to Bristol railway, opened in 1841 and soon to become the Great Western Railway, laid wrought-iron bridge-section rails on longitudinal baulks, still in essence largely a wooden railway. Wooden railways continued to be built for safety reasons, such as in explosives works, or where timber was easily available and iron was not, such as logging systems. The 'pole road', much used in timber-harvesting in the south-eastern USA, must rank as the most rudimentary form of railway ever devised, yet well into the twentieth century, steam and petrol locomotives hauled bogies along rails made of unmarketable logs laid directly on the ground.[11] Wooden railways lingered on in the Australian outback and in coal-pits in the Forest of Dean until very recent times indeed, though they have now effectively vanished.[12]

The steam locomotive itself became one of these time-expired technologies. Its immediate demise was much heralded in the 1920s, though it is only now that it is finally reaching the end of its commercial life.[13] It could be replaced by other forms of locomotive, diesel or electric, without having to make a fundamental change to the system, which is why other railway technologies from this early period have proved even more durable. These include the concept of an integrated network, of public control, a standard gauge, articulated rolling stock, even the idea of a train of vehicles and of dedicated passenger facilities. The railway, even now, is to a great extent founded on the technology that evolved from 1750 to 1850 which then led to the first regional and national networks.

The iron railway truly made the modern world. It would redefine warfare, sustain empires and deepen commitment to a fossil economy. Its effect on national economies, industrial development and globalisation is beyond calculation.[14] It changed employment practice and would also profoundly affect less obvious matters such as diet and health, as well as ethnic and gender relationships.[15] Total route mileage has declined since the twentieth century, but there are still about 1,000,000 km of railway in the world, carrying 10,000 billion freight tonne-km and 3,000 billion passenger-km annually.[16] Freight services now connect Zhejiang in China with Madrid in Spain. Sleek bullet-shaped trains powered by electricity running on rails of steel can reach speeds of over 400 km an hour. However, if we are to understand how the railway has shaped the world we inhabit, we need to return to its transformation from wooden waggonway to 'iron road'.

1

Trade, transport and coal 1767–1815

For coals on rails preceded, in Britain, all else on rails.[1]

In January 1768 Joseph Banks (1743–1820), making his way slowly back to London from a visit to his friend Thomas Pennant of Downing Hall in north-east Wales, came across a railway in Shropshire. It connected the Quaker iron-furnaces at Ketley with those at Horsehay and at Coalbrookdale by the River Severn, making it about four miles long. Banks described it in detail, as he did all the many other industrial developments he saw on his travels, noting that it was 'the most extensive one I believe in this Part of the Kingdom'. It used a double wooden rail, with sacrificial strips, which could be replaced more frequently, along the top made of the finest heart of oak, but they soon wore out, 'so much so that they have begun at the dale to make the upper barrs of cast Iron & have thoughts of continuing it all their ways'.[2] These conferred a further advantage, which he did not mention, in that they also enabled heavier loads to be carried than on a wooden railway, still more than in a cart on a made-up road.[3] An area that had already seen many innovations in metallurgy had now created the iron railway.

Banks, only 24 years old yet already a fellow of the Royal Society, had made his name as a natural philosopher and botanist with an expedition to Newfoundland and Labrador. The year after his journey from Wales to London he would accompany Captain Cook on the *Endeavour* to observe the transit of Venus, thereby calculating the earth's distance from the sun, and to map the Australian coast. Banks embodied Enlightenment rationalism and restless scientific enquiry, and introduced eighteenth-century Britain to the wonders of the wider world. It seems appropriate that he should also have been the first to describe an 'iron road'.

His journey home from Wales lay along Watling Street, a route that had been in use since prehistory, and the manner in which he travelled would have been familiar to its Roman wayfarers – slowly, on horseback, staying overnight at inns. In truth, there was little that would have surprised a visitor from the ancient world in the way that goods and people moved about England, or any other country, in the middle of the eighteenth century. Wooden ships and boats plied great continental rivers and inland seas, and were moved by muscle, currents or wind. Dutch, Italian and French engineers were once again building artificial canals, as the Romans had done and as the Chinese had continued to do for two thousand years. Human beings – merchants, trappers, the enslaved – and livestock made their own way along trackways and trails. In sub-Saharan Africa, goods were carried by head porterage. Animal-drawn carts running on made-up roads had been long used over much of the Eurasian land mass, and could now be found in European colonies in North America and South Africa. In pre-Columbian America the absence of large domesticated animals limited the use of wheels to children's toys; goods were moved by rafts and by pack llamas, until the conquistadors arrived with their cavalry and their baggage trains.

In Europe, both economic growth and warfare were leading to better techniques in horse breeding, as well as in agricultural productivity to feed both people and draught animals. New vehicle components such as steel springs accompanied road-surfacing techniques associated with the engineer Pierre-Marie-Jérôme Trésaguet which were later to be developed by Thomas Telford and John Loudon McAdam into what in English was called a 'metalled road' and in French (and in German and Russian) a *chaussée*.[4] The Seven Years War (1756–63) had left France with debts it could not afford, but gave victorious Britain an enlarged sea-based empire which brought it to industrial 'take off', the point where technical innovation and economic growth become mutually self-supporting. Very shortly after hostilities were brought to a close by the Treaty of Paris, the spinning jenny made its appearance; James Watt conceived the idea of the separate condenser for the Newcomen steam engine; and the Scottish philosopher and economist Adam Smith began work on what became *The Wealth of Nations*, making the case for low-tariff trade between nations and with their colonies.

Britannia had now come to rule the waves, its sea-based empire providing the raw materials for the industries which were transforming the global economy. Cart roads connected the cotton growers of Gujarat with weaving villages along the Indian coast, and the sugar planters and enslaved workers of Jamaica with wharves on the island's inland rivers. Another such industry lay in Britain itself: mining for coal, which had its own unique transport needs.

Britain produced more coal than any other country in the world, and by the mid-eighteenth century was rapidly developing into an 'energy economy'.[5] By far the biggest suppliers were the pits of the north-east of England, but there were also productive areas from Ayr to Fife in Scotland, in Lancashire, Yorkshire, Cumberland, Nottinghamshire, Staffordshire, Warwickshire, Shropshire and in north-east and south-east Wales, as well as a small coalfield in County Tyrone in Ireland. Between the mid-seventeenth century and 1800, production rose sixtyfold. London's need for fuel was voracious and growing; from being home to about 50,000 people in the 1520s, by 1700 England's metropolis rivalled Paris for the distinction of being the most populous city in Europe, with half a million inhabitants, a figure which doubled in the course of the eighteenth century to make it one of the largest in the world, on a par with Beijing and Edo (Tokyo). London's new townhouses were designed around the burning of coal for warmth and cooking.[6] Coal was also increasingly being used in industries as varied as metallurgy, lime burning, glassmaking, brewing, brickmaking and salt boiling.[7] British collieries were not only the largest but also the most heavily mechanised in the world.[8] They were themselves becoming consumers of their own output, as steam pumping and winding engines became more common.[9]

Coal has a low value-to-weight ratio and incurs high movement costs, but because it provided a very effective form of energy, it could generate a high return on investment for those who could mine it – if it could be transported in sufficient quantity. Overland transport of this heavy but compact and valuable fossil from pithead to navigable water or to a point of use therefore became a matter of the utmost significance, to the extent that central government – the Westminster Parliament and the Privy Council, as well as the Irish Parliament until its abolition in 1800 – took an active interest in the trade and were prepared to intervene when problems arose.[10]

Coal owners had several options, depending on topography, on output and on the level of capital at their disposal. In some impoverished and out-of-the-way places, packhorses were still used, carrying two or three hundredweight at a time. Cart roads had a greater capacity, and artificial waterways could carry more still. Collieries at Worsley in Lancashire were badly served by their tidal rivers, leading their owner, the Duke of Bridgewater, to construct a canal system directly from the coalface to their principal markets in Manchester. Collieries in south-west Wales and in County Tyrone also came to be served by canals. However, the railway not only ultimately prevailed over all the other systems but also gave the industrial age its definitive transport technology.

It is not entirely true to say that 'coals on rails preceded, in Britain, all else on rails', since railways did not first appear in the coal industry.¹¹ Nor were they invented in the British Isles. In the 1560s experienced miners from the free imperial city of Augsburg had installed at a copper mine in Caldbeck in Cumberland a system which used a small hand-propelled vehicle known as a *hund*, running on plain wheels, with a vertical guide-pin fitting into the very narrow gap between plank-like wooden rails, such as had been commonly used in the mining fields of central Europe since at least the late medieval period.¹² So far as is known, this was the first railway in Britain, but it was not long before the coal industry also saw the value of this continental European invention.¹³ The first wooden colliery railway may have been at work in the Severn Gorge in Shropshire before the end of the sixteenth century. Though this area's output was far smaller than the north-eastern coalfield, it was the scene of many significant technological innovations and an important element of the economy of western England, benefitting from access to markets by means of the river and of Watling Street.¹⁴

By 1604 a railway had been laid in the Nottinghamshire coalfield by the entrepreneur Huntingdon Beaumont (*c.* 1560–1624), who also introduced them to the north-east of England – one on the coast at Bedlington, which was laid around 1608, and at Cowpen and Bebside, probably at much the same time.¹⁵ In 1621 the first recorded railway to the River Tyne was built, from pits in Whickham parish. After the Restoration of 1660 they became common.

At some very early stage, their basic technology underwent a significant change, anticipating the modern railway in that guidance was no longer provided by a pin between the rails but by a flange on the wheels – a projecting rim which interfaces with the side of the rail, just as the wheel's rim is supported by the rail's surface. On overland wooden railways serving English collieries, this enabled a broader gauge and horse traction, and hence higher capacity, whereas the *hund* was only effective for hand-propulsion over short distances underground.¹⁶

Some important variations also became evident early on, with two distinct types of wooden railway evolving within the coalfields of

1. An inclined plane to the River Severn, sketched by a Swedish visitor in 1754.

Britain and Ireland. The Shropshire type was adopted in north-east Wales and in parts of south-east Wales and Lancashire. The other came from the north-east of England and spread to Cumberland, Yorkshire, County Tyrone, parts of lowland Scotland and Wales around Swansea and Wrexham. Both were to have a lasting impact on the way that the successor technology of the iron railway developed, with consequences that are clear to this day.

The typical Shropshire system was a compact one. The gauge (the distance between the rails, normally measured from the inside faces) was narrow, not more than 3' 9" (1140 mm), reflecting the fact that many were initially devised to run directly from the coalface along a level tunnel into the open air, then down to the River Severn.[17] Some involved inclined planes down the valley sides; these worked on the 'self-acting' principle, whereby the downward traffic enabled the empty waggons to be hauled upwards. The Swedish metallurgist Reinhold Rücker Angerstein (1718–60) sketched one during his visit to Coalbrookdale in the Severn Gorge in October 1754, showing a single track with a passing loop to a small jetty.[18]

Waggons were generally small in capacity, between 14 cwt (711 kg) and 60 cwt (3048 kg). These systems were often not heavily engineered and might have very short lives, 'of the mushroom breed, only the produce of one night'.[19] Others were more substantial and lasted longer. In 1729 iron wheels were used at Coalbrookdale; their use soon spread.[20] The establishment of new coke-fired furnaces in the 1750s and the expansion of mining brought longer systems into being, including the one Banks saw from Ketley (near Watling Street) to Coalbrookdale Wharf on the Severn. A few years later, the Quaker minister and industrialist Abiah Darby stated that the Company had 20 miles of cast-iron 'road'.[21] They were sometimes called 'waggonways', or 'insets' or 'footrails', but also 'railed ways' or (as Joseph Banks called them) 'railways'.[22]

In the great coalfield around Newcastle-on-Tyne, the wooden 'waggonways', or sometimes simply 'ways', were long, high-capacity overland systems requiring significant investment. The main coalfield was compact – barely more than 15 miles east to west and 12 north to south – but coal was increasingly having to be dug further inland and deeper, in higher ground rising up to 1,000 feet away from the coast, up the steep-sided Tyne and Wear valleys – impossible terrain for a canal, and a long haul by road, but within the reach of wooden railways running from the pitheads to the rivers. The Tyne was navigable 10 miles upstream from Newcastle, the Wear as far inland as Lambton Castle. Coal was transferred from waggonways either into river keelboats or directly into the sea-going collier ships themselves by means of 'staithes', substantial timber structures whose scale was dictated by the larger waggons, by the size of the vessels they were loading and by the fact that they had to accommodate a tidal range.[23]

Waggonways also carried timber for pit props which was landed from Scandinavia. The Chopwell way and the Tanfield way carried lead from the north Pennines, the East Winlaton way and its successor the Western way carried billets of Swedish steel inland to Cowley Ironworks.[24]

These waggonways were all built and operated by colliery owners, either as individuals or as partnerships, and were routed to their advantage, avoiding as much as possible land held by others and the consequent need to pay for way leaves and royalties. The 'lords of coal' clashed repeatedly with each other, and also tended to be at odds with the Newcastle merchants. The Tyne and the Wear trades were bitter rivals, and powerful regional dynasties who had never accepted the Reformation had little common ground with the Whig descendants of Puritan families. Not for nothing were coal exports described as 'A fighting trade run in a shallow tempestuous stream', and as we shall see, this has some bearing on the ways both in which the wooden waggonway did not need to change its form significantly for many years and in how it ultimately began to do so at the very end of the eighteenth century.[25]

Central to their evolution were the colliery 'viewers', a formidable caste of individuals who were responsible as engineers for the working of the pits themselves and for the building and operation of waggonways and staithes, but who also took on other roles, as negotiators and political enforcers. They often worked as independent consultants, and so were much in demand in other coalfields, a state of affairs which explains the often rapid spread of particular technologies. Viewers might be the younger sons of gentlemen or they might be pitmen who had shown particular aptitude. They had to win the confidence of workmen and to combine respectful deference to their employers with frankness, and they needed to show good judgement in commercial and technical matters.[26]

The waggonways they constructed were extremely well adapted for the Newcastle trade. They were built to a broader gauge than their Shropshire counterparts, invariably between 3' 10" and 5' (1168–1524 mm), a reflection of the optimal axle length on a single vehicle pulled by a single horse, from which the current UK, continental European and USA standard gauge of 4' 8½" (1435 mm) derives. Typical practice was for the wooden rails to be fixed to wooden sleepers, and the space between and around them filled with stone ballast. Outside the rails, ballast was laid nearly to the top of the rail to form a path for the waggonmen to follow. The sleepers were 4 to 8 inches (101.6–203.2 mm) square and longer than the gauge of the way. The rails were 4 to 5 inches wide and high, with the width usually slightly exceeding the height. They might be 6 foot (1828 mm) long, pegged to the sleepers every 18 inches with treenails of young oak 6 to 7 inches (152–177 mm) long. Rail joints were made above a sleeper, with adjoining rails merely butting up to each other. If a rail became damaged it could be turned over. From the mid-eighteenth century the practice evolved of laying a 'double way', whereby a second rail was fixed to the top of the first. This allowed more ballast to be laid above the sleepers, thereby protecting them from the horses' hooves, and had the advantage that the joints between the running surface did not have to be above a sleeper, which made for a more durable system.[27]

The waggons used on these systems had a much greater carrying capacity than the Shropshire types, typically between 37 cwt (1880 kg) and 50 cwt (2540 kg), and were known as 'chaldron waggons' after an archaic term for a measure of coal (53 cwt in the north-east of England from 1694).[28] James Fittler's 1783 engraving shows a typical operation, the single waggon making its way down to the Tyne at Newcastle by gravity, with the horse acting as a brake, under the control of one man.

The volume of traffic was such that some Tyneside systems had been built, or rebuilt, to double- or even triple-track, on which loaded

TRADE, TRANSPORT AND COAL 1767–1815

2. A 'run' to the Tyne on the Parkmore waggonway, depicted by James Fittler in 1783.

waggons made their way down a 'main way', laid wherever possible on a regular falling alignment, and horses pulled the empties back along a 'bye way', which might not necessarily run parallel to it but wandered around to find an easier route.[29] In practice, the maximum gradient was 1 in 10, which was neither too steep for a horse hauling an empty chaldron waggon up a 'pull', nor for keeping a loaded one in check on a 'run'. By accepting these ferocious gradients, and some civil engineering on a massive scale, waggonway wrights could choose a route which placed the coal owner at an advantage over the way leave owner.[30]

The coalfield had long been a place of wonder and curiosity to visitors. The pioneering French metallurgist Gabriel Jars (1732–69), who came to Newcastle in 1765, stated that the waggonways were costly to build but soon paid for themselves by the ease of moving coal at all times of the year. He explained how the formation was prepared on as constant a gradient as possible towards the river, and earthworks constructed as necessary, with wooden bridges across small side valleys. Wooden sleepers were then laid down and wooden rails dowelled into them to a gauge generally of about four *'pieds'*. The waggons varied in size but all followed the same basic pattern, hopper-shaped, with an opening floor, running on four flanged wheels, of which the two at the river end were made of cast iron and were larger in diameter than the wooden discs at the rear in order to keep the floor level. Axles were made of iron and fixed to the wheels. A brake lever operated on one of these wheels. A single horse pulled a single such waggon or acted as a brake on downhill sections, where derailments were not unknown. Sharp changes in alignment were carried out by a turntable. Routes varied in length. Some collieries were very near the Tyne, but others lay 7 or 8 miles away ('three French leagues').[31] Jars remarks:

> The longest way [*chemin*] belongs to a very rich company, in which Milord Bute, a former minister, is one of the principal parties. This company is not only the owner of the *Royalty* of several collieries, but it also leases very many such all along the road. It works coal on a very extensive scale, as the road is almost always covered with waggons. This company does not spare any expense in developing its operations, and the market for its output.[32]

Coal from Bute's collieries was carried on several 'ways', though Jars is probably referring here to the Tanfield, from South Moor to the Tyne at Redheugh. This included a single 120 foot-span (36.6 m) stone arch carrying the double-track way over the Causey Burn, hailed by a local newspaper on its completion in 1727 as 'the greatest Piece of Work that has been done in England since the Time of the Romans'.[33]

In 1726 some of the most powerful coal-owning families – the Bowes, Liddells, Ords and Montagues – formed a cartel which came to be known as 'The Grand Allies', creating a regional monopoly by not only working coal on their own manors but also by buying up land and royalties, and managing way leaves to ensure their own rights of way to navigable water and to deny them to others. Jars's visit came at the end of 40 years of rising prices and steady output, the golden age of these coal owners and of the wooden Newcastle waggonways. The level of profits they generated may have discouraged innovation, or restricted it to the collieries themselves, and the rationing of the London market may have delayed the exhaustion of some coal seams. Ultimately, the Grand Allies' monopoly and profits were eroded by their rivals in the Tyne basin and along the River Wear, and the price of coal fell in 1767. A negotiated agreement, the 'Limitations of the Vend' in 1771, restored some stability to the trade at a time of renewed expansion of its markets. The Lambton waggonway, which ran to staithes on the Wear, was built in 1770 and was the last entirely wooden way serving a large area and running to the highest point which could be served by river keels. The need to transport ever greater tonnages of coal made it clear that the days of the wooden waggonway were numbered.

The coming of the iron railway

Iron components had been used to build and operate railways before 1767. Thin wrought-iron strips beaten out by a blacksmith on an anvil had been laid on the top of wooden rails on some systems since perhaps the early eighteenth century, and iron axles, wheel treads and entire wheels began to be used not long after.[34] There was no one moment when the iron railway made its appearance, but a process of gradual evolution took place. Between 1786 and 1815 a growing understanding of beam theory and of the properties of cast iron made it possible to produce brittle but serviceable rails in this medium, rather than simply an iron element to be laid on wood such as Banks had observed at Coalbrookdale. These took two very different basic forms. Some were 'edge rails' for flanged wheels, which were initially no more than untutored attempts to reproduce the form of a squared timber but which grew in sophistication over this period. Others were L-section plate rails for flangeless wheels, an iron version of the existing wooden flanged barrow-ways which were used underground in some collieries. These two types were completely incompatible with each other, which mattered little for as long as railways were short-haul unconnected handling systems, just as differences in gauge were still of little account but became problematic in later years.[35]

In some places, these technologies converged with another, mechanical traction in the form of both fixed and 'locomotive' (self-propelling) steam engines. Fixed engines, for winding waggons by rope, were used with some success, but locomotives proved more challenging, as they had to be sufficiently heavy to gain traction but not so much so as to break the rails or plates on which they ran. Not until 1815 did the steam locomotive become an entirely practical and useful technology, after about 12 years of experiment.

These changes are generally ascribed to the growing demand for coal, a fall in the price of iron and a rise in the price of wood as a consequence of the Revolutionary and Napoleonic wars (1793–1815). Warfare also increased the cost of horses and of fodder, which not only meant that animal traction had to be used more effectively but also prompted the search for mechanical alternatives.

Remarkably, none of these innovations took place in the northeast of England coalfield. The first all-iron rails were laid in collieries near Merthyr Tydfil in South Wales and at Sheffield in Yorkshire in 1786–7, and though they were known on Tyneside a few years later, their use spread in the context of the iron industry, as well as of the collieries that served it, and above all the growing network of canals. Many hundreds of miles of canal feeder railway came to be built, even into the 1830s. Earlier ones were typically a wooden waggonway, later ones an iron road. The form the rails and other track components took is described in the following chapter, and the railways themselves in chapter 3. The first use of steam to haul railway vehicles was in Shropshire in 1793, in the form of fixed engines on railed inclined planes functioning in place of locks on canals. Experimental use of railway locomotives began in South Wales in

1804, to the design of the Cornishman Richard Trevithick and with the capital and encouragement of the ironmaster Samuel Homfray. Though further trials took place in the north-east of England shortly afterwards, regular use of locomotives only began in Leeds in 1812, operating on the rack system patented by the viewer John Blenkinsop, to haul coal from Middleton Colliery. The use of these rack locomotives spread to the Lancashire coalfield in January 1813, to the north-east in September of that year and to South Wales by 1814. They are described in chapter 4.

The rapid spread of all these technologies is itself remarkable. The movement of experts between different coalfields ensured that new developments soon became common knowledge, but a more general 'spirit of association' was also abroad, often rational and utilitarian in its ambitions and sometimes, but not always, associated with radical politics and freethinking. Experiment and enquiry depended on sharing and discussing results: a polymath of the Enlightenment like Joseph Banks could benefit from membership of the Royal Society (founded 1660) and of the Society of Arts (1754), or of the similar institutions which flourished in Edinburgh and Dublin as well as in and around Birmingham and Shropshire, Sheffield and Newcastle. Some of these were formal organisations with meeting rooms and membership rolls; others, like the Birmingham Lunar Society, were informal associations of like-minded individuals.[36] This went with a belief in 'improvement', a word which meant many different things to different people but which conveyed gradual and cumulative betterment through more productive farming and manufacture. Since this was a time when warfare had driven up the price of many basic commodities and made the use of horses very costly, it also encompassed more effective transport, whether by road, canal or railway. For many, 'improvement' was also supposed to reconcile the working poor to the ways things were, once the French Revolution had shown the ruling classes of Britain and Ireland how inefficient farming and poor transport threatened the food supply on which the social order depended. Sympathy for radical change and for free thought gave way to cautious reform: if the poor had to be taught their responsibilities as dutiful subjects, servants and Christians, they also deserved to be properly fed. The Board of Agriculture and Internal Improvement, founded in 1793, was a voluntary association, but it had government backing. Its pioneering president, Sir John Sinclair (1754–1835), introduced the word 'statistics' to the English language, sought to measure the 'quantum of happiness' of the population and studied the possibilities that railways had to offer.[37]

It might seem remarkable that technical developments were least evident in Newcastle and the north-east coalfield, a region that lacked neither capital nor expertise, but it had invested very heavily in its wooden waggonways, and at this stage mostly chose to adapt rather than replace. There were as yet few foundries in the region, and cast-iron components had to be brought from elsewhere. So changes to the waggonways were gradual. In 1793, instead of the time-honoured system of running one waggon with one horse, the Western way began to couple two waggons together, preventing them from running away by a 'long brake'.[38] Running a train in the Shropshire manner might have seemed an obvious way of increasing capacity, but for the fact that this would have required an unfeasibly large number of horses to haul waggons up the steep 'pulls' or to control them on a downward 'run'. Where demand was high and changes had to be made, it sometimes became more cost-effective either to realign an existing system or to build a completely new one, but in both cases to lay iron rails.

In 1798, for the first time, a colliery railway in the north-east of England was laid with cast-iron edge rails, on stone sleepers. It was only 700 yards long and connected the Walker pits east of Newcastle with an existing wooden waggonway and hence to staithes on the Tyne. It was the work of Thomas Barnes (1765–1801), one of the ablest of the region's viewers, and may have been based on recent practice on a canal tributary railway in Leicestershire.[39] Outside the region, but on what was effectively a Yorkshire colony of the north-east, iron rails were laid alongside the wooden way on the main route of the Middleton Railway at Leeds between 1799 and 1808 to form a double track, the heavier loaded waggons running on the iron, the lighter empties returning on the wood.[40]

Using rails that were compatible with the existing networks, and with their waggons, gave coal owners and viewers the option of making a costly change slowly. In 1807 it was recorded that these same Walker pits were 'gradually giving up their wood Roads and introducing Iron, and they make them piece by piece, keeping up the old Waggons, and running them partly on wood and partly Iron'.[41]

Iron rails and iron–wood hybrids were also being laid on new systems. One of the earliest was laid in 1805 from Isabella pit to Heworth Colliery staithes on the Tyne as part of the Washington waggonway. A circuitous waggonway from Sheriff Hill which also led to Heworth, only built in the 1790s, was replaced in 1806.[42] In 1808–09, pits at Coxlodge and Kenton could export from deep-water staithes on the Tyne at Wallsend by a new railway using iron straps on wooden rails which connected with an older unconverted waggonway down to the river.[43] It was the work of the colliery tenant, the accomplished civil and mechanical engineer William Chapman (1749–1832). Chapman was unusual among his peers in that his background was not confined to mining – he had been a mariner and a canal and drainage engineer, and had worked in Ireland and in Yorkshire. He had recently taken out a patent for a 'drop' whereby loaded waggons could be lowered vertically from the end of the staithe into a ship's hold, avoiding breakage of the coal and reducing handling costs. Improved rope-making techniques, another of Chapman's areas of expertise, was crucial to the operation not only of his 'drop' but also of inclined planes, colliery winding and of shipping. This system was applied at Benwell staithes on the north bank of the Tyne, four miles upstream from Newcastle, where Barnes had formerly held sway.[44]

DROPS AT WALLSEND.

3. 'Drops' reduced handling costs and avoided breaking the coal.

Other systems followed. A new iron railway was laid in 1810 to replace the Felling waggonway, running parallel to it.[45] The same year, it could be proudly announced that one horse had pulled eight waggons of coal on the inaugural journey of an iron railway constructed by Simon Temple's Harton Coal Company to the Tyne at South Shields, where an inclined plane led them directly to the collier's deck.[46] A waggonway wright of the old generation might now think of himself as a 'mechanic', working additionally in iron and no longer exclusively in wood.[47]

Steep sections were problematic. Iron rails made it easier for horses on a 'pull' but more difficult on a 'run', where the load was now in greater danger of running away. The solution lay in building a new formation, as near level as possible, where horses could pull trains of several waggons coupled together, but with much steeper gradients in between, where rope haulage could be used. This particular technology – the inclined plane – was another Shropshire practice, which duly made its appearance in the north-east of England. An Edinburgh advocate familiar with the Newcastle trade was awarded a patent for them in 1750, but they do not seem to have come into use.[48] The first was built in 1797, at Benwell staithes, and was also the work of Thomas Barnes.[49]

Barnes's incline was self-acting, as most later ones were, since the topography of the coalfield generally favoured the flow of traffic towards the two rivers, but loads sometimes had to be hauled up the brow of a hill, either by uncoupling a train into separate waggons and hauling them up one by one by horse, or by using a fixed steam engine to wind them up in batches. The inauguration of one such engine on Birtley Fell (Black Fell) on the morning of 17 May 1809 drew a crowd allegedly 10,000 strong, who watched four waggons from Bewicke Main pit being wound up the two inclined planes. In the evening, six men pushed the waggons as far as the staithes on the Tyne 'in order to prove the excellence of the level railway'.[50]

In 1812 the landowner and lawyer John Douthwaite Nesham built an iron railway and an inclined plane to connect his pits at Newbottle with a deep-water staithe at the mouth of the River Wear, superseding earlier wooden waggonways which had reached the river miles inland at Penshaw. A detailed watercolour plan and elevation which he commissioned five years later, measuring 10 foot by nearly 3 foot (305 x 91 cm), now preserved in the Sunderland Museum, shows that it was little different from its predecessor systems in many respects, with branches from the various pitheads all converging on a 'main way', and horses each leading a single waggon, yet it ran on iron rails – perhaps aptly, it is described as a 'waggon rail way'. The document includes a pen-and-ink drawing of the Wearmouth iron bridge of 1796, implicitly contrasted with a cartouche of the untamed upper river.[51] Well might Nesham present his iron road as another symbol of regional 'improvement', since it served seagoing colliers, cutting out the need for the river keelboats which had previously carried his output from Penshaw down the River Wear.[52]

This might have proved the railway's undoing since in March 1815 keelmen fearful of losing income pulled down the wooden bridge carrying the railway across Galley's Gill and set fire to the Wearmouth staithes, including its inclined plane.[53] This was only the start of a series of misfortunes: an explosion in one of the pits claimed the lives of 57 men on 2 June the same year, and on 31 July 1815 an experiment with a locomotive to the design of William Brunton which pushed itself along on legs came to an explosive end, claiming over a dozen lives.[54] Thereafter Nesham made further changes involving both self-acting and powered inclined planes until in 1822 his collieries and railway were acquired by the Lambton interests.[55]

Nesham's locomotive was not the first to earn its keep in the region, a distinction which belongs to a machine purchased for Coxlodge Colliery, similar to those already operating near Leeds and in Lancashire.[56] The section of line destined to receive it, already recently relaid with iron rails, now had to be prepared for the rack system with its projecting teeth on one side of the track. At around 1pm on 2 September 1813 'the new invention' set off from the Jubilee pit at Coxlodge Colliery with 16 waggons, 'in the presence of a vast concourse of spectators', after which the gentlemen repaired to the nearby race-track for lunch, 'when the afternoon was spent in the most agreeable and convivial manner'.[57]

The Coxlodge locomotive and its two later companions fell victim not to mechanical disaster like Nesham's, but to an intense managerial disagreement, when changes in personnel led to accusations of incompetence and wrongful dismissal. The steep section had to be adapted for cable haulage, other realignments were put in place and a new, more powerful locomotive purchased.[58] Matters did not end there, but the Kenton and Coxlodge nevertheless had a long working life, carrying coal until 1884.[59]

More successful in the immediate term was the rebuilding of the isolated Wylam waggonway with plate rails, in 1808. This was carried out with steam traction in mind, though it was several years before it

THE JUBILEE PIT, COXLODGE COLLIERY.

4. Coxlodge Colliery ran locomotive-hauled trains to deep-water staithes on the Tyne from 1813.

actually saw a locomotive – perhaps in March 1813 – and even then it was not until the summer of the following year that it went into regular use. Experiments here established the relationship between locomotive weight and the ability of a smooth wheel to pull a load on a smooth rail.[60]

These events are set out in greater detail in chapter 4. However, even more significant in the long term was the construction of the locomotive *Blücher* for the Killingworth Colliery railway in 1814, to a basic design which had its weaknesses but which proved capable of improvement. It was the work of the engine-wright George Stephenson (1781–1848), one of the outstanding figures of the Industrial Revolution and often hailed as the 'father of railways'. A collier's son, still barely literate but charismatic, skilled and persuasive, he was already on an annual salary of £100 from the Grand Allies, with the use of a horse to carry out consultancy work at other pits and on behalf of other coal owners. He and his son Robert (1803–59) between them, more than any other engineers and technicians, created the iron railway, above all by identifying and fostering

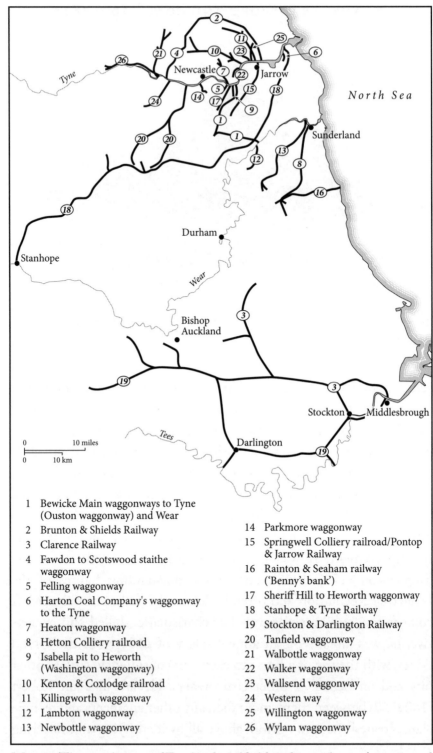

Map 1. The north-east of England coalfield and its railways (mentioned in the text).

a system in which all the components – track, motive power and rolling stock – were able to operate harmoniously.

That they were ultimately able to do this was because at this stage the north-eastern coalfield chose not to invest heavily in completely new systems, even to the point of sticking with wooden rails in many cases, and otherwise remained faithful both to the edge rail system and to a gauge of between 4 and 5 foot (1219–1524 mm). The Wylam plateway was the only significant departure from type in the region. When new technologies became available after 1815 – improved forms of iron edge rail and more effective locomotive traction – the tried and tested elements which had served the region well since the seventeenth century could be combined with them to create an effective new system, triumphantly vindicated with the opening of the Stephensons' Stockton and Darlington Railway in the southern part of the coalfield in 1825, and soon widely adopted not only in the United Kingdom but also in France and the USA.[61]

In the meantime, very many new railways were being built in Britain to connect with the growing network of canals. Some served collieries, others served different industries entirely and very many of them were built as newfangled plateways, using the L-section flanged rails, an aberration in railway history but which formed the dominant technology for a generation. Their development was to a great extent independent of practice in the north-east of England. The next chapter turns to these canal railways and to other systems using the same technology.

2

'Rails best adapted to the road'
Cast-iron rails and their alternatives in Britain 1767–1832

It is also an important matter to determine the description or form of rails best adapted to the road.[1]

The single most fundamental development in railway history was the change from wooden to iron rails. This chapter charts the engineering of the first stage in that process, the adoption of track components made from cast iron and the forms they took, from when they first appeared in the form of straps in 1767, through the all-iron rails introduced in 1787 up to their last significant installation in 1832, as well as the contemporary use of wooden and stone elements. These made it possible not only to improve existing systems, such as the coal carriers of the north-east of England, but also to build new ones serving other industries and economies, often in association with canals and turnpike roads, and so coming under some measure of public control. They also made mechanical traction a possibility. They reflect the emergence of cast iron as the dominant material for any high-performance construction during the 'iron revolution' of the period 1750 to 1850, which was in many ways the motor of the Industrial Revolution.

This chapter concentrates on the track system itself, leaving the development and fortunes of these systems, and their emulation in continental Europe and the USA, to be discussed subsequently. The second stage, the successor technology of wrought ('malleable' or 'bar') iron rails from 1808 onwards, is set out in chapter 6. These could carry heavier loads, could be bent into curves and could be produced in longer lengths. They ultimately offered speed, and made the mainline passenger railway possible.

Rails and sleepers undoubtedly lack the stirring emotional appeal of the steam locomotive, but they are the most fundamental aspect of railway engineering. Whatever they are made of, whether wood, stone, cast or wrought iron and more latterly steel, or any combination of these, they are the elements that support the vehicle, and it is the rails' interface with the wheels which provides the guidance, by means of a flange – a protecting rim – either on the wheel itself or on the rail. As chapter 1 described, a remarkable development of this cast-iron era was that many railways were laid with flat plate rails with a tall vertical flange, a fundamental if short-lived departure from mainstream edge-railway practice.[2]

The choice between plate and edge rails seems to have depended on professional and personal connections, individual preference and good salesmanship as much as on any perceived technical superiority. Some regional peculiarities are evident: north-west Wales, for instance, was mainly edge-rail territory, and it was a family connection with this part of the world that led to the construction of one of the few edge railways built in Cheshire, where otherwise plateways predominated.[3] Practice in Ireland and Scotland was mixed.[4] The English Midlands, where the influence of the engineer Benjamin Outram was strong, were mainly plateway territory, as was South Wales, to the extent that a number of its early railroads were relaid as plateways very shortly after they were constructed, at Outram's urging. A detailed description would be very long, but there were in addition many design variations within the two basic types.[5] Because rails were such an important item of expenditure, promoters were anxious to ensure that they were spending their money wisely and that they were indeed investing in designs 'best adapted to the road'. The Plymouth and Dartmoor, for instance, a 20-mile plateway completed in 1823, received tenders from four local foundries, three in South Wales, one in the West Midlands and two London merchants. One of the Welsh applicants enclosed a letter recommending that it be built as an edge railway.[6] Many market towns now had foundries which might be entrusted with smaller contracts, even if making castings for the whole of the track might be beyond them. Timber and stone components could often be resourced locally.

Rails require additional components which transfer the load to the 'ballast', a layer generally made up of broken stone which also facilitates drainage, and to the earthwork formation below, and which keep them the right distance apart. On the railways described here, these were often wooden sleepers extending from rail to rail, also called 'ties' or 'crossties' in the United States, just as they had been in the waggonway era and were to be again on modern railways, but they could also be made of cut stone, either a single embedded block laid under each rail or, less commonly, extending under both sets. Fastenings that we now refer to as 'chairs', secured to the sleepers or blocks by pins set into wooden plugs, supported the rails, though a variant type known as a 'sill' extending from one rail to another was also used in this period. Rails could simply slot into the chairs or sills, or could be tightened with an iron or wooden wedge.

All of these together constitute the 'track' of a railway, also called the 'permanent way'. The track needs to function effectively in conjunction with the vehicles that run on it, according to their weight, their wheels and how frequently they operate. If these vehicles are hauled by horses or by cables, the motive power has no impact on the track, but a steam locomotive pulling its load by means of friction ('adhesion') between its wheels and the rails will inflict additional stresses on it. Cast-iron rails could be made strong enough to bear the weight of a locomotive, but they were a poor fit for each other.

Also crucial to the effective operation of a railway of any type was, and is, the question of gauge. In the period under discussion, this was not necessarily measured between the inside faces of the rails but occasionally from the centres, and, in the case of plateways (where the rail flange was nearly always on the inside), from the angles formed by two sections, thus ⌐ ⌐. The gauge of a railway affected its construction costs and its carrying capacity – the narrower the gauge, the smaller the earthworks on which it ran, and hence the cheaper it was to build, but the less it could carry. Particular types of load required different gauges. A compact mineral which required careful packing, such as North Wales slate, suited small waggons and a narrower gauge, whereas coal or limestone, both of which had to be shovelled into waggons, did better on a broader gauge. With no need for systems to interconnect, gauges and rail types could be of any kind.[7]

Another important factor is the gradient on which the track is laid. Unlike wooden waggonways where a single horse could pull an empty chaldron waggon uphill from the staithes to the pithead, cast-iron railways with their trains of waggons required level or near-level formations wherever possible.

Cast-iron components

Production of cast iron was limited until the successful use of coke for smelting, first achieved in 1709 but not exploited to the full for another 60 years until costs began to fall, at a time when both timber and horses were becoming more expensive. A blast furnace needs 12 tons of coal, ironstone and limestone to produce 1 ton of cast iron, with the result that British output was dominated by the English West Midlands and by south-east Wales, where these raw materials were found together.

It is a brittle material but because it is cast by pouring it into a mould of sand in a foundry, it can be produced in a variety of complicated shapes. A rail can incorporate the means by which it is attached to other track elements and also by which it is connected end to end with other rails. Cast-iron components are strong in compression but weak in tension and cannot be bent, which meant that, on anything other than an absolutely straight alignment, the alternatives were either to invest in a complex set of castings of different size and shape or to accept a rough-and-ready system involving 'threepenny bit' curves. Cast-iron rails were rarely longer than 7 foot (2.1 m), which meant that wheels were constantly passing over joints.

Cast-iron straps

The railway which Joseph Banks saw at Coalbrookdale in 1768 involved cast-iron straps laid on wooden rails and held in place by nails through projecting lugs – a revolutionary step indeed, though they were crudely made.[8] Effective rather than sophisticated, they were adopted in 1776 by John Morris, of Landore near Swansea, 'for wheeling Coal on in my Collieries' as he put it, and in 1793 at Mancot in north-east Wales.[9] A late and large-scale use of this technology was on the Budweis–Linz railway in Bohemia and Austria, laid between 1827 and 1832, where, in the absence of any rolling mills in the empire, they were favoured above wrought-iron straps (chapter 10).

Cast-iron plate rails

God bliss the man wi' peace and plenty
That furst invented metal plates!
Draw out his years to five times twenty,
Then slide him through the heevenly gates.[10]

'RAILS BEST ADAPTED TO THE ROAD'

The cast-iron L-section plate rail for flangeless wheels came to be the dominant form of railway technology from the 1790s to the mid-1820s – an aberration in railway history perhaps, but an effective system wherever speeds did not need to be high. Hundreds of miles of these 'plateways' were laid in many parts of the United Kingdom, most notably in Wales and the border country, above and below ground, and on both internal and cross-country systems. Only the north-east of England remained immune, other than an isolated system connecting Wylam Colliery with the Tyne, which was built in anticipation of locomotive operation.

They first date from 1787, when John Curr (*c*. 1756–1823), viewer of the Duke of Norfolk's Sheffield collieries, devised and brought into use L-section cast plates for flangeless wheels for use underground, an iron version of existing wooden flanged barrow-ways.[11] They proved to be a considerably more cost-effective way of moving coal from the face to the foot of a shaft, requiring no horses and fewer men, which is why a Tyneside collier hoped that he would live to be a hundred and then enter heaven. In 1788 two such plateways were built by Joseph Butler, one from iron mines and collieries to his furnace at Wingerworth in Derbyshire, the other to carry coke from the nearby Lings Colliery to the furnaces and ironworks.[12] Another was an inclined plane which carried tub boats on the Ketley Canal in Shropshire (see chapters 3 and 4), also opened in 1788.[13] However, it was in collieries that they were most widely adopted. The viewer John Buddle Senior (d. 1809) reported favourably on them, and they were adopted at the Middleton Colliery in Leeds (1790), South Wales (by 1792), in the Shropshire coalfield (1793) and then in Cumberland, Scotland and the north-east of England.[14] In 1797 Curr self-published *The Coal Viewer, and Engine Builder's Practical Companion* after it had been in draft for several years; it describes these plate rails as well as giving a detailed account of rope haulage, of improvements in rope-making and of the potential of the fixed steam engine for pumping and winding – all of them technologies which would become increasing important in railway building. It was dedicated to his noble patron.[15] Curr claimed that 'no work of the like kind has hitherto been published'.[16] In its detailed and specific suggestions for constructing new types of railway and associated new technologies, this was certainly true. It differed from the more general observations offered by visitors like Angerstein and Jars, discussed in chapter 1, and was the first of a variety of 'how to' books which touched on the subject.

Curr's plates were used underground or on internal systems, and most were built to very narrow gauges – between 1' 8" and 2' (508–610 mm).[17] In South Wales, plates to Curr's pattern were soon being cast with the ends convex and concave, lapping them so that only a single nail held both, and incorporating a downward-projecting sleeve or cone around the nail hole to attach it more firmly to the sleeper.

Larger plates were soon being used on the new overland systems which were being laid as tributaries to the growing network of canals, as well as others in which the influence of canal engineers is also apparent. Their spread owed much to the advocacy of Benjamin Outram (1764–1805), who had trained as a surveyor but was now turning himself into a civil engineer and an ironmaster. In 1789 he had begun building the Cromford Canal in Derbyshire, and in 1790 he established the partnership ultimately known as the Butterley Company.[18] Outram developed a type of plate rail which was markedly heavier than Curr's, for a broader gauge and for use on overland railways of some length, characterised by raised feet, rounded lugs on the inner side for stability and a notch at each end for spiking down. He guaranteed those that he cast and laid himself.[19] His first systems, from Crich limestone quarries to the Cromford Canal and the Little Eaton gangway to the Derby Canal, were 3' 6" (1067 mm) gauge. The rails he designed for the Peak Forest Railway between 1794 and 1796 became the Outram standard, 42 lb (19 kg) rails to 4' 2" (1270 mm) gauge laid on stone blocks rather than wooden sleepers, which thereafter became the preferred option for heavy-duty rails. His approach to construction was extremely thorough.[20] As Dr Michael Lewis observes:

> Outram, more than any other man, determined the shape of industrial railways for a full thirty years. He was responsible, by direct advice, or by example, for countless plateways throughout the Midlands. When he was consulted about the Monmouthshire Canal railways in 1799, he recommended a wholesale change-over from edge to plate rail, and practically every other railway-owner in South Wales followed his advice. Even after his death in 1805, his disciples carried on the work, building new railways by the score and adapting old ones, until by the 1820s, south of County Durham, the edge rail was very much the exception, the plate rail the rule.[21]

These systems served the iron and coal industries of South Wales for many years to come. The Merthyr plateway on which Trevithick's locomotive performed in 1804 was one such; so was the very extensive system in Monmouthshire leading down the canal at Crumlin and to the Severn Estuary at Newport.[22] They could work effectively with locomotives as long as speeds were low and the driving wheels were sprung.[23]

Though this was a time when other industrialising countries were looking to the United Kingdom for technical inspiration, plateways seem to have found little favour elsewhere, though they could be found on a Shropshire-inspired canal lift in Silesia built in 1806, at Alexandrovsk cannon-works between St Petersburg and the White Sea, underground at collieries in Liège in 1810 and in New South Wales in 1832, on temporary and construction railways in Sweden and at collieries and iron mines in Alleghany County in Maryland.[24]

Their weakness lay in the fact that the plate rails are effectively flat brittle beams, supporting transverse loads, with the result that they frequently failed in tension at mid-point on the lower side when waggons, or more particularly locomotives, ran along them. This problem could be minimised by casting them in shapes which increased the amount of iron here, either a rib on the underside or a hogged-profile flange, and by making sure that they were laid on a solid and well-built formation. Damaged plates could in any case easily be taken out and replaced. A further inherent weakness in the design was the friction of the tall flange on the wheels when trains ran round curves; this increased resistance, which could be lessened by 'dishing' the wheels inwards so they ran in the angle rather than against the flange.[25] It did not matter particularly that traffic could at most proceed at a walking pace, but their efficiency was drastically reduced by misalignment or by dirt or debris on the surface.[26]

Cast-iron edge rails

The introduction of cast-iron edge rails, as distinct from cast-iron straps for wooden rails, was a significant advance in railway technology, even though many early attempts were no more than untutored efforts to reproduce in metal what had already been done in wood.[27] On present evidence, it seems highly likely that they were first used in Wales, with England following shortly after, and it is probable that they preceded the L-section plates for flangeless wheels described above.

Such 'railroads' first appeared either in the lower Swansea valley or in one of the ironworks between Merthyr Tydfil and Blaenavon in the 1780s – the exact dates are unrecorded. By 1791, all-iron rails were being cast at Dowlais Ironworks near Merthyr to a design by Thomas Dadford senior, the Glamorganshire Canal contractor, with a male–female joint at each end, for fixing to transverse wooden sleepers. Modified versions were adopted by Dadford's son, Thomas junior, for railroads built for the Monmouthshire Canal Company (from 1792) and for the Brecknock and Abergavenny Canal (from

1793), with bulbous lugs to accommodate the spikes. It became the most common form of rail in the heads of the South Wales valleys until 1796, when plateways came to be adopted. Similar section rails were used on the Cwm Clydach railroad near Swansea, engineered by John Dadford, as well as on the Somerset Coal Canal in 1795, at Walker Colliery in the north-east of England in 1798 and on the Lake Lock railroad near Wakefield in 1798–9.[28] Some of these were cast with integral feet to attach them to the sleepers, others were used in conjunction with separate chairs.

As with plate rails, tendencies to break under load could be obviated by the design of the casting, either by widening the rail below the running surface or by deepening it to form a 'fish-belly'. An oval section, probably designed by Thomas Dadford, for use with double-flanged wheels, was used on the Penrhyn slate quarry railroad in North Wales in 1800–1, which was emulated elsewhere in the region and on a colliery system at Congleton in Cheshire where there were family links with Wales.[29] Other forms of cast-iron rails of varying degrees of sophistication came to be used, such as an inverted T-section which was to be found on railways serving the Montgomeryshire Canal and at the Mostyn Colliery at Point of Ayr in Flintshire, north-east Wales.[30]

Both cast-iron straps and rails were more widely adopted in continental Europe than plate rails, just as purely wooden railways on the English model were also to be found. In 1786 either cast- or wrought-iron straps and wooden rails were laid to connect a colliery to the Ruhr, becoming the model for a number of other coal carriers in the area. Cast-iron straps were added to an existing Shropshire-inspired waggonway carrying silver ore at Clausthal in the Harz mountains around 1806, and spread elsewhere in the Harz, which was at that time part of the kingdom of Hanover, and so under the same crown as Britain. Wooden railways had been installed in Silesian collieries before the end of the eighteenth century, and cast-iron straps and rails made their appearance there soon after. The French ordnance factory at l'Indret at the mouth of the Loire began in 1778 or 1779 with wooden rails to which cast-iron straps were soon added, as did the Le Creusot Foundry at Saône-et-Loire a few years later.

The Stephenson and Losh rail

As the demand for coal grew, and with locomotives under development, there was still every need to devise more effective forms of cast-iron track for the edge railways of the north-east of England.

In 1815 William Losh (1770–1861), a manufacturer of chemicals, invited George Stephenson to devote two days a week to the Newcastle ironworks in which he was a senior partner, for which he was to receive £100 a year. The following year the two men were jointly granted a patent which encompassed three separate inventions – a steam locomotive spring, described as being 'kinder' to the track; an improved wheel using malleable iron instead of cast iron; and, more importantly, an improved type of cast-iron rail involving half-lapped jointing through which a horizontal pin was passed transversely to join the rails together and also to fasten them to a more sophisticated cast-iron chair, the lap resting on the apex of the curve forming the bottom of the chair.[31] These were more stable than the butt joints used up until then.

Each of these three inventions was designed to make the best use of iron components and reflects Stephenson's grasp of the all-important relationship between the rolling stock, locomotive and rail. Losh rails were laid on new or improved systems in the north-east of England, where his influence was evident. They replaced wooden rails on the Killingworth waggonway by the end of 1818 and were used on the Wallsend waggonway by 1821.[32] The first all-steam railway, serving Hetton Colliery, was built with them in 1821–2.[33] Though most of the route of the Stockton and Darlington Railway was laid with the new wrought-iron rails in 1825 (chapter 6), a considerable part used Losh rails.[34] The Wylam plateway was

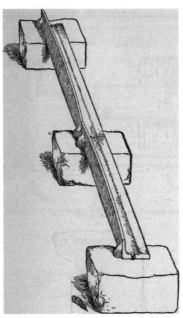

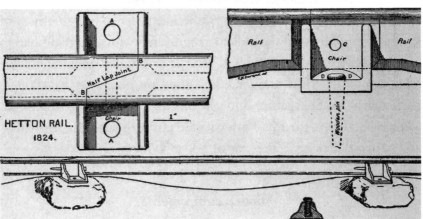

5a&b. Cast-iron plate rails came in many shapes and sizes but all had in common that the flange is on the rail, not the wheel. The Losh–Stephenson patent represented a considerable improvement on earlier cast-iron edge rails.

probably relaid with them in 1828–30.³⁵ The first public railway in France, from collieries at Saint-Étienne to Andrézieux on the River Loire, made use of them, and the cast-iron edge rails sent out to Newcastle Colliery in New South Wales in 1830, the first iron railway in Oceania, were a variation on this design.³⁶ Their last significant use was on the Cromford and High Peak Railway, completed in 1832, where they were installed by the Butterley Ironworks and by Josias Jessop (1781–1826), son of the engineer William Jessop (1745–1814).³⁷

Chairs and sills

'Chairs', also sometimes known in this period as 'saddles' or 'pedestals', the means by which cast-iron rails were fixed to wooden or stone sleepers and blocks, were themselves castings, as were sills, the combined sleeper-chairs used in some locations to bed the track in the ballast and to keep the rails to gauge. Chairs for Losh rails had to be cast to incorporate the holes for the horizontal pins. Sills were first adopted in 1794 and were also used on plateways by 1805–6, particularly in Shropshire and the north-east Wales coalfield. In some cases the crossbar had a downward curve to bury it in the ballast and to avoid getting in the way of horses' hooves.³⁸

Wooden components

The all-wooden waggonways in the north-east of England described in the previous chapter were on the verge of becoming design-expired by the 1790s, but were still being built. The Lambton, built in 1770, was the last major wooden 'way', but a branch put in to the D pit between 1789 and 1791 was laid with wooden rails.³⁹ Others continued to operate unchanged for many years to come, even in some cases into the 1840s.⁴⁰ Where traffic was light or the cost of replacement prohibitive, wooden rails remained a rational choice. As a short-term solution, George Stephenson was prepared to run locomotives on wooden rails, where their main problems were friction on curves and adhesion on any sort of gradient in wet weather.[41] The very earliest systems built in North America, from 1795 onwards, were entirely wooden.[42] So was the first railway in Australia, laid in a colliery at Newcastle in the Colony of New South Wales sometime between 1824 and 1829.[43] As noted in the Introduction, purely wooden rails lasted long and late in some circumstances. By adding a cast-iron strap along the running surface, such as Joseph Banks observed at Coalbrookdale, rolling resistance could be significantly reduced, and later wrought iron, even steel, could be used in this way.[44]

Wooden sleepers had been universal on the old generation of waggonways, and were also often to be found in conjunction with cast-iron rails and with hybrids, though stone blocks and sleepers and cast-iron sills were also widely used (see below). The illustrations to Curr's book for instance show plates slotting into wooden dovetailed sleepers, or attached to sleepers by spikes driven through holes in projecting lugs.[45]

Stone components

Stone components have been used in British railway building since records began. Broken stone had been used as ballast in conjunction with wooden rails, though other materials might also suffice, such as furnace slag, clinker or cinders.[46] An innovation of the cast-iron period was the use of shaped stone blocks to support rails instead of wooden sleepers, first noted in South Wales in 1792 and used on a plateway from Crich limestone quarries in Derbyshire to the Cromford Canal a year or two later.[47] This was also the work of Benjamin Outram, who recommended that they be 'not less than 8, nor more than 12 inches in thickness ... their shape not material, so as they have a flat bottom to rest upon, and a small portion of their upper surface level'.[48] It is no surprise that stone blocks became near universal on railways for many years to come, even after cast iron had given way to wrought iron in the 1830s. In some circumstances, stone sleepers extending under both sets of rail were to be found. Slate slabs lent themselves to this use.

Iron railways and roads developed in close association, and the modern distinction between the two would not necessarily have been drawn at the time. A number of railways from the cast-iron era were operated like a turnpike, sometimes to the extent of sharing a formation (chapter 3), and some canal acts permitted tributary railways or 'stone roads', alternative but related technologies. In 1793, for instance, the engineer William Jessop, then best known for his work on canals, considered whether the owner of Penrhyn slate quarry in North Wales should build a 'stone road on a regular declivity ... to be used with Carriages with Cast Iron Cylindrical Wheels' to the harbour. Jessop preferred a 2' 6" (762 mm) gauge plateway, on the grounds that, though a stone road would be cheaper, it would entail a cost in transhipping slate from whatever vehicles were used in the quarry itself.[49]

Other proposals were more in the nature of a technical hybrid. A fanciful scheme had been put to the Society of Arts by Richard Lovell Edgeworth (1744–1817) 'about the year 1768'. Edgeworth, an Anglo-Irishman whose wealth and social standing insulated him from having to turn his ideas into practical reality, was acquainted with Joseph Banks from a series of weekly meetings of 'men of real science and distinguished merit' in London coffee houses and had surely listened to him discuss his visit to Coalbrookdale.[50] Edgeworth proposed iron railways running along public roads for trains of four or five baggage waggons, which won him a gold medal from the Society for the Encouragement of Arts and Manufactures but no more, and thereafter he experimented with wooden or iron-strapped railways for carrying limestone and peat. Then in 1802 he came up with an idea for an experimental four-track railway to be laid on the main roads from London for a distance of 10 or 12 miles, the inner ones to carry waggons, the outer ones for wheeled platforms on which chaises, carriages and stagecoaches could be placed

'RAILS BEST ADAPTED TO THE ROAD'

and either drawn by a single horse or by fixed steam engines. Where gradients were steep, new alignments would have to be constructed.[51]

Nothing came of any of these schemes, but attempts to combine the technical advantages of the railway with the easy practicality of the road are evident in a document which the House of Commons ordered to be printed in May 1808. Its *Report from the Committee on the Highways of the Kingdom* showed how thinking about roads and railways was beginning to overlap. What exercised the committee most was the optimal width of a road vehicle wheel, as narrow ones tended to damage the surface. Evidence was heard from a number of people, including William Jessop and Edgeworth. A 'Country Gentleman' observed that a 'stone railway' required a groove to accommodate the wheel, as a raised notch in the stone would be destroyed by the wheel rim. Mr Adam Walker, of Hayes in Middlesex, advocated a 'single railway', in which one set of wheels ran on flat cast-iron plates about a foot wide, and the other ran on the ordinary road surface, until money could be found to lay a second set of plates. This would serve carts, coaches, chaises and gigs, as waggons and pack horses would, he suggested, run on the road. A raised bank by the side would serve as a flange and stop vehicles from straying. Others raised the possibility of laying cast-iron or stone railways along the sides of main roads.[52] In Scotland, cast-iron plates were laid on an ordinary road between Glasgow and Port Dundas in 1816.[53]

A number of stoneways were built in England with two sunken wheel tracks, and a path for draught animals between them. Some ran on dedicated routes. One such connected the West India Docks with London and ran along the Commercial Road. It was completed in 1830 and was 7 foot wide with a horse-path 4 foot wide in the middle, enabling it to support, and perhaps offer minimal guidance to, vehicles of very different sizes.[54] Other proposals never came to fruition. In 1833–4 there was talk of introducing stoneways on the London to Holyhead Road, again perhaps with some minimal form of guidance, for steam road coaches, following a trial observed by Telford and other distinguished engineers. Proponents also urged the laying of 'Granite Tramways, not off, but on Common Roads'.[55] Investors did not bite.[56]

Remarkably, several railways were built in which stone was used for the rails themselves. Carved stone rut-ways may have been the earliest form of railway in antiquity; they made a brief reappearance in the 1820s, with and without iron elements.[57] Three significant stone railways were built. One was in the United Kingdom, in effect a stone plateway which was built in 1820 to carry granite from quarries on Dartmoor to the Stover Canal, following the award of a contract to supply stone for the new London Bridge. It ran eight and a half miles from Haytor and Holwell quarries down to the Stover Canal, a descent of some 1,300 feet (396 m). The rails were cut from granite blocks, an upstand providing the guidance. At points, pivoted iron tongues guided the waggons.[58]

The other two were in the United States, and both used wrought-iron strips laid on the stones. The Granite Railway, so called, at Quincy, Massachusetts, connected a quarry with a river wharf from 1826, and is described in the following chapter, whereas the Baltimore and Ohio, which opened its first section in 1830, belongs with the other first-generation main lines discussed in chapter 11, though both bear witness to the way in which stone track components were regarded as a useful technology in a country where imported iron was very expensive.

Monorails

A number of monorail systems were constructed in this period, though with little success. They are more commonly associated with the late nineteenth and early twentieth centuries, and are a much-vaunted technology which has rarely proved itself to be particularly effective. Their claimed advantages are that they can run on pillars or columns without expensive earthworks and that the use of just one rail will reduce friction. In practice, maintenance is costly and most systems rely on additional guide rails. At a time when railway technology was going through such a period of experimentation, it is not surprising that monorails should have been patented, nor even that the first English patentee should have been the Bavarian Joseph von Baader, a restless and enquiring individual who had studied at Göttingen and Edinburgh, and who in 1790 had spent six months as manager of an ironworks in Wigan. He secured the patent on a return visit to England in 1815 and published a depiction of his proposed system in 1822. The vehicles were to be supported above rail level, running on vertical wheels on a 24-inch (610 mm) wide cast plate supported on piles sunk into the ground, and guided by horizontal wheels on either side of a downward flange. He claimed it could be powered by a 'nimble-footed dog', by cranks or levers or even by sail.[59]

The other type of monorail is one in which the vehicles are suspended below, like panniers on a packhorse, and was first patented in 1821 by Henry Robinson Palmer, engineer to London Docks. A single rail of timber covered by a cast- or wrought-iron plate supported a double-flanged wheel and was itself carried on pillars, which Palmer recommended be of cast iron, at intervals of 9 or 10 foot. Like Baader, Palmer promised great things for his system, claiming that it would enable a horse to pull 20 tons (20321 kg) along it, rather than 11 on an edge railway and 6 on a plateway, on the basis of trials he had conducted for Thomas Telford. He may have thought that halving the number of rails and wheels would reduce rolling resistance accordingly. Two were built in England, one to carry provisions at the Royal Victualling Yard in Deptford, and the other connecting lime-kilns at Cheshunt in Hertfordshire with the River Lea. Other proposals were entertained and some short-lived systems were built in Hungary and the USA (chapter 3) but, like the stoneway, they did not offer a better alternative to the more conventional form of cast-iron railway, still less to the wrought-iron systems which would come in their stead.[60]

Demise of cast iron

In February 1828 the American engineer Horatio Allen was given a guided tour of Liverpool docks, and noted the change in the railway systems which served them:

> ... railways are laid down, crossing the work in many directions. Upon these ways the material is taken out by means of horse power. The rails formerly made use of were of the tram kind, 3 feet long, clumsily made; at present they are using the trams where they have [sic], but all new rails are of the edge species.[61]

The 'tram kind' – plate rails – would evidently no longer serve. Nor would cast iron generally; though their design had benefitted from a culture of scientific experiment and innovation, they were still essentially a loose fit technology, with joints between each rail every few feet and serious friction problems, and they needed to be very well laid and maintained to work at full capacity. Their widespread adoption had made possible public railway systems and locomotive traction and had led to technology transfer from the United Kingdom to Europe and to the USA, but by 1825 they were no longer 'best adapted to the road' and would need to give way to newer alternatives on the main lines that were then being planned in order to ensure the next stage of development.

They by no means immediately disappeared. Some railways from this period still made do with their cast-iron rails for years to come, and either converted when they had to or reached the end of their working lives without ever having made the change (chapter 3).

Wrought iron had a higher initial cost and poorer resistance to corrosion than cast iron, but it was tougher and allowed for longer rails and a smoother ride.[62] Between its first introduction in 1808 and 1825, when it proved itself on the Stockton and Darlington Railway (chapters 6 and 9), it became evident that this would complete the change from wooden way to iron road.

3

Canal feeders, quarry railways and construction sites

... these damned tram-roads— there's mischief in them!'
(words attributed to the Duke of Bridgewater)[1]

The canals and turnpikes built in Britain and Ireland in the late eighteenth and early nineteenth centuries formed the first transport networks capable of meeting the needs of a minerals-based industrial economy. Though they ultimately found themselves subordinated to the growing mainline rail network from the 1830s onwards, in the meantime they created many feeder railways which connected them with industrial sites, and which made their own contributions to the evolving technology of the 'iron road'. These interconnected systems enabled the growth of coalfields which lacked easy access to the sea, as well as linking iron, lead and copper mines with their forges and furnaces, thereby also increasing the demand for fuel.[2] They typically made use of one form or another of the cast-iron rails described in chapter 2 as well as of technologies such as inclined planes, which are discussed here and in chapter 4. Similar systems were also devised to transport other mineral products to tidal water, or directly to a point of use – limestone from a quarry to a blast furnace or a chemical plant or for use as an agricultural fertiliser for instance – or to carry out internal movement in a manufactory, ironworks, dockyard, mine, quarry, farm or peat bog or on a construction site. Some carried general goods and passengers in the manner of a turnpike road. As cast iron came down in price, and as foundry techniques improved, it became evident that railways such as these could more expeditiously carry out tasks which had previously relied on wheelbarrows, pack animals, the carrier's horse and cart and canal boats.

Some of these railways were only a few yards long, from a wharf to a nearby warehouse or a limekiln.[3] The ones which served quarries or construction work might be laid and relaid on a regular basis. Others, however, were substantially built for long-term operation and might connect one market town with another, or might be double-track high-capacity mineral carriers. Some were purely private systems – 'industrial railways' in later parlance – whereas others were publicly funded. All together, these made up the great bulk of the railways constructed in the United Kingdom between 1767 and 1832, not only in absolute numbers but also in terms of overall route length, which eventually ran into many hundreds of miles.

As chapter 2 sets out, railways, canals and roads developed in very close association with each other. Their relative merits would be the subject of vigorous discussion for many years, not only in Britain and Ireland but also in continental Europe and in North America. The plans for road-rail hybrids set out in the previous chapter were more discussed than implemented, but canals and railways developed in very similar ways, described below. Though there was no shortage of opinions on the subject, it became clear that a single horse could pull 8 tons (8128 kg) on level iron rails and 3 tons on a wooden 'way', 50 tons by canal boat, 30 tons by river boat and 2 tons on a macadam road but only one-eighth of a ton with a pack on its back.[4] Furthermore, canals did not dry up in hot weather, as natural rivers were apt to do, and were less affected by frosts, so their advantages might seem obvious – but they were also very much more costly to build than the alternatives.

Neither canals nor turnpikes were new by the mid-eighteenth century. The Romans had built artificial waterways for irrigation and drainage as well as navigation, and many British rivers had been 'improved' since medieval times to permit the passage of industrial goods; these made up most of the 1,400 miles or so of navigable waterway which Britain possessed by 1760. The difference was the level of investment in new systems which by the end of the eighteenth century already connected all the principal estuary navigations of southern Britain and led to a 'canal mania' which lasted until 1805. When the Birmingham and Liverpool Junction Canal was completed in 1835 (the last significant expansion), the network of waterways in Britain extended to about 4,000 miles. Ireland, well endowed with lakes, rivers and watercourses, and where canal building continued longer, by mid-century had about 840 miles.[5]

Canals worked well in central England and in the Irish midlands but in hilly country they needed locks, which were slow, caused bottlenecks and used up water. Vertical lifts were quicker but required substantial and costly earthworks, and did not solve the problem of water loss if loads needed to be counterbalanced. One solution was to revive a much earlier technology, carrying canal vessels on inclined planes.[6] Other schemes involved carrying railway waggons in canal boats, and even the construction of wheeled canal barges. These are described below.

Similarly, the British road network had seen progressive improvement over hundreds of years, until by the late eighteenth century it entered on a golden age. Routes for coaching and carrying were maintained by turnpike trusts and were often state-funded. Journey times for those rich enough to afford stagecoach travel were far swifter by 1800 than they had been even 50 years before, and carts, waggons and vans were now bringing goods even to remote villages. Few railways acted as feeders to turnpikes, but some in Wales and the borders, the Sirhowy, the Grosmont and the Brecon Forest shared their formation with roads, and proposals were more plentiful still.[7]

Though edge railroads for use with flanged wheels and plate-ways using L-section rails and flangeless wheels were incompatible systems, and though they were built to many different gauges and for different purposes, these railways together form one broad recognisable type, very different in conception to the colliery 'ways' to which the north-east of England and its outposts remained faithful.

The most extensive systems were to be found in south-east Wales. Some of the longest connected ironworks and collieries at the heads of the valleys with navigable water. The most famous, because it was the scene of Trevithick's locomotive experiments in 1804, ran from Merthyr Tydfil to the Glamorganshire Canal at Abercynon, though the Rhymney, Sirhowy and Ebbw valleys were served by a much more extensive plateway network feeding the Monmouthshire Canal as well as the tidal harbour at Newport.[8] Others ran from the Brecknock and Abergavenny Canal through the Welsh border country – the Hay and the Kington to limestone quarries in Radnorshire, the Grosmont and the Hereford to the River Wye. They carried timber, corn and cider as well as coal, minerals and iron. These made up about 120 route miles.[9] Add to them the many shorter yet technically varied systems which brought limestone and coal to the furnaces themselves, the dense networks of track within each ironworks and those which operated underground, and an 1824 estimate that 500 miles of railway operated in this region becomes plausible.[10] It was the first part of the world where railways might be said to have formed the built environment through which they passed. The sudden rise of its iron industry had created a need not only for better transport links but also for housing. With little in the way of pre-existing roads and in a landscape of fields and farms, homes for miners, puddlers and rollermen and their families often had to be built along the course of the plateways through these narrow and winding valleys. Some newcomers settled into model villages provided

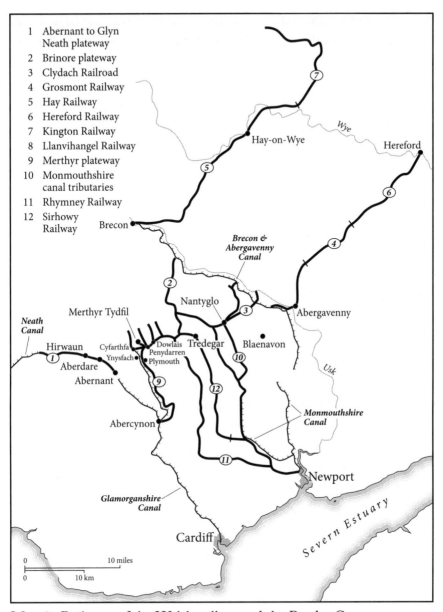

Map 2. Railways of the Welsh valleys and the Border Country.

by sympathetic industrialists, but most had no choice but to make do with squalid hovels and cabins, and to seek oblivion in lineside drinking dens.

The transport needs of this region were pressing but it also had capital and rapidly growing engineering expertise. For about 30 years, its railways were exceptionally innovative. They were the first to run over iron bridges (1793), to make use of locomotives (1804), to operate regular passenger services in purpose-built vehicles (1807) and to develop articulated rolling stock (1821). They were early in securing public funding for their role as common carriers, handling a variety of goods in the public interest, all essential preconditions for the future development of an integrated transport network.[11]

Yet this area did not become the cradle of the modern railway. That distinction belongs to the north-east of England colliery 'way', which adopted some of the technologies and the legal measures which had shown promise in South Wales, enabling Stephenson and his contemporaries to transform it into the classic nineteenth-century railway. Once the Stockton and Darlington showed the way forward in 1825, even short-haul mineral railways tended to be variations on the Stephenson system. The slate carriers of north-west Wales illustrate this change. The first significant systems here were railroads to the very narrow gauge of 2 foot (610 mm) – the Penrhyn of 1801 and the Dinorwic of 1825 – so a proposal to build a 3' 6" (1067 mm) gauge plateway from the Nantlle slate quarries to the sea at Caernarfon represented a break with regional practice. The plate rails were landed on the quay but never laid, and the system adopted represented a more significant departure from local practice. The Stephensons advised, and the railway's Liverpool bankers insisted, that wrought-iron rails and stone block sleepers be substituted, though they failed to persuade the committee of management to adopt their favoured gauge of 4' 8" (1422 mm).[12] This same track technology was adopted on the nearby Ffestiniog Railway, built between 1832 and 1836, but in conjunction with the 2' gauge, by now the slate industry 'standard', which was ultimately adopted from the 1860s onwards for a very wide variety of industrial, public and military railway systems all over the world.[13]

Funding and legislation

Some of these railways were built by individuals, partnerships or by unincorporated companies as private ventures along negotiated way leaves wherever they crossed anyone else's land, as in earlier years, but many were built by joint stock companies incorporated by Act of Parliament with right of compulsory purchase, since this carried with it limited shareholder liability.[14]

This involved preparing a private bill ('for the particular interest or benefit of any person or persons') subject to scrutiny by a select committee of Members of Parliament hearing evidence from promoters and supporters and those who were opposed. British and Irish legislators were landed gentlemen, broadly sympathetic to 'improvement' (chapter 1), and prepared to interfere with property rights through compulsory purchase if there was evident public benefit. This process was significant for the later development of the railway network.[15]

Most canal acts had clauses enabling the construction of feeder railways. The Stourbridge (1776), Dudley (1776) and the Birmingham and Fazeley (1785) canals permitted the construction of 'one or more Railway or Railways' to mineral workings within 1,000 yards of the canal, whereas later acts increased the distance to three miles or more, or laid down no limit.[16] Typically these rights were granted to the canal proprietors themselves or to the 'owners of any lands containing mines, minerals or quarries, or the proprietors, lessees or occupiers of any iron furnaces' within a specified distance of a canal, who could build 'any railways or roads ... over the lands of any other persons for the purpose of conveying the produce of their mines or quarries to the canals or railways of the company', provided that the proprietors had not, within three months of the interested parties' application, started to build that railway or road themselves. It had to be open to the public, and the promoters were limited in the charges they could levy on others using it. What these clauses did not stipulate was that the road or railway had to reach the canal at any particular location. One consequence was that the ironmasters of Merthyr Tydfil in South Wales could promote a plateway to run alongside the Glamorganshire Canal as far as an intermediate point at Abercynon, a distance of nearly 10 miles, in order to deprive their neighbour and rival Samuel Homfray of revenue.[17]

The 'Bubble Act' of 1720 had forbidden the creation of companies without royal charter or act of parliament, but a turnpike could be operated under a trust deed, and this was the arrangement under which the Lake Lock Rail Road Company in Yorkshire was organised. Formed in 1796, its 128 shares were purchased variously by a lawyer, a banker, a doctor, a clergyman, a merchant and a widow, and its three-mile route from the Outwood collieries to the Aire and Calder Navigation opened in 1798.[18] The next stage was authorising railways by their own acts of parliament, their shares traded on primary and secondary markets. The first was the Surrey Iron Railway in 1801, a plateway which began its operations down the Wandle Valley to the Thames the following year. Its share certificates showed the goddess Abundantia with a horn of plenty and surveyors' tools, holding a wheel and a plate rail, as well as a cherub pulling a waggon.[19] Shares were bought by the operators of oil and calico mills along the route, including James Perry, the radical proprietor of *The Morning Chronicle*, a newspaper in which he extolled the national possibilities of the iron road.[20] The bankers were Henry Hoare of Mitcham, of Hoare's Bank, and John Brickwood of Croydon, and the brokers were the brothers Abraham, Asher and Benjamin Goldsmid. Despite incorporation by statute, investment by local industrialists, strong publicity and the backing of an Anglo-Sephardi bullion house, it made very little money.[21]

Neither did the next statutory railway, the Carmarthenshire Rail Road. This was a plateway supposed to connect Llanelli docks with Castell y Garreg limestone quarries, but which went no further than the collieries and iron mines of Mynydd Mawr, several miles short.[22] The first public railway in Scotland was the Kilmarnock and Troon in Ayrshire, operational from 1811, a much more successful venture. All but three of the shares were held by the owner of the collieries it served, William Henry Cavendish-Scott-Bentinck, fourth Duke of Portland (1768–1854), and his daughter. Perhaps the decision to apply for incorporation by statute reflected lukewarm support from other local landowners.[23] By contrast, the social base of the Gloucester and Cheltenham's subscription list was wider, though still socially distinguished – two peers, a baronet, five clergymen (one a Doctor of Divinity), thirteen 'esquires' and four 'misters'.[24] Regional elites such as this had provided the membership of turnpike trusts but, because railways and canals were costlier, any access to London capital was helpful. Opening ceremonies demonstrated how private wealth was being ventured for the public good. Well-accoutred horses would set off with their inaugural train to the peal of church bells or a fusillade of cannon; gifts of fuel might be made to the deserving poor, and subscribers would sit down to a dinner where the king's health, the union of church and state and the success of the new enterprise were besought in claret and with much enthusiastic cheering.[25] The iron road, in other words, was becoming part of the Hanoverian world, and a conscious symbol of 'improvement' (chapter 1). The Gloucester and Cheltenham's seal represented in allegorical form the Leckhampton freestone quarries, the River Severn and the plateway itself.[26] The Monmouth Railway went so far as to adorn its share certificates with a bogus coat of arms.[27]

After 1817 Parliament became even more directly involved, when Exchequer funds became available under the Public Works Loans Act to support projects that would create employment. These included the Hay, Grosmont and Llanvihangel in the Welsh borderland, a colliery plateway at Llansamlet near Swansea and the Plymouth and Dartmoor, as well as canals, harbours, drainage, roads and bridges. Government aid nevertheless stopped well short of the central planning and public financing for navigable rivers and canals which had long been a feature of the French economy.[28]

Operation

Where such a railway served only its owning concern, operations were comparatively uncomplicated. Horses might be owned outright or belong to the journeymen and contractors responsible for day-to-day running, or be hired in by the day from a local farm. On a public system, where there was a legal obligation to transport whatever might be offered, normal practice was for those who wished to make use of it – the 'carriers' – to provide their own motive power and rolling stock, to employ hauliers, generally men but sometimes women, and to run trains as and when they saw fit, just as carts and carriages might operate on a turnpike. On a single-track system, passing loops were provided at regular visible intervals, and published rules or agreed conventions supposedly governed precedence in occupying the next section, often the first to reach an intermediate post set into the ground at a halfway point. Punch-ups between hauliers were inevitably common. On a 500-yard tunnel on the Bullo Pill plateway down to the Severn, the last team in each direction had to carry a tree branch as acknowledgment that the way was now clear and sound a horn so that those at the opposite end would know that this was so.[29]

Tolls were fixed by each railway's act, so traffic was recorded by clerks and weighbridge-men, and a few labourers would keep the track in good order. In at least one case, managing such a railway was the start of a prosperous career: German Wheatcroft (1773–1841) ran the Peak Forest Railway from the limestone quarries at Dove Holes in Derbyshire to the canal terminus from when it opened in

1794 until 1809, before establishing a carrying business serving London, Bristol, Bath and Nottingham, by road, water and rail. We will meet him again, operating a passenger service on the Cromford and High Peak Railway from 1833 (chapter 10).[30]

Engineers

The labourers, stonemasons, blacksmiths, foundrymen and carpenters who made these systems possible formed the common stock from which better-known engineers emerged, at a time when huge infrastructure projects were being undertaken across Britain and Ireland – docks, harbours, roads, bridges, canals and navigations – and when both scientific and practical skills were increasingly acknowledged and respected.[31] Thomas Telford (1757–1834) used construction railways on the Pontcysyllte Aqueduct and the Menai Suspension Bridge, and on a very significant scale indeed on the Caledonian Canal contract, across the Great Glen of Scotland, but never built a railway as a transport system in its own right. John Rennie (1761–1821), one of the leading civil engineers of the period, used plateways on construction projects at Bell Rock lighthouse off the coast of Angus, at Howth, Holyhead, Kingstown (Dún Laoghaire) and Plymouth harbours and in the building of London Docks, though he also built a railroad from limestone quarries at Caldon Low in Staffordshire to the Leek branch of the Grand Trunk Canal in 1802–4.[32] His protégé William Stuart (1773–1854) surveyed the Plymouth and Dartmoor plateway to carry granite from upland quarries to the breakwater of which he was the engineer from 1811 until his death. William Jessop (1745–1814), whose principal works were canals and docks, recommended that a proposed canal from Croydon to the Thames at Wandsworth be built as a railway, because most of the water resources of its route down the Wandle Valley were already used by industry. His views prevailed, and he constructed the Surrey Iron Railway in 1802–3. He also built the Kilmarnock and

Troon Railway, from 1808 to 1811. However, Rennie and Jessop were so much in demand that even where they were nominally responsible for a transport project, the man on the ground often had to make the decisions; it was William Cartwright, the resident engineer, who resolved on an intermediate plateway and inclined planes instead of a costly aqueduct to carry the Lancaster Canal over the Ribble Valley.[33]

Another 'man on the ground' was Roger Hopkins of Llangyfelach near Swansea; he and his father Evan and brother David not only surveyed but also built canals, roads and railways all over south-east Wales. Roger showed that he could rescue a railway project that had run into difficulties when he worked on the Severn and Wye, which may have led to his appointment as assistant on the Plymouth and Dartmoor when it became clear that it needed a thorough redesign.[34] A less proficient surveyor-engineer was George Overton who worked mostly in South Wales, though he also planned an unbuilt route for the Stockton and Darlington in 1818.[35]

Some of these men we have already encountered. Thomas Dadford Senior (1730–1809) was particularly associated with the canals of the Black Country and the 'Grand Cross' connecting the Thames, Severn, Humber and Mersey as well as with the South Wales valleys, but also surveyed the Penrhyn quarry railroad. His elder son Thomas (c. 1761–1801) worked with his father on the Stourbridge and Neath canals and the River Trent, and on his own account on the Monmouthshire Canal, to which he contributed several tributaries. The younger brother John (c. 1769–c. 1800) built the Clydach railroad.

As chapter 2 points out, a formidable influence on the development of these railways was Benjamin Outram, who began his working life as a land surveyor. He joined Jessop's team on the Cromford Canal at the age of 24, and with the company's help was able to acquire the Butterley estate, the start of his ironworking business, in partnership with Jessop and with John Wright, a Nottingham banker. This began trading in 1790. The first railway he constructed, just over a mile long, was a plateway to carry limestone from quarries at Crich to the Cromford Canal to supply the ironworks. He went on to build plateways from the Denby collieries to the Erewash Canal at Little Eaton, the Peak Forest, to serve collieries at Ashby-de-la-Zouche, and in the Forest of Dean and South Wales. He was skilled at identifying ways in which iron components could be used, not only in aqueducts but in the particular form of plate rail he devised. His business and family connections show how tight-knit the world of the engineers and savants in this age of Enlightenment had become. His father-in-law was the Scottish political economist and writer James Anderson of Hermiston (1739–1808), inventor of the Scotch plough, friend of Jeremy Bentham and (ultimately) inspiration to Karl Marx. Not only was Jessop a partner in the Butterley Ironworks, but he was connected by both business interest and marriage to the Hodgkinson family of Overton Hall in Derbyshire, themselves related to Sir Joseph Banks. John Hodgkinson (1773–1861) was working for Outram by 1796, before supervising the construction of the Ashby-de-la-Zouche rail systems three years later.[36] He did little canal work after 1808, but became engineer to the Sirhowy, to the Gloucester and Cheltenham, and then to the plateways in the Welsh border country.

Another whose background combined the scientific and the practical was William Reynolds, the Shropshire ironmaster, a Quaker on friendly terms with the Birmingham 'Lunar men' (see chapter 2). As a young man he had learnt from Professor Joseph Black, James Watt's Glasgow mentor, how to base technical decisions on the close observation of data and on experiment, and had his own laboratory and geological collections. He was the first person to apply steam power to railway operations (chapter 4), and was directly involved in the design and construction of railed canal inclined planes at Coalbrookdale, which were widely discussed and imitated in France and Prussia.[37]

An altogether less practical man was Robert Fulton (1765–1815), who also busied himself with ideas for canal inclines but never built one. His background and career were different from his contemporaries, and his influence was only indirect. He was, first of all, an American, from Pennsylvania, and he initially came to England to make a career for himself as an artist. Finding insufficient encouragement, he turned to engineering, and by 1793 was designing tub boat canals with inclined planes instead of locks, for which he obtained a patent the following year.[38] This was when he was living in Manchester, having won a contract to excavate a cutting on the Peak Forest Canal, despite having no experience at all. He had evidently talked his neighbour, the socialist Robert Owen, into financing the development not only of his ideas for inclined planes but also for a digging machine, which came to nothing. He disappeared from the canal company's records in September 1795, without having completed the work.[39] Fulton might then perhaps have considered himself fortunate in securing the patronage of the Third Earl Stanhope (1753–1816), with whom he had been in correspondence for a number of years, who not only shared his radical politics and scientific interests but was also anxious to build a canal through his extensive properties in Devon from the English Channel to the Bristol Channel.[40] The Earl sponsored the publication in 1796 of Fulton's *Treatise on the Improvement of Canal Navigation*, with its slightly eerie engravings showing tub boats raised by counterweights or by a waterwheel, in a benign but barely populated landscape.[41]

Sir John Sinclair, the President of the Board of Agriculture and Internal Improvement (chapter 1), showed interest in his proposals but stopped short of giving them political support. Portraits of Fulton suggest a man who cultivated an air of romantic detachment, which may explain both his talent for latching on to potentially helpful patrons, and the fact that none stayed with him for long.

John Brunton (1812–99) was the last significant engineer associated with any of these railways. In 1832, fresh out of his time at

Hayle Foundry in Cornwall, he was asked by his father to take on the post of resident engineer to extend a plateway through the Forest of Brecon to the Swansea Canal, a challenging position for a young man with no word of Welsh. He persevered, stayed two years and went on to make a name for himself on the London and Birmingham Railway; he then enjoyed a distinguished career building railways in India, before developing intra-urban street tramways in Milan, Karachi and Oxford.[42]

Engineering

Gradients had to be kept light. Outram recommended a gradient of no more than 1/100 for railways carrying traffic predominantly in one direction. The Penrhyn quarry railroad was built on a gradient of 1/96, and the Kilmarnock and Troon about 1/660, enabling a horse to draw at least three waggons each of 13 cwt (660 kg) and containing about 33 cwt (1676 kg) of coal at about 3 mph, and to return without difficulty with the empties.[43] Topography did not always permit such engineering, and the Peak Forest Railway had to be built with gradients as steep as 1/60, as well as a substantial inclined plane near Chapel-en-le-Frith.[44] There were limits beyond which such a railway could not be operated: Stuart's initial survey for the Plymouth and Dartmoor anticipated a gradient of 1/36, which was what led to the appointment of 'an engineer practically acquainted with railways', in the person of Roger Hopkins, to sort matters out.[45]

Hopkins' native Wales certainly offered some engineering challenges. The Brinore plateway, constructed in 1814–15 by Overton under an eight-mile clause, had to cling to the slopes of Dyffryn Crawnon to reach the Brecknock and Abergavenny Canal.[46] Even those which followed the broader Rhymney, Sirhowy and Ebbw valleys required very sharp curves. Earthworks might need to be on a significant scale – an embankment on the Carmarthenshire Railway contained 40,000 cubic yards (30,572 m^3) of spoil.[47] Bridges included a 10-arch viaduct at Blaenavon in about 1788, and the 'long bridge' at Risca, a fine red pennant sandstone double-track viaduct with 32 arches, built in 1802–5 by John Hodgkinson to enable the Sirhowy plateway to cross the Ebbw River. At Bassaleg near Newport, George Overton's 4-arch stone viaduct of 1826 still carries trains over the same river and a nearby road. Pont y Cafnau at Merthyr Tydfil is the world's oldest known iron railway bridge, built in 1793 to carry a 3-foot gauge plateway from limestone quarries to the ironworks. As originally constructed, it was on three levels, the middle and the highest carrying water channels. The designer was Watkin George, the chief engineer at Cyfarthfa Ironworks, whose initial training as a carpenter is evident in the construction.[48] At Robertstown in Aberdare there survives an iron bridge of 1811 over the River Cynon, cast at the nearby Abernant Ironworks, to connect Hirwaun Ironworks with the canal.

Civil engineering features in other parts of Britain were less striking. The viaduct built around 1800 over the River Cegin on the Penrhyn Quarry railroad, even William Jessop's viaduct of 1812 on the Kilmarnock and Troon, owe more to late eighteenth-century bridge-building technique than to the more advanced practice of the nineteenth. Josias Jessop constructed a 5-arch viaduct in 1819 for the Mansfield and Pinxton railway, which survives, though reconstructed for locomotive traffic in 1847, as a coursed squared stone structure with ashlar dressings.[49] Wooden bridges were common. Cartwright's trestle viaduct which took the Lancaster Canal plateway over the Ribble was a workaday substitute for Rennie and Jessop's intended lofty stone aqueduct, but served its purpose and lasted for 160 years.

Formations were generally single track, though there were many exceptions. The Kilmarnock and Troon was double track from the start, and was well provided with crossovers and sidings. Others were rebuilt as traffic grew – the Peak Forest Railway had been doubled by 1803, apart from where the track went through a tunnel and under a bridge.[50] By 1830 the first three miles of the Ruabon Brook system from the Llangollen Canal wharf at Trefor in north-east Wales were described as a 'double railway'.[51]

The need to keep gradients as shallow as possible on horse-worked sections meant that inclined planes were a vital component of these systems, and they also came to be incorporated into some canals in the form of railed tub boat lifts instead of locks. The first was on the Tyrone Navigation in Ireland, following a grant from the Dublin parliament to build a canal branch from Drumglass Colliery. The French-Italian engineer Davis Ducart (*c.* 1735–*c.* 1785) began work in 1768 on a surface canal where height differences were to be overcome by inclined planes ('dry hurries' or 'dry wherries' locally) with rollers fitted on the ramps. Jessop, who had been called in to advise, suggested counterbalancing the boats and installing horse gins to haul them out of the upper pound, but they could not be made to work, and the rollers were replaced by railed cradles carrying the boats. In 1777 the first boats arrived at the basin amid scenes of 'mild enthusiasm'.[52]

The first British example was built by William Reynolds on the east Shropshire network in 1788, and proved more successful. It was self-acting and double-track, connecting his Ketley Canal to his furnaces below. It appears as the background to his portrait by a fashionable artist commissioned by the South Wales ironmaster Richard Crawshay.[53] Reynolds returned the compliment by preparing a drawing of an inclined plane at Crawshay's Cyfarthfa Ironworks in 1797 which used a similar arrangement, again perhaps Ketley-inspired, in the form of a railed traveller carriage permanently attached to a rope on which another railed vehicle rather than a boat could be placed and secured.[54] Traveller inclined planes such as this came to be particularly associated with the slate industry of North Wales, but they were also extensively used on the Stanhope and Tyne Railway (see chapter 9) and they remained an effective technology for many years to come.[55]

On the Shropshire Canal itself, Reynolds also constructed tub boat inclined lifts at Wrockwardine Wood, at Windmill Farm and at the Hay, Coalport. These were fully operational and equipped with steam engines by 1793, the first use of steam to move railed vehicles anywhere in the world (see chapter 4). At nearby Trench in 1794, Reynolds built another inclined lift to connect the Shrewsbury Canal to the tub-boat system, self-acting but also with a steam engine to pull the cradles over the top of the plane. Another was built at Hugh's Bridge on a branch connecting the Lilleshall limestone quarries with the Donnington Wood Canal in 1796, again steam-powered, replacing a box hoist.[56]

Reynolds's work inspired a self-acting boat lift incorporated into the vast underground and surface canal system initially engineered by James Brindley for Francis Egerton, the Third Duke of Bridgewater (1736–1803). Since 1761 this had directly connected the coalface at Worsley in Lancashire with the towns of Salford and Manchester. However, the highest of its three underground levels had no outlet at the surface and only connected with the main canal by shafts, down which the coal had to be lowered. To replace them, work began in September 1795 on a lift driven under a stratum of sandstone, three miles from the entrance. It was completed in October 1797. An account written in French by the eccentric Eighth Earl of Bridgewater and published in 1812 ascribed the whole undertaking to the genius of the Third Duke.[57] He was certainly a hands-on industrialist to a degree uncommon in the annals of the Hanoverian peerage, chaffing his employees on Worsley bank in his shabby, snuff-stained clothes, but conception and design belonged to his able estate manager John Gilbert (1724–95), who did not live to see it completed. Gilbert owned collieries at Donington Wood in Coalbrookdale, where Reynolds operated a forge, and his inspiration came from having constructed the first part of the Shropshire tub boat canal network.[58] Robert Fulton also claimed, unconvincingly, to have worked on its construction.[59]

Both Worsley and the Shropshire Canal in turn influenced the design of lifts on the Klodnitz Canal in Silesia, the first major transfer of railway technology from Britain to another country. This was built in 1806 to transport coal from Zabrze to the ironworks at Gleiwitz (Gliwice), an important and recently developed metallurgical centre which adopted English technology and employed an English engineer, John Baildon. The Bridgewater Canal may have inspired the existing underground canal in the colliery, which it was now proposed to extend. In order to accommodate the fall in surface level to Gleiwitz, two self-acting railed lifts were designed, which initially anticipated a Shropshire Canal-type system whereby boats would be hauled out of a dock on and off a cradle, later changed to a guillotine lock system as at Worsley. A Shropshire touch was a cradle with a lower pair of wheels larger than the upper to allow it to remain horizontal as the upper lock was emptied.[60] A possible source of information was the Spanish-Canarian polymath Agustín Betancourt y Molina (1758–1824), who studied the Shropshire Canal lifts in the mid-1790s, and drew up detailed plans.[61] He grasped the possibilities they offered, above all that they did not waste water as locks did, an important consideration where the route lay across a watershed, but he did not publish his views for another 10 years, diminishing any immediate impact they might have had.[62] Vienna also sent an official delegation to Shropshire in 1795, including an engineering officer, Sebastian von Maillard (1746–1822), who already had a published description of steam power to his name, and who had been charged with an ambitious plan to design a canal from the Neustadt collieries to the capital, and even to link the Danube with the Adriatic.[63] It never completed its report, though drawings were lodged in the imperial archive.[64] Even after Waterloo, the Shropshire boat lifts still exerted a fascination: the French engineer and political economist Joseph Dutens described them in his *Mémoires sur les travaux publics de l'Angleterre* in 1819.[65]

Robert Fulton, now enjoying the protection of Earl Stanhope in London, at a prudent distance from the incomplete cutting on the Peak Forest Canal, acknowledged 'the genius of Mr. William Reynolds's when he published his own plans for a network of low-cost narrow canals in 1796. He proposed wheeled boats to negotiate both the water channel and the railed lifts, which might be self-acting, counterweighted or powered by a waterwheel.[66] William Chapman (1749–1832), the versatile civil and mechanical engineer whom we met in chapter 1, was unconvinced – or, as he tactfully put it, had 'objections to the universality of the system proposed by Mr Fulton'. He suggested either an eight-wheeled boat cradle on rails, with equalising beams to accommodate changes in gradient at the foot and summit of the lifts, or loading railway waggons into canal boats to avoid the cost of transferring goods.[67]

Fulton's career only blossomed when he went to Paris in 1797 to offer his services to the governing Directory, where he developed his interest in the submarines and steamboats for which he is best known.[68] He may have inspired an inclined lift and a vertical lift on the Canal du Creusot between 1801 and 1806.[69] In Britain, his influence was mainly felt in the West Country, perhaps reflecting continued interest shown by Earl Stanhope and his family. James Green (1781–1849) adopted his system of wheeled tub boats and railed lifts on the Bude, Torrington (Rolle), Grand Western and Kidwelly and Llanelly canals between 1823 and 1836.[70] A branch of the Tavistock Canal carried tub boats in cradles to a slate quarry.[71]

These were all short railed sections on what was otherwise a waterway. A proposed canal connecting Lord Penrhyn's slate quarries with the sea at Abercegin in north-west Wales would have adopted intermediate railed inclined planes, but it was a sign of the times that when this transport link was constructed, in 1800–1, it was a railroad throughout, using a variation on the edge rails favoured by Thomas Dadford but in which the hand of the agent of the estate, Benjamin Wyatt (1745–1818), is also evident.[72]

Ten years after Reynolds's pioneering use of steam in 1793, other systems installed fixed engines to operate inclined planes, at Penwortham and Avenham on the Lancaster Canal plateway, at Glynneath in Wales, on the construction of London Docks and on a ballast bank at

Willington Quay, on the north bank of the Tyne. These and other forms of traction are discussed in chapter 4. They were never common. A late (1832–4) example was the inclined plane at Ynysgedwyn, which gave the Brecon Forest plateway access to the Swansea Canal and was designed to be self-acting but was also equipped with a winding engine for loads which needed to be hauled against the gradient. This was built by Neath Abbey Ironworks, designed by William Brunton, builder of the locomotive which pushed itself along on legs (chapters 1 and 4) and installed by his son John.[73]

Locomotives were even more of a rarity. Trevithick's successful trials between Merthyr Tydfil and Abercynon in 1804 demonstrated their practicality, but subsequent developments took place in the north-east of England. There was some further use of locomotives in South Wales and in the Forest of Dean from 1814, as well as briefly on the Kilmarnock and Troon.[74] It was only from 1829 that the South Wales valleys saw regular use of steam locomotives, beginning with two from Robert Stephenson & Co. of Newcastle, one for Bute Ironworks, the other for Samuel George Homfray of Tredegar in the Sirhowy Valley, to haul iron to Newport. Homfray, as a little boy of eight, had probably seen the 1804 trials, and came to favour steam traction; in any event, he set his engineer, Thomas Ellis, to building further locomotives to the basic Stephenson design at Tredegar's own workshops.[75] Other industrialists and engineers in the region followed his example, including the Neath Abbey Ironworks near Swansea, which built the first of its remarkable series of locomotives for plateways and industrial lines in 1830 (chapter 8).[76]

Stables and depots

The means by which loads were moved on these and other systems is described in chapters 4 (to 1815), 7 and 8 (1815–34). Suffice it to say here that the near-universal use of horses left its mark in the provision of stables. Those on the Haytor quarries were probably typical of shorter lines, where horses were accommodated at the canal interchange, next to a forge and a public house, all of which were managed as one family business.[77] Where a railway was particularly long or was built with intermediate inclined planes, horses did not work the entire length, and stables would be provided at suitable points. The Monmouth Cap public house and its stables on the border between Wales and England was where the Grosmont Railway met the Hereford Railway and was the limit of travel for teams of horses in either direction – 10.5 miles from Abergavenny, 12 from Hereford.[78] On the Peak Forest, stables were provided at the Bugsworth Basin and at both the foot and the summit of the Chapel-en-le-Frith inclined plane, where there was also a carpenters' and blacksmiths' shop and a house for the supervisor. These were substantial buildings, constructed of local gritstone.[79] However, by far the most impressive was the Palladian quadrangle with a rail-served central yard built by Sir Josiah John Guest in 1820 for the horses at his Dowlais Ironworks, with its dense internal system. Its architecture echoed the stable block associated with a country house, and the same courtyard principle was to be found at the Cnewr depot on the Brecon Forest, built the following year, which incorporated lime sheds, wool stores and staff dwellings as well as provision for horses.[80]

Traffic

Some of these railways were purely internal systems, moving whatever was needed within an industrial undertaking or a construction site.[81] The simplest traffic pattern is exemplified by the 'cob' sea defence on the Glaslyn River in North Wales, where two railways dumped stone from quarries at each end until their embankments met in the middle of the estuary, and the united railway was then retained for maintenance purposes.[82] Others served a mineral export area, carrying a single commodity – coal, limestone or slate – in one direction to navigable water, transporting in the other direction only what was needed

CANAL FEEDERS, QUARRY RAILWAYS & CONSTRUCTION SITES

to keep a colliery, a mine or a quarry operational, such as pit props or machine parts. Furnaces in South Wales needed to export pig iron and bar iron down the valleys but also depended on railways to feed them coal, iron ore and limestone from moorland outcrops.[83]

Others developed more complex traffic patterns, requiring cranes, tipplers, ore bins and warehouses at interchange points with canals, tidal water and roads, depending on the goods they were called upon to carry.[84] The first of the several railways built to serve the Llangollen Canal initially only carried coal; a nearby pit provided the ceremonial run of waggons to the basin at Trefor on the opening day in 1805. As it was extended, it began to transport iron ore, cast- and wrought-iron products, limestone and clay.[85] The Surrey Iron Railway, opened in 1802–3, carried not only limestone, chalk, flint and fullers' earth from Merstham to the Thames at Wandsworth but also coal, and linseed for the oil mills of the Wandle Valley, in the other direction.[86] Government enclosure in 1808 of the Forest of Dean on the Gloucestershire–Monmouthshire border to secure timber for ship building led to the building of three plateways, the Severn and Wye, the Bullo Pill and the Monmouth, which carried iron ore, coal and stone to water.[87] Other royal forests were sold outright, releasing into private hands a number of upland areas with permission to build and develop railways. The Brecon Forest plateway was unusual in that traffic moved in both directions, to the Swansea Canal and to the Usk Valley turnpike – transporting limestone, wool and bituminous and anthracite coal.[88] Though the Plymouth and Dartmoor was built to carry granite blocks to the harbour, its promoter, the courtier Sir Thomas Tyrwhitt, hoped that it might also export peat, mineral ores, flax and hemp from the Forest of Dartmoor, and bring back coal, manure, timber, groceries, furniture and plants – all the essentials of 'improvement'.[89]

In April 1807 the first known public railway passenger service was inaugurated, enabling tourists to enjoy the beauties of Swansea Bay along a plateway opened the previous year to carry limestone from the Mumbles to the copper smelters, and coal and manure in the opposite direction.[90] Such services were soon found on railways in the border country, Scotland and the English West Country. Some carriers offered pleasure carriages for hire. Well-heeled people could now make railway journeys in order to enjoy attractive scenery, and humbler folk could travel by train to buy and sell – both Dr Griffiths' plateway from the Rhondda to Pontypridd and the rail section of the Somerset Coal Canal to Radstock were used by women taking farm produce to market. A proposal to transport prisoners to and from Princetown jail on the Plymouth and Dartmoor came to nothing.[91]

Rolling stock

The varied types of traffic carried by these systems led to the adoption of novel types of vehicle. The waggonway hopper-type chaldron for coal seems to have been unknown, and if anything the design ancestor of the rolling stock used on these railways were the smaller box waggons which evolved in Shropshire. Most were built of wood, with iron wheels, axles and drawbars, to carry between one and three tons. Coal, coke, quarried stone, burnt lime and manure needed enclosed sides, while slate could be carried in lath-sided crates. Where distribution relied on different transport systems over a fairly restricted area, there was, as engineers soon realised, every reason to use an early form of containerisation, whereby a box could be craned off one system and transferred to another, to avoid the cost of transhipment. Outram recommended such a scheme in 1792 to transport coal to Derby. Jessop suggested that only the section from Derby to Little Eaton be built as a canal, the remaining section from the collieries as a plateway, and that the waggons could be drawn onto the boats, then hauled through Derby on their own wheels. In the event, detachable boxes were used.[92] Outram may have been inspired by a slightly earlier system at a furnace in Ankerbold in Derbyshire.[93] The use of railed cradles to carry tub boats on canal lifts was also a form of containerisation, but Chapman's notion of carrying rail vehicles in water craft was actually put into effect by John Rennie during the construction of Plymouth Breakwater, begun in 1812, where flat waggons carrying blocks of stone were run or hoisted onto double-deck ferries, of which there were 10, each capable of carrying 20 waggons.[94] End-tipping waggons for waste rock were noted at Penrhyn slate quarry in the 1820s, and may have been introduced when the internal rail system was brought into use in 1801.[95] Thomas Telford used such waggons together with side-tippers, long-wheelbase box waggons, and flats with concave bases to build the Caledonian Canal. This massive project went on from 1805 to 1822.[96]

Carriers offering passenger services for wealthy tourists built specially designed vehicles, including long-wheelbase carriages on the Sirhowy, hauled at 6 or 7 mph, and on the Oystermouth.[97] William Chapman suggested in 1813 that 'long carriages, properly constructed, and placed on two different sets of Wheels, viz. 8 in all, may take 30 or 40 people with their articles to market'.[98] A 'market caravan' on the Plymouth and Dartmoor had fireplaces to keep passengers warm, and there was also an open carriage with an awning. The Kilmarnock and Troon had a coach called 'the Caledonia', another called 'the boat', then one variously described as 'an enormous Gypsy caravan', 'the Czar's winter sledge' and a 'Brobdingnagian diligence'.[99] On other systems, humbler passengers rode on unconverted waggons, perhaps for the price of some beer money to the haulier or some other acknowledgement, or paid a fare to travel in a coal waggon which had been brushed out, and had planks inserted to serve as seats.[100]

Another innovation was the use of articulated vehicles such as are now practically universal on railways throughout the world, equipped with 'bogie' swivelling undercarriages pivoted under the main frame at each end, enabling them to carry longer loads and negotiate sharper curves, and spreading the load on brittle track. They may have originated in the 1797 suggestion by William Chapman (see chapter 1) that railed cradles carrying barges on canal inclined planes be equipped with equalising beams to accommodate changes in gradient at summit and foot. He realised that, by enabling them to swivel on a vertical pivot instead, they could accommodate changes in lateral alignment; this was reflected in a patent he took out in 1812 (chapter 4).[101] However, it was the need to transport long loads that led to the widespread adoption of bogie vehicles. The introduction of rolling mills in the iron industry of South Wales from 1807 onwards meant that wrought-iron bars had to be taken on the tightly curved plateways down the valleys. Initially, long-wheelbase four-wheelers were used, but it became common to couple together two or three smaller vehicles equipped with a swivelling bolster, either by chain or drawbar, and to use the load itself as the frame. The next stage was Tredegar Ironworks' use of bogies permanently attached to a frame by bolts or kingpins to carry coal to Newport for export in 1821, and their wider adoption elsewhere in the region.[102]

Stone-carrying railways may have been going through a similar evolutionary process. The first train on the Haytor quarries in Devon in 1820 was, in the words of a local poet, a 'twelve-wheeled car' pulled by no less than 19 horses – unless the versifier, never having seen a train before, saw three waggons as one. The Severn and Wye Railway made provision for large blocks of stone to be carried on more than one vehicle in its byelaws of 1811. Its Bicslade branch remained in use into the 1950s, one of the last plateways to see service in Britain, running eight-wheel flatbeds in which the load was taken through iron drawbars in the bogies themselves and a connecting link between them, in which there was a limited amount of play to allow for curves. The plateway transporting stone to build Kingstown Harbour in Ireland may have been running such a vehicle in 1831.[103]

Continental Europe and North America

Continental European engineers who visited Britain went home better informed about feeder railways as well as about canal building and boat lifts. After his visit to Shropshire, Sebastian von Maillard reconfigured the Vienna–Neustadt waterway on English lines and contemplated a plateway feeder from the Semmering Pass, in the hope that this might complete the intended link between the Austrian heartlands and the sea at Trieste.[104] However, nothing came of this scheme, nor of an 1816 proposal for a 128-mile-long railway from the Zabrze coalfield in Silesia to the River Oder.[105] Railways built in emulation of British prototypes were numerous but were no more than short internal handling systems, just as colliery 'ways' were built in Prussia and the Ruhr at much the same time (chapter 1). Inclined planes were installed at Falkenberg (Sokolec) in Silesia within a blast-furnace complex by 1800, and at Bad Gastein in what was then (briefly) the Electorate of Salzburg in 1803–4, on wooden rails to lower gold ore and to haul miners to their work, as well as any particularly intrepid tourists who wished to save themselves a long climb.[106] Books and plans circulated, but there was also much toing and froing of mining and metallurgical experts between Britain, the German-speaking lands and France in this period.[107] Hanover controlled the silver mines at Clausthal and elsewhere in the Upper Harz, and since its Prince-Elector was also King of Great Britain, personnel links between the two countries were easy to facilitate. Iron rails first appeared here in the 1770s. More than anyone else, William Wilkinson (1744–1808) of Bersham Ironworks near Wrexham in north-east Wales passed on British metallurgical practice to continental Europe. In 1777 he accepted an invitation from the French Government to modernise cannon production at l'Indret near Nantes; then five years later he established the new Fonderie Royale at Le Creusot in Burgundy. He made sure that both were equipped with railways, perhaps based on those at Bersham or Coalbrookdale. He later made his way to Silesia, where he may have had a hand in designing a short colliery system from Königsgrube (Kopalnia Król) to Königshutte (Królewska Huta, Chorzów).[108] A plateway was installed underground at a colliery at Liège in 1810.[109]

Catherine the Great appointed experts from the Carron ironworks in Scotland to advise on the organisation of Alexandrovsk

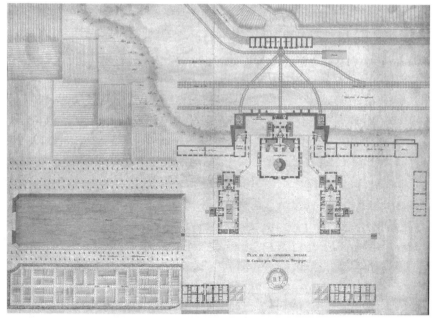

6. The ironmaster William Wilkinson installed railways at the French Fonderie Royale at Le Creusot.

cannon works in 1786.[110] Russia's well-organised and productive iron industry had made some use of wooden railways and inclined planes since the early eighteenth century, but by 1806 the iron road had even reached southern Siberia, at Zmeinogorsk ironworks.[111] The temporary and construction plateways built in Sweden (chapter 2) were meant to give service at times of year when neither snow sledges nor boats were able to move materials.[112] Two short-lived monorails were built in Hungary, one to the canal at Ödenburg (Sopron) and the other to supply Pesth with building materials.[113]

Short wooden railways had been built in the United States as early as 1795, but the first attested iron rails on the continent were in British North America, on inclined planes laid in 1823 or shortly afterwards to raise stone to extend the Cape Diamond citadel above Quebec.[114] These were built at the instigation of Major General Elias

Walker Durnford (1774–1850), who had been educated at the Royal Military Academy in Woolwich before being gazetted to the Royal Engineers, and who had built canals to the Rideau and Ottawa rivers.[115] British influence is less easy to trace in the Quincy Granite Railway (chapter 2), constructed in Massachusetts in 1826 by a civic group building the Bunker Hill monument, north of Boston.[116] It ran from Bunker Hill quarry, Quincy, some four miles to a wharf on the Neponset River at Milton, from which barges carried granite blocks to the construction site. Rails at first were of wood and then of granite, both topped by iron straps. A self-acting inclined plane was added in 1831. On the level sections, oxen and horses hauled waggons which looked like farm carts, until in 1828 its engineer Gridley Bryant (1792-1870) began building smaller four-wheelers and discovered that stones of up to 10 tons could be carried on a timber bridge spanning two or even four of these.[117] The Quincy may have been inspired by a similar system in Britain, or it may represent an emerging American design tradition.[118] By this time the first proposals were being made in the USA for longer railways, publicly funded, and it was already clear that the Stephenson design model would be adopted, or a variant of it. The plate rails used at collieries and iron mines in Alleghany County in Maryland (chapter 2) may reflect Welsh influence but were no doubt cast locally, and thought was given to building the Ithaca and Owego in New York State as a plateway.[119] Two short monorails inspired by Palmer's patent systems (chapter 3) were built in the USA, one in an amusement park, the other for an unspecified purpose. Abundant American supplies of timber for the supports might have made them seem an attractive proposition.[120]

Demise and aftermath

The north-east of England design tradition developed wooden colliery 'ways' into a unified transport system offering efficiency, capacity and ultimately even speed over long distances, and at a national level. By 1834 it was clear that it would shortly supersede Britain's canals and turnpikes and turn them into secondary networks, and this would inevitably also be the fate of their feeder railways. Furthermore, the cast-iron rails of which most of them were built, particularly plateways, were an evolutionary dead end. Yet they made their own contribution to the development of the iron road by lowering transport costs and enabling industries to develop. Above all, they established the principle that railways could be funded and regulated in the public interest, as common carriers, even in some cases to the extent of running passenger carriages. The greater variety of services they offered and goods they carried was in turn reflected in new forms of rolling stock and in experiments with locomotive traction which were adopted elsewhere.

More to the point, from about 1790 to 1825 they were simply the only option available to promoters of any overland transport system which was expected to carry a significant payload – wherever branch canals were going to be too costly, and roads would require too many horses. Since modernisation meant costly replacement of rails and

7. The Little Eaton gangway closed in 1908, little changed from when it opened in 1795.

wheel sets, as well as easing curves and updating facilities, some of these systems remained in use, largely unaltered, well into the modern railway era, especially where they served canals which still returned a profit, and the goods they carried – limestone, coal – did not call for speed. The process by which the plateways down the Rhumney, Sirhowy and Ebbw valleys to Newport in Wales were converted into Stephenson railways was long and painful, lasting from 1850 to 1865.[121] A few others lasted into the twentieth century, regarded either as anachronistic embarrassments by the concerns which had inherited them, or as picturesque survivors, but faithfully performing the service for which they were built.[122]

4

'Art has supplied the place of horses'
Traction 1767–1815

. . . art has supplied the place of horses . . .[1]

. . . machinery of the most scientific and perfect description yet invented . . .[2]

As cast-iron rails took over from wood, more effective types of traction also became possible, whether on the colliery 'ways' described in chapter 1 or the canal feeders, quarry and construction railways discussed in the last chapter. Engineers were able to make better use of animal power and of gravity but could also now devise novel forms of mechanical haulage, both fixed and self-propelling. By the time of Waterloo, and with the coming of peace, an initial experimental phase had led to one of the most remarkable technologies of the Industrial Revolution, perhaps even its symbol, the steam railway locomotive.

Animal power

Short-haul movements could be powered by human muscle. By the late eighteenth and early nineteenth centuries this often meant women and girls as well as men and boys. Female employment in British collieries and in ironstone, lead and copper mines increased significantly in this period. In these circumstances, an iron railway required less physically punishing labour than a sled running on the bare floor or a wheeled tram on a barrow-way, but it was still hard and unpleasant work. Increasingly, public opinion was becoming outraged at the way women and children were labouring in conditions where they had no educational opportunities and where they were liable to abuse, and the practice was ended in 1842.[3] Workers in other industries pushed waggons on internal systems, though human muscle power was rarely used for longer journeys. The six men who pushed four waggons from Bewicke Main pit to the staithes on the Tyne in 1809 (chapter 1), for instance, did so to demonstrate 'the excellence of the level railway', not because this was a cost-effective way of moving coal.[4]

Another way in which physical strength was put to moving railed vehicles was by operating winches on short sections, such as at the Leeds staithe on the Middleton railway, or to haul waggons off ferry boats during the construction of Plymouth Breakwater.[5]

For journeys of any length, animals of the horse family had probably hauled railway vehicles since the late sixteenth century. Their use on overland railway systems remained near universal well after practical steam locomotives became available by 1815, and persisted long into the twentieth century. It has not ended even now.[6] The colliery viewer Nicholas Wood (1796–1865) set out their advantages in his *Practical Treatise on Rail-Roads*:

> Any description of this species of power would be quite superfluous. Of all quadrupeds the horse is best adapted for use as a moving power, especially in the way that his muscular action can be employed. In dragging carriages upon a Rail-road, we can always adopt the line of draught to the direction of his muscular force, so that the greatest effect is thrown upon his line of

traction. When a horse makes an effort to drag a carriage, he bends his body forward, and throws so much of his weight upon the collar, as is required to overcome the resistance of the carriage. And the muscular force of his legs is employed to keep up this action, and to move his body forward. The effort then resolves itself into two parts – that of the action of the load, and that required to urge his own body forward. No very satisfactory experiments had yet been made to ascertain the precise rates of each, or what proportion the constant exertion which a horse was capable of bestowing upon the load, bore to his own weight.[7]

Though Wood criticised the lack of scientific evaluation, experiments had in fact taken place, as described in chapter 3, leading to a consensus that a horse could pull 8 tons (8128 kg) on an iron railway laid on absolutely level ground, compared to about 3 on wooden rails, though its capacity would be severely restricted by any sort of gradient.[8] A good horse worked most effectively at a steady pace of 2.5 to 3 mph and could draw about 100 tons for 19 or 20 miles in a nine-hour day, with one day's rest in every seven, but it might cost £40 to buy, with depreciation at £2 a year, and would need 8 shillings (40p) weekly for keep. It might work for five years but could fall ill, and would consume annually the produce of five acres of land, at a time when the cost of fodder was increasing.[9] Horses offered a reliable and well-understood form of traction. That they were slow hardly mattered; but they were also becoming very costly indeed.

How they were used depended on the type and length of railway, and the traffic it carried. On the wooden ways of the north-east of England, they had typically hauled one waggon each, occasionally in conjunction with another horse on steep sections, until the day in 1793 when the Western Way began to couple two waggons together to run in a train.[10] With iron rails on shallower or favourable gradients, trains could be made longer still, as with the eight waggons pulled by a single horse to the Tyne at South Shields in 1810 (chapter 1).[11] If a railway had a short steep section on an otherwise undemanding formation, there was the time-consuming option of uncoupling the waggons and drawing them up the slope one by one.[12]

For light loads horses worked singly, while for heavier traffic they worked in teams. The 1801 Penrhyn railroad's 24-waggon slate trains, each waggon carrying 15 cwt (762 kg), were initially hauled by two horses but a few years later it was remarked 'three are now more prudently employed'.[13] Where rolling resistance was greater, for instance on a plateway, more still might be harnessed. Four horses were used to haul 20 waggons carrying a total of 45 tons of limestone down the steeply graded Peak Forest, though the 19 horses which hauled the 'twelve wheeled car' and its granite load along the stone rails from the Haytor quarries to the Stover Canal in Devon (chapter 3) surely set a record in railway history.[14] On most systems, they were normally harnessed to the vehicles they were pulling by chains kept apart with a spreader, though the train horse might be placed in shafts on the first waggon, in order that it might be able to contribute to breaking.[15] Otherwise, band brakes could be made to apply to the broad wheel rims of edge rail systems but not narrow plateway wheels, where a bar ('sprag') was inserted between the spokes and jammed against the vehicle body.

In the course of the eighteenth century, horses and ponies also increasingly came to be used for haulage in collieries underground.[16] In 1815 Percy Main Colliery near the Tyne, where John Buddle was viewer, found work for 203 'vigorous horses', each drawing three waggons along iron rails at a brisk trot.[17] The large and technically advanced colliery at Whitehaven on the Lancashire coast required 313 horses in 1781, and still employed almost as many after it entered the iron railway age, because, although they could then be used more efficiently, output was growing and the working faces were becoming further from the shafts.[18] By 1816 its James pit had an underground stable for 60 horses.[19] There were 104 horses working on the Traeth

Mawr sea-defence 'cob' in North Wales in 1810, and so desperate was its promoter, the radical MP William Madocks, to keep the work going that he even agreed that his curricle mares and a wall-eyed pony 'might go to the waggons'.[20] On Telford's Caledonian Canal, a wooden stable was built for the railway horses which could be moved as the work progressed.[21]

Girls and women often led horses, both underground and on the surface, employment which does not seem to have perturbed respectable opinion as much as requiring them actually to propel waggons themselves did.[22]

Horses were also used to turn whims and capstans operating inclined planes, just as humans occasionally were called upon to operate winches (see above). These were used on the Penrhyn railroad (chapter 2), and it is possible they were found in quarries and on construction sites in this period where differences in height and the need to uphaul loads might have required them, though evidence is lacking. They were used as a temporary measure on the Shropshire Canal inclined planes before the steam engines were installed (see below) and on the construction railways at Cape Diamond citadel at Quebec in 1823 (chapter 3).[23] Their capacity can never have been great, but they came to be used even on quite substantial systems in the USA and the United Kingdom in later years (chapters 7 and 10).

The only other animals used to pull trains were oxen, which are slower than horses but can pull heavier loads for longer periods and eat less. They were used on some wooden waggonways in South Wales, both above and below ground, very occasionally on colliery 'ways' around Newcastle, and were permitted on some Shropshire railways.[24] Their only known use on an iron road was a plateway at Landore, near Swansea, in conjunction with a horse or two in front, in the 1820s. An ox called Hunter was long recalled in the neighbourhood for his immense strength, as he could haul 12 waggons.[25]

Gravity

Gravity is an impellent force rather than a means of traction but it was one that could be exploited by a skilful engineer. Mineral railways leading to navigable water could be built on a downward alignment, easing the work of the horses or locomotives, so long as the gradient was not problematic for the return journey with the empties. The other way in which gravity was used, as will be clear from previous chapters, was in the operation of self-acting inclined planes, which were known from at least the mid-eighteenth century in Shropshire, and which spread very rapidly in the 1790s.[26]

This technology influenced not only railway building as the iron road spread throughout the United Kingdom but also the construction of canals. William Reynolds's 1788 Ketley lift involved a cradle on rails which awaited the boat in one of a pair of locks at the summit; the gates would open and the cradle would begin its 73-foot descent, hauling up its counterpart and its boat on the parallel track. The Duke of Bridgewater's underground lift at Worsley Colliery was self-acting, 151 yards long, with two parallel locks at the summit. Plate rails were laid on a shallow gradient in the locks, steepening to 1/4 on the main incline and ending in the water of the lower canal.[27] A 2.5 inch (6.35 cm) hemp rope was used, wound on to drums 4' 11" (150 cm) diameter. A brick construction between the lines of rail supported the roof of the chamber in which the plane was built.[28]

These more considered principles were also applied to self-acting inclined planes on iron railways. Outram's 1796 inclined plane at Chapel-en-le-Frith on the Peak Forest Railway is a case in point, built on a concave or 'catenary' profile with the gradient at the top twice as steep as at the foot, to assist both initial acceleration of the 'gang' of waggons and final deceleration, and employing a continuous chain to avoid the effect of a single rope adding to the weight, and hence the velocity, of the descending load as it paid out.[29]

8. Outram's Peak Forest inclined plane followed a concave profile to assist acceleration and deceleration.

This may have been the first inclined plane to be built with such a profile, but it was soon followed by others, on the Somerset Coal Canal (1801), on the Lancaster Canal plateway (1803–4) and on the Caldon Low Railway (1804).[30] They were sometimes rebuilt in a more scientific form, such as on the Middleton Railway near Leeds, where in 1813 iron rails, and locomotive traction on the near-level sections, required that an existing rope-worked section had to be altered, and a substantial sheave mechanism installed in a brick housing.[31]

Mechanical haulage

The first machine to haul railway vehicles long predates the iron railway era. This was a waterwheel which operated a wooden mine system in Sweden by the end of the seventeenth century.[32] As chapter 3 notes, Fulton's 1796 *Treatise* showed how they could be put to operating inclined planes, but a waterwheel had already been installed to wind a short-railed lift on the Hadley Canal in Massachusetts. The proprietors were proud enough to show it on their seal, in which a boat is carried on a cradle, perhaps reflecting Shropshire influence.[33] Possibly Joseph Gainschnigg's waterwheel-powered inclined plane of 1803–4 at Bad Gastein was also based on British practice.[34]

In Britain their use was initially restricted to Devon, where the mining engineer John Taylor put in a 40-foot wheel at Wheal

9. Robert Fulton advocated using waterwheels to raise loads on inclined planes.

Friendship mine in 1807–8, and a similar one at Wheal Crebor in 1812, in both cases to wind an inclined shaft. In any region with a heavy year-round rainfall over a wide upland area, they were an eminently practical technology, and were later introduced on a significant scale in the slate industry of North Wales.[35]

However, they were only practical for very short operations and were unsuitable for places where it did not rain as much as Wales or Devon. With the rising cost of horses and fodder, the advantages of steam traction became increasingly apparent, whether fixed engines operating waggons by means of ropes or chains or 'locomotives', self-propelling machines which hauled them in trains. Steam was, in a sense, a well-established technology. A patent awarded to Thomas Savery in 1698 had applied it to raising water, and the principle of the piston and cylinder had been introduced by Thomas Newcomen in 1712, but the crucial development was the harnessing of high-pressure or 'strong' steam towards the very end of the eighteenth century, by which time perhaps as many as 3,000 steam engines had already been erected in Great Britain and Ireland, as well as others in continental Europe and in the United States. Newcomen's slow-moving, low-pressure atmospheric engines had revolutionised the operation of collieries and mines, and were even experimentally used in river boats, but for railways they were not an option. The more compact double-acting Boulton and Watt rotary engines which evolved in the 1780s could be put to winding shafts and powering watercraft, dredgers, rolling mills or agricultural machinery, but once boiler construction had been improved in the following decade, higher pressures and greater power meant that designs could be reconfigured and components reduced to sizes suitable for transport purposes. New types of engine which would all ultimately be put to railway use were being built by James Sadler (1753–1828), Adam Heslop (1759–1826), Henry Maudslay (1771–1831) and above all by Richard Trevithick (1771–1833).

Reynolds's Ketley incline had been innovative but could only haul loads uphill if there were a heavier weight descending.[36] When

Reynolds carried out a survey for a canal to connect the Oakengates collieries with the River Severn, he concluded that the wastage of water from locking would be prohibitive, and after much hesitation and a public competition, he and the other shareholders resolved on a modification of the Ketley system, but with fixed steam engines on its three inclined planes, at Donnington Wood, Windmill Farm and Hay. Instead of locks at the summit as at Ketley, reverse railed slopes were constructed into docks permanently kept in water, and the cradles were equipped with overlapping wheels which ran on ledges on the docksides to maintain them in a horizontal position. The engines were used to draw boats and cradles out of the docks and to haul up the main incline if necessary. All three were built to a hybrid design by Reynold's protégé, Adam Heslop.[37] These were the first locations in the world where railed vehicles were moved by steam. They were operational by 1793.[38]

These were, of course, only short railed sections on what was otherwise an artificial waterway. So in a sense was the next location where railway vehicles were moved by steam: a plateway across the Ribble Valley, at Walton Summit, completed in 1803 to connect the northern and southern ends of the Lancaster Canal. Its three inclined planes were each equipped with a high-pressure 6-horsepower 13-inch cylinder engine costing £350 and made by Summerfield and Atkinson, a local foundry which offered 'patent steam engines', and which also built the waggons. The first was installed in May of that year.[39] In June, a 6-horsepower steam engine was installed on a plateway incline to haul spoil on the construction of London Docks.[40]

It was a fixed steam engine which provided George Stephenson's introduction to railway work. For a while after his marriage in November 1802 he operated the engine at the foot of the inclined plane which hauled waggons to the summit of the ballast bank at Willington Quay, on the north bank of the Tyne, about six miles below Newcastle, performing the humdrum but vital task of keeping England's main artery for fossil fuel free for trade. This huge mound was formed by the accumulation of earth, chalk and Thames mud carried in the holds of collier ships. He would have been an employee of the Corporation of Newcastle. Nothing is known of the engine itself, through by now foundries and workshops were springing up along the Tyne with the capacity to build such a machine.[41]

Wales followed shortly afterwards. The Abernant Ironworks supplied an engine at a cost of £775 for a plateway at Glynneath which became fully operational in November 1805, connecting it and the Aberdare Ironworks with Neath Canal.[42] It was built to a patent taken out by Richard Trevithick, the most able of a talented group of Cornish engineers who had already made a significant contribution to steam power. His native region's tin and copper mines needed pumping and winding machinery to reach ever greater depths, and now sustained a network of foundries and smithies which made it less dependent on established manufacturers in Birmingham and Shropshire than it had been. With his cousin the mine manager and financier Andrew Vivian (1759–1842), Trevithick patented the high-pressure steam engine for stationary and locomotive use in 1802. He came to Wales at the bidding of Samuel Homfray of Penydarren Ironworks the following year, a visit which led to the construction of the first operational railway locomotive, discussed below. So it was not surprising that Samuel's brother Jeremiah, a partner in Abernant Ironworks, should have opted for a Trevithick design for Glynneath.

It was another few years before fixed steam engines came to be found on railways in the north-east of England.[43] They appeared where wooden waggonways were either being rebuilt with iron rails or were giving way to completely new systems – where once there would have been a 'pull' against the gradient for a horse and its one chaldron, a fixed engine was now required. The first, perhaps a single vertical cylinder condensing beam engine, appeared around 1805 on a waggonway from Bewicke Main (Urpeth) Colliery to the Wear at Fatfield, which involved a haul against the gradient up the slopes of Black Fell.[44] This was realigned to the Tyne in 1809, which was when the alleged crowd

'ART HAS SUPPLIED THE PLACE OF HORSES'

of 10,000 came to witness its completion (chapter 1), though it is not clear if a new engine was installed.[45]

Fixed engines and iron rails transformed colliery 'ways' into higher capacity systems, and ushered in an era of 'steam and rope' in the north-east of England which lasted until the mid-twentieth century.[46] Locomotives made less of an immediate impact; nevertheless, between 1802 and 1815, they evolved from an experimental technology into a practical and cost-effective means of moving a payload along an iron railway – in some limited circumstances.

The challenge of constructing a self-propelling machine had long appealed to visionaries and to enthusiasts. Richard Lovell Edgeworth (chapters 2 and 3) had conceived of the notion in the 1760s when he urged the building of a national iron road, and had discussed it with his 'Lunar' friends in Birmingham.[47] Edgeworth always brimmed with ideas, but he was far from being a practical man, and initial experiments were carried out by Trevithick's Scottish next-door neighbour in Redruth, William Murdoch (1754–1839), who was employed by Boulton and Watt as their engine erector in Cornwall. In 1784 he constructed a model steam carriage which did no more than run around his living room floor, but it is the first recorded example in Great Britain of a machine moving around completely under its own power. It followed the steam *fardier* ('dray') to haul artillery built by the army officer Nicolas-Joseph Cugnot (1725–1804) in Paris in 1769–70. Whether Edgeworth, Trevithick and Murdoch were aware of the French experiments is not clear, but in 1802 Trevithick took out with Vivian a patent for the 'Construction of steam engines to drive steam carriages and other purposes'.

He had by then already built a road locomotive, which he named *Puffing Devil*. Its first outing was on Christmas Eve 1801, when it successfully carried several men up Camborne Hill in Cornwall and then on to the nearby village of Beacon, but three days later it broke down after passing over a gully in the road. Abandoned with a fire still in it while the crew repaired to a public house for a roast goose and drinks, it overheated and was destroyed. A London road carriage was hardly more successful in 1803: with Trevithick firing, Vivian steered carelessly, crashing it into a paling fence 'to the great annoyance of the owner'. The engine was sold to operate a hoop-rolling mill.[48]

The first attempt to build a railway locomotive was at Coalbrookdale in Shropshire, and once again the hand of William Reynolds is evident. In August 1802 Trevithick recorded that 'the Dale Co have begun a carriage at their own cost for the realroads [*sic*]'.[49] The design was probably his, but there is nothing to suggest that he supervised its construction. On 2 May 1803 he added that 'there has been no further trial at the Dale'.[50] It may have been abandoned because Reynolds died on 10 June at the age of 45 after several months of illness. Fears about high-pressure steam and the dangers of a boiler explosion may also have been factors – on 8 September of that year a Trevithick pumping engine exploded at Greenwich, killing four men. It is probable that the locomotive was assembled, at least in part, but there is nothing to indicate that it ever ran.[51] Like *Puffing Devil*, it had a vertical cylinder, which measured 4 inches by 3 foot (10 x 91 cm).[52]

A well-known drawing in the Science Museum, the 'Tram Engine', dated December 1803, which shows a 3-foot gauge locomotive, was for long taken to represent Reynold's Coalbrookdale locomotive, but it is far more likely that this was a design for a yard shunter for the Tredegar Ironworks in South Wales, part of Samuel Homfray's empire. Tredegar's furnaces were newly in blast, their feeder plateways either under construction or already operational, and Homfray may well have thought that they needed a locomotive to keep the supplies of coal and limestone moving. There is no evidence that it was ever built, but at Penydarren Ironworks, Rees Jones the fitter (1776–1862) was already putting together a locomotive for Homfray, under Trevithick's nominal supervision.

It was this machine which performed the first attested run of a steam railway locomotive, on 21 February 1804, when it hauled a loaded train along the 9.5 mile plateway built a year or two before from Merthyr Tydfil to the Glamorgan Canal at Abercynon.[53] Samuel Homfray's neighbour and rival, Richard Crawshay of Cyfarthfa Ironworks, did not believe that it could do the same work as the horses, so a wager of 500 guineas was placed with Richard Hill of the neighbouring Plymouth Ironworks. Homfray was an aggressive individual who had fallen out with Crawshay before, over the Glamorganshire Canal and, it was rumoured, for dishonouring his daughter. The sum wagered was considerable. King George III's subjects were happy to gamble on more or less anything, so it was no surprise to find a locomotive being made the subject of a bet. But its operations also constituted a scientific trial: observers included the Cornish engineer and mathematician Davies Giddy, who was a future president of the Royal Society, and George Overton, who had engineered the plateway to Abercynon.

The locomotive's exact form is unknown, despite many depictions over the years of its supposed appearance. It worked by the adhesion of wheels on the plate rails. Only a horizontal cylinder would have enabled it to pass through the tunnel at Plymouth Ironworks, and this was probably set into the boiler, like the road carriage Trevithick had built in London the previous year. It probably operated the four wheels by gearing from a trombone-style piston, slide valve and connecting rod. Though it may have had a flywheel, this cannot have been large, and its tall chimney would have been hinged, again because of the tunnel confines. As well as the smoke, the chimney carried the exhaust from the cylinder in order to excite the fire, an arrangement which would become important with the development of multitube boilers a generation later.[54] Reports mention its tractability, suggesting that it was built with the piston and crosshead at the opposite end from the fireplace, which would make it easier to operate. Trevithick made the entire 19-mile return journey with the locomotive on his own at one point, which would have required him to fire it, adjust the damper, maintain the water level, open and shut the regulator and adjust the valve gear, tasks which would have been much more difficult if he was also having to dodge the piston and rods. It weighed about 4½ tons empty, 5 tons with water in the boiler, which was probably cast iron. It is likely that its fuel and water were carried in a dedicated vehicle, a 'tender', coupled to it, which became a common feature of locomotive technology.

Whether the bet was ever honoured we do not know, but the operation of the locomotive was on the whole a remarkable success. It proved quite capable of hauling a load over a considerable distance, one not exceeded for more than 20 years; it coped with a gradient, and could even reach a speed of 16 mph. Trevithick was delighted, writing to Davis Giddy: 'The public until now called me a scheming fellow, but now their tone is much altered'.[55] For the rest of his life, he harked back to the Penydarren trials as evidence that his design had proved itself. However, it had a fundamental weakness, and this was the friction it produced on the cast-iron plate rails, and its tendency to break them. By July 1804 it had been put to driving a forge hammer, and did not run on rails again thereafter. It may well have been Trevithick's intention to show it off as a multipurpose machine capable of demonstrating the various uses of high-pressure steam, even if primarily intended as a locomotive. It served out its days as a winding engine on an inclined plane, before being broken up as late as 1859.[56]

An account of the trials appeared in the *Cambrian* newspaper on 25 February 1804, tucked away after a description of how some new and powerful Trevithick fixed engines had been installed at a colliery in Llanelli and demonstrated in London; this was copied in several London and provincial papers over the next few weeks.[57] The article refers to the 'long expected' trial witnessed by many hundreds drawn by 'invincible curiosity', and prophesied that 'the use of horses in the kingdom will be very considerably reduced, and the machine, in the hands of present proprietors, will be made use of in a thousand instances never yet thought of for an engine'. This also found its way into the March edition of the *Universal Magazine*, a London monthly which published a wide range of material and which had a wide and informed readership.[58] The *Royal Cornwall Gazette* was less sanguine about the future of the 'steam "engine" cart' ('It is somewhat premature to form any conjectures to which so useful a machine may be applied') but was 'happy to be enabled to give a correct account of our countryman's success in this new and wonderful discovery, especially as we understand it has met with opposition from many rival mechanics'.[59] In February 1806 Rees's *Cyclopaedia* briefly alluded to the trials in its article on canals.[60] *Nicholson's Journal*, the first monthly scientific journal in Great Britain, published an article by Davies Giddy on the effect of a steam exhaust on a fire and referred to the Penydarren locomotive. Both these publications were widely read and respected, particularly by individuals whose taste ran to philosophic radicalism and scientific reason. Giddy's article was reprinted in French in Geneva and Brussels, in German in Saxony and Bavaria, and in the USA in 1813.[61]

Just possibly, the Penydarren locomotive was not the only one in South Wales at this time. Whether or not the 'Tram engine' was constructed, and whether this was indeed intended for Tredegar, later traditions refer to a Trevithick locomotive running along the plateway from Aberdare to Hirwaun Ironworks, perhaps a misremembered or garbled reference either to events in nearby Merthyr or to the Trevithick fixed engine which wound the Glynneath inclined plane on this system, but significantly these locations also formed part of the Homfray empire. A locomotive suggestion which fell on deaf ears came from Edward Martin, an engineer associated with a proposed transport link between Swansea and Oystermouth, who wrote on 6 March 1804 to the *Cambrian* that if 'Mr Trevethick's very ingenious machine is brought to that perfection, which some persons are confident it will be, and which, from the very liberal patronage it has received from Mr Homfray, of Pendarran, there is every reason to hope for', then it would be wiser to build it as a railway than a canal. A plateway was duly built, but for horse traction.[62] Other possible Trevithick locomotives from this period were a machine for the West India Docks in London to pump water for firefighting, and act as a steam crane, and one for Plymouth Breakwater which could break rock and work a crane. Both were intended to be self-propelling but it is not clear if either was intended to run on rails, or was ever built.[63]

Trevithick's locomotive ideas were, however, soon taken up in the north-east of England. In January 1805 he noted 'I expect there are some of the travelling engines at work in Newcastle'. Whether this was true, and what hand he had in them, remains unknown; however, a document dated 3 October 1804 which emerged many years later at the Gateshead Park Ironworks, on the opposite bank of the Tyne, shows a 'Waggon Engine', a locomotive for a 5-foot gauge wooden waggonway with a 7-foot flywheel and a horizontal cylinder set at the opposite end to the chimney. The same source also yielded undated elevations and a plan for a similar locomotive but with a shorter boiler, larger wheels and gearing of 3:2, also intended for a 5-foot gauge 'way'.

One locomotive at least was indeed built at Gateshead at this time, and was the work of Trevithick's agent, John Whinfield, with the assistance or supervision of John Steel, who may have previously worked at Penydarren. According to Trevithick's son, 'a temporary way was laid down at the works "to let the quality see her run"', but no buyer was found, and the engine spent the next 55 years powering the foundry blower. The 5-foot gauge strongly suggests that the plans were prepared, and the locomotive built, to interest Christopher Blackett, the owner of Wylam Colliery, but, if so, it was rejected as unsuitable. Blackett's caution is understandable, as a locomotive would have been of no use on the Wylam's wooden rails, but the idea remained with him.[64]

The final Trevithick patent locomotive was *Catch Me Who Can*, constructed in 1808 by Rastrick and Hazledine of Bridgnorth in

10. Trevithick's *Catch Me Who Can* was the first named steam locomotive.

Shropshire, who had built several stationary and dredging engines for him, and who were working on his abortive attempt to build a tunnel under the Thames. His original idea was to run it at Newmarket races 'against any mare, horse or gelding that may be produced at the next October meeting' for a wager of £10,000, claiming that it would attain a speed of 20 mph. One again, he placed his trust in both scientific evaluation and high-rolling gamblers, though the stakes were now nearly 20 times higher than at Penydarren.

In the meantime, he was persuaded 'by several distinguished personages' to offer it as a public circus in London.[65] It was demonstrated between 8 July and 18 September 1808 on a specially created showground on what is now the site of University College, the first railway locomotive to operate in the capital, and the first one to be given a name. Its design derived from that of his steam dredger, a horizontally mounted boiler on four wheels with a vertical cylinder, piston and connecting rod driving one pair of the wheels. Trevithick charged one shilling for entry to the circular track within a paling fence on which the 'racing steam engine' pulled a passenger carriage. This was built over soft ground, and heavy rain caused subsidence in the rails. They broke and the locomotive became derailed. Public interest waned and very soon the 'steam circus' came to an end. The 'extraordinary wager ... so novel in the sporting world' was scheduled for 21 and 22 September but did not take place, and *Catch Me Who Can* never went to Newmarket.[66]

That was the end of the matter so far as Trevithick was concerned. When Blackett raised the matter of a locomotive again in 1809, after rebuilding the Wylam waggonway with plate rails, Trevithick loftily replied 'that he was engaged in other pursuits, and having declined the business, he could render no assistance'.[67] Trevithick did indeed develop other interests over the next few years, with little success, then spent from 1816 to 1827 in Peru, as a mining engineer and consultant. A touchy and combative individual to the last, he lacked the business and personal skills that George and Robert Stephenson had in abundance which would have ensured take-up and general acceptance of the technologies he brought into being. Though he made the locomotive a reality, ultimately all he had was a patent, with no foundry or manufactory behind him. When he died in 1833, on the brink of the railway age, he had played no part in locomotive development for 25 years. Yet in later life he always harked back to Penydarren as proof that such machines could function effectively, and that a smooth wheel could indeed grip a smooth rail and pull a load.

So they could; but relying on a combination of friction and weight to start a train was problematic on brittle cast-iron rails, and over the next few years other engineers experimented with nonadhesion systems. These included cogged rails and wheels, an on-board steam winch which hauled itself along a chain and machines that pushed themselves along on legs.

The first of these was moderately successful. On 1 October 1808 the 25-year-old John Blenkinsop took over management of the Middleton estate just south of Leeds on behalf of its owner, Charles John Brandling, where the collieries had been connected to staithes in the town by a waggonway since the middle of the previous century. Some time before he arrived, changes had already been made, involving the substitution of cast iron for wooden rails and of a self-acting inclined plane for the 'run' halfway along its length. Faced with the rising cost of horse traction, Blenkinsop concluded that a locomotive was needed, and grasped that the problem lay in constructing one strong enough to pull a commercial load along the short level sections without damaging the track. In 1811 he secured a patent for a rack rail and wheel, in which a cog engaged with projections on the side of one rail.[68]

The basic design was the work of Matthew Murray, a Tynesider like both Brandling and Blenkinsop, who had established a foundry at Holbeck, on the outskirts of Leeds. Detailed descriptions and diagrams published in the Parisian *Bulletin de la Société d'Encouragement pour l'Industrie Nationale* in April 1815 confirm an oval boiler with a flue passing through it, with the fire laid on a grid at one end and the gases passing to a chimney at the other. Two four-way plug cocks, coupled by a rod, controlled the steam supply from the boiler to the two vertical inset cylinders, which exhausted into the atmosphere through a central pipe. Each piston rod was controlled by two vertical guides, and power was transferred by connecting rods and crankshafts to the cogged wheel, on the left-hand side of the engine. The boiler was cast in two halves, each incorporating a cylinder set into the top, and bolted together across their central flange, effectively two Trevithick-types combined.[69]

On 24 June 1812 it was loaded at the foundry, carted to the coal staithes, put in steam and driven to the top of Hunslet Moor, a distance of 1.5 miles, returning with 14 waggons loaded with coal and spectators, the first time that a locomotive was put to a train as part of an ordinary operation, not as a trial or speculative display. It was soon named *Salamanca* after the Duke of Wellington's recent victorious battle but in Leeds it was known as the 'machine'. Others followed: the *Prince Regent*, *Marquis Wellesley* and *Lord Wellington*.[70]

The first Middleton 'machine' was very quickly emulated at Orrell Colliery in Lancashire. The *Walking Horse* was built at the Haigh Foundry in Wigan by the engineer Robert Daglish (*c.* 1779–1865), who had trained as a coal viewer in the north-east and then at Leeds and had managed the foundry on behalf of the Earl of Balcarres since 1804, producing pumping, winding and blast engines to a high standard. From January 1813 his locomotive was pulling waggons to the Leeds and Liverpool Canal, saving his employers £500 a year compared to the cost of horses. Two others followed.[71]

On 2 September 1813 another such locomotive was set to work in the north-east of England, on the railway connecting the Kenton and Coxlodge collieries to the Tyne at Wallsend, where the viewer, John Watson, knew Blenkinsop. A second locomotive was at work here by late 1814, but, as chapter 1 describes, their operation became a matter of disagreement; a new agent was condemned not only for being 'no judge of hay', and inflicting distemper on the horses, but also for redeploying the locomotives on a previously horse-worked section where the gradient went against the load (which presumably also had to be relaid), and putting inexperienced men in charge of them. By May 1815, they were out of use and disappeared from the

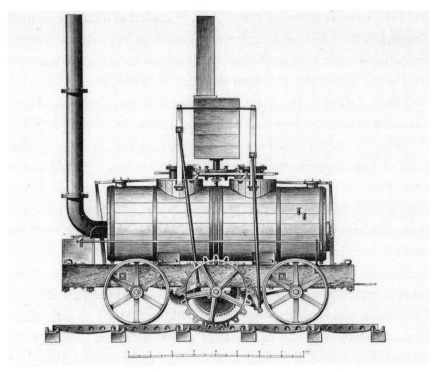

11. *Salamanca* was the first locomotive ever put to ordinary revenue-earning service, in 1812.

record. The steep section was converted to cable haulage, other realignments were put in place and a third, more powerful locomotive purchased – changes which emphasise that the evolution of the iron road was not simply a question of substituting a newer and better technology for an old one so much as a constant re-evaluation of both innovative and established elements.[72] Another locomotive was built, perhaps in-house, in around 1814 to haul pig iron from the Nantyglo furnaces to the Brecon and Abergavenny Canal in South Wales.[73]

Locomotives of this type were the first to appear in continental Europe. One operated between Murébure Colliery at Horloz, near Liège, to the River Meuse.[74] According to one source, it was put to work in 1814, another suggests 1817.[75] It may therefore predate the description and diagrams of the Leeds 'machine' published in Paris. In

1814–15, two Prussian engineers, C.H. Eckardt and Johann Friedrich Krigar, saw the Leeds and the Kenton–Coxlodge locomotives at work. Immediately on returning, Krigar set about constructing two locomotives at the Royal Prussian Iron Foundry in Berlin, one to transport coal from Zabrze to the ironworks at Königshütte (Królewska Huta, Chorzów) in Silesia, one for the Saarland collieries. On 9 July 1816 the Silesian engine was paraded in a park in the Prussian capital before being dispatched. On arrival on 23 October, it turned out to have been built to the wrong gauge – to three Prussian feet, rather than to three Silesian feet.[76] Workmanship was found to be poor and it was underpowered. The cylinders were replaced, but the machine only ever earned its keep powering a pump. The other fared no better. Again, indifferent workmanship became evident; it could barely be made to move when it was trialled on a railway to the Saar River at Bauerwald, and the parts were finally sold in 1837. The centralised Prussian state was not as effective as experienced Tynesiders in delivering a workable steam locomotive; neither could Berlin as yet challenge Holbeck or Wigan in quality control and technical proficiency.[77]

Though there were variations between them, the Murray–Blenkinsop designs may almost be called the first 'class' of locomotives to be constructed. When their workmanship was sound, they were clearly very effective, and some of them led long and useful lives. Accidents took place on the Middleton, some of them gruesome; one claimed the life of a 13-year old boy, John Bruce, in January 1813. He had been in the habit of running alongside the waggons, when he tripped and fell under them, making him the first person to be killed by a locomotive-hauled train. Then 1818 saw the death of the first railwayman to be killed by a locomotive: George Hutchinson's body was blown from the footplate of *Salamanca* when the boiler exploded, to be found in a field 100 yards away. He had been screwing down the safety valves to operate the locomotive well above what Blenkinsop considered to be the maximum safe working pressure of 55 PSI.

Another boiler explosion took place on 12 February 1834, when James Hewitt, who had driven the locomotives from the beginning, was 'literally blown to atoms'. By this time the old 'machines' were nearing the end of their lives. Traffic was falling off, and by the following year the first railway in the world to operate steam locomotives commercially had replaced them with a stationary engine.[78] The Orrell machines lasted longer, until about 1850.[79]

Locomotives hauling themselves along a chain proved not to be an option. They were the brainchild of William Chapman (chapters 1 and 3), who believed they might offer an inexpensive mechanical system whereby 'no alteration is required in the waggonways, whether of wood or iron'. The patent he took out with his brother Edward, with financial help from Buddle, specified a powered drum operating on a chain secured at each end of the railway, as well as articulated subframes for the wheels in order to spread the load and minimise damage to the rails, a development of his earlier ideas for railed canal cradles.[80]

Three locomotives were built under this patent but only one made use of both the drum-and-chain system and the articulated wheel sets. This was constructed by the Butterley Foundry in Derbyshire for the Heaton Colliery waggonway to the Tyne in 1813. A later source claims that it only worked for a few weeks in October of that year, that the chain broke constantly, and that it was a costly failure.[81] The next two were both adhesion locomotives, eight-wheelers on subframes, one of which was trialled on the Lambton waggonway on 21 December 1814, the other built for Whitehaven Colliery in Lancashire in 1816. Phineas Crowther of the Ouseburn Foundry in Newcastle built the Lambton locomotive and supplied components to Whitehaven, other components being produced locally by Taylor Swainson (1761–1839), the colliery's brilliant but alcoholic engineer. In 1818 it was adapted to wind an inclined plane and to operate a pump.[82] Chapman may also have inspired some unsuccessful trials on the Severn and Wye and on the Sirhowy between 1814 and 1816.[83] These were not the end of Chapman's career as a locomotive designer, and his use of articulation is now widespread, but he seems wisely to have abandoned chain haulage in favour of adhesion early on.

Two locomotives were constructed between 1813 and 1814 which pushed themselves on legs. These were the work of William Brunton (1777–1851), a Scotsman, the son of a watch and clock maker, who came to work as senior engineer at Butterley from Boulton and Watt's Soho manufactory in Birmingham in 1808. In 1813 he constructed a locomotive as a private venture which was trialled on the railway at the nearby Crich limestone quarry, which formed part of the Butterley empire. The locomotive 'may then be inspected by any gentleman desirous of seeing it'. It performed satisfactorily, but thereafter disappears from the record. He built a larger two-cylinder version for Nesham's Newbottle waggonway in the north-east of England the following year, which operated successfully up a gradient of 1/36. Brunton was permitted to enlarge the boiler, as initial concerns about weight had proved groundless, but it exploded on its first working day on 31 July 1815, killing over a dozen people and scalding and maiming many more, the first multiple fatality railway accident, and one which followed a recent underground disaster at one of Nesham's pits (chapter 1).[84] Brunton's reputation was severely damaged, and he remained deeply distressed by what had happened for the rest of his life, though he went on to enjoy a distinguished career.[85]

Wylam Colliery

Christopher Blackett, the owner of Wylam Colliery, had rejected the proffered Gateshead locomotive in 1805, but was clearly minded to improve his rail link to the staithes at Lemington on the Tyne, and to that end rebuilt it as a plateway in 1808. This opened up the possibility of locomotive traction and, as we have seen, he raised the matter with

Trevithick, only to receive a frosty reply. Blackett, however, was fortunate in that he had on his payroll several remarkable individuals who were to make the adhesion railway locomotive a practical technology. These were William Hedley (1779–1843, the viewer), Timothy Hackworth (1786–1850, the foreman blacksmith), and Thomas Waters of Gateshead.

Hedley had been appointed viewer in 1805 when he may have seen the Gateshead locomotive.[86] It was he who supervised the rebuilding of the waggonway. Hackworth had been born at Wylam and succeeded his father once he had completed his apprenticeship. Waters had been 'well acquainted' with the Gateshead locomotive.[87] Together they began a series of experiments with a man-powered test carriage to ascertain the practicality of smooth wheels gripping smooth cast-iron plates; this led to a patent, and the subsequent modification of the carriage to accommodate a boiler and single cylinder with flywheel, resulting in a locomotive known as *Black Billy*. An account of its trial run captures the mixture of fear and wonderment that these machines inspired.

> On its first trip, in returning after accomplishing about half the distance, it turned restive, declining to move backwards or forwards. The alternative resorted to by its bold and heroic constructor, Tom Waters, to surmount the difficulty was extra weighting of the safety valve and a peremptory order to the fireman to 'fire up'. Whereupon the group of onlookers, leaving the others to their fate, retired to a safe distance to await expected explosion. However, the engine made a further advance, and after a similar act of daring at 'Street House', Waters brought his locomotive successfully through its first trip, landing safely at the starting point, Wylam.[88]

Trials took place from four o' clock in the morning, before regular work began, and on Saturdays.[89] On balance it seems likely that the test carriage was under construction by November 1812, that *Black Billy* was being assembled in March 1813 and was operational by early 1814, and that two-cylinder engines were working by the following year. It is not clear how many there were of these, but it was probably three.[90] One locomotive was named *Lady Mary*, but it now seems clear that this is the one that has always popularly been known as *Puffing Billy*, and that the unofficial name has taken precedence.[91] All the Wylam locomotives were referred to as 'dillies', a contraction of 'diligence', a horse-drawn road vehicle, so not surprisingly another became simply *Wylam Dilly*. The colliery 'caller' would go around Wylam waking up the engine men with the cry of 'Get up to the dilly!'[92]

The two vertical cylinders on *Puffing Billy* and *Wylam Dilly* imparted motion to the driving wheels by 'grasshopper' beams pivoted on top of the boiler, which gave an insect-like appearance in operation. Little is known about the initial form these locomotives took. A sketch of 1815, and hence the earliest representation, shows a six-wheeler, coupled to a single-axle tender carrying a water barrel.[93] By 1825, when Nicholas Woods depicted them, they had become tall, ungainly machines, running on two Chapman-type subframes, both of which carried four wheels. Perhaps the original three-axle configuration had damaged the track. At any rate, they operated as plateway locomotives until 1828–30, when edge rails were substituted, which is when they probably assumed the form the two survivors retained until they were taken out of service in the 1860s. One was preserved in the Science Museum and the other in the National Museum of Scotland.

Long-lived and comparatively successful though *Puffing Billy* and *Wylam Dilly* were, the design path for the steam locomotive only became evident after George Stephenson was appointed enginewright in 1812 at Killingworth Colliery, north of Newcastle. How this happened is for a later chapter, but Stephenson's new home was barely six miles away from Coxlodge, where the Murray–Blenkinsops were put to work, and about 15 from Wylam. His first locomotive for the Killingworth Colliery railway in 1814, popularly but unofficially known as *Blücher*, followed the Blenkinsop design in having two vertical cylinders set in line with the boiler top, and the Hedley–Hackworth design in being built for adhesion running.[94] Over the next 11 years George Stephenson set to work with a tight-knit group of blacksmiths, machinists, engine fitters and foundrymen at Killingworth to develop this initial design. Cornwall, Shropshire, South Wales, Lancashire and Yorkshire had all contributed to steam locomotion, but the initiative now lay with the north-east of England.

'... a great improvement upon the horse for drawing coals'[95]

By the time of Waterloo, fixed engines winding railed vehicles by rope, cable or chain on inclined planes were an accepted technology. By contrast, few locomotives were yet at work in revenue service – maybe four Blenkinsops at Leeds, three on the Kenton and Coxlodge, three others at Orrell, one at Nantyglo, possibly one at Murébure, as well as perhaps three Hedleys at Wylam. George Stephenson went on to build more in the 'Killingworth' series after *Blücher* (chapter 8). Eleven years of experiment had established that adhesion working by steam locomotive was practical and that it was indeed 'a great improvement upon the horse for drawing coals', above all because it was cheaper. However, a fundamental problem remained: existing track-systems were unsuitable, since rails strong enough for locomotives were prohibitively expensive. Peace in Europe and North America would soon make the iron road a matter of interest not only to the 'lords of coal' and to ironmasters but also to the merchants and landowners who were looking for regional and national systems, and who would ultimately require quicker turn-rounds and hence faster speeds. In these circumstances, mechanical traction would have to be further refined, and harmonised with new forms of track.

5

War and peace 1814–1834

Between March 1814 and June 1815 diplomacy brought to an end the wars that had raged for many years over Europe and North America, a process sealed by military victory over Napoleon at Waterloo. To many, it seemed that peace had merely restored old borders and old elites, yet in truth nothing would ever be the same again. The forces of nationalism and democracy had both been unleashed, and the nature of warfare had evolved. All but the most rigid of conservatives now considered that monarchs ruled not by divine right but by ensuring the happiness and prosperity of their subjects. States would need to industrialise and accept some of the social changes that this implied, including more effective transport.

A consequence was that, over the following 20 years, the way that railways were built and equipped, and the purposes to which they were put, changed dramatically. Mechanical traction became both more common and increasingly sophisticated, new types of rail were introduced and dedicated passenger stations were built, some of them on a lavish scale. Longer and more ambitious systems came to be constructed, not only in the United Kingdom but across continental Europe and in the USA, to match the still-growing network of canals and turnpikes, at a time when railway technology itself was evolving very rapidly.[1] Some were built to carry coal, but on a far more ambitious scale than previously, or were constructed with a perception of national trade in mind. Others still were manifestly 'main lines' as we understand the phrase now, running on a heavily engineered double-track formation, carrying goods and people at something like speed, and over a long distance. These are described in the chapters that follow.

This was a multifaceted process. It was clear that the iron railway as it had evolved in Britain in its various forms offered a robust and potentially useful way of moving goods and people, but it also became evident that it was not necessarily suitable for other cultures and environments, even where the function was similar, unless it were significantly adapted.[2] The systems installed in France, Prussia and the USA to connect collieries to navigable water, for instance, did not follow British models slavishly.

Even so, the United Kingdom led the way. Though small, it commanded the Atlantic economy. It had retained its wealth-generating West Indian colonies from its pre-1776 American empire, and had consolidated its hold on India, enabling it to import raw cotton on a significant scale, the staple of the emerging industrial economy. It was itself exceptionally rich in natural resources, above all iron and coal. It had a well-organised business sector and a very effective banking system, dominated by the Rothschilds, which had made it victorious in war as well as successful in peacetime commerce. Its patent system was an effective way of safeguarding intellectual property rights and rewarding invention.[3] In addition, it already had long experience of constructing costly transport systems, in the form of docks, harbours, canals and turnpikes. It was now poised to deliver the world's first fully fledged minerals-based industrial revolution, and to consolidate the huge maritime empire on which it would depend.

After 1814, when its inconclusive war with Britain came to an end, the United States would also become immeasurably strengthened and enlarged, not yet interested in a global role but anxious to connect its eastern seaboard with its interior, by crossing the mountains which barred the way to the major rivers which flowed into the Gulf of Mexico. Though the young republic had no shortage of visionaries calling for industrialisation and for transport links to develop its inland empire, its democracy was agrarian, its population was small and dispersed, the relationship between Americans of African, Caucasian and indigenous heritage unresolved, skilled workers few, capital in short supply and its mineral reserves largely unexploited. What it did have in abundance was timber, which could be used both as a fuel and as a construction material.

France, still the most powerful country in continental Europe, lost the territories it had acquired since the Revolution, including the province of Liège, with its rich coal deposits, to the kingdom of the Netherlands, and the barely developed coalfields of the Saarland to Prussia. It now had to look elsewhere to fuel its industrial revolution, above all to the mineral basin between the Rhône and the Loire around Saint-Étienne in the Massif Central.

The Saarland was not Prussia's only source of coal, as under Frederick the Great it had occupied mineral-rich Silesia in 1742. Many idealistic nationalists looked forward to the day when this energetic and modernising kingdom would form the nucleus of a German state, one which would require efficient means of internal communication to bind it together. In the meantime it vied with slow-moving Austria for supremacy in the new Deutscher Bund ('German Confederation') established after the Holy Roman Empire had been swiftly dissolved following the battle of Austerlitz in 1805. Austria shared a border with another land-based empire which had survived the Napoleonic period but which was nevertheless weakening irrecoverably. This was Ottoman Turkey, no longer able to control its Christian subjects along the Danube, which was now declared free for navigation. Flourishing river trade might one day enable the double eagle of Vienna to supplant the crescent of the Prophet on its lower reaches.

Russia had been saved from Napoleon not only by its vast size and harsh climate but also by its poor transport. When the *Grande Armée* invaded after the June harvest of 1812, its supply carts had struggled on the narrow dirt roads; when it retreated during the *rasputitsa*, the time of mud and rain, they became hopelessly bogged down. In winter, only sledges were practicable on many of the roads, and canals and rivers froze. Goods might take two years to travel from Astrakhan and the Lower Volga to St Petersburg. For this very reason, the Russian elite had maintained an interest in new forms of transport since the time of Peter the Great. Alexander I (r. 1801–25) continued a policy of building *chaussées* and further canals to promote long-distance trade, but his country lacked capital. He himself was becoming increasingly conservative after 1815, suspicious of outside influences and prone to intellectual and spiritual excitements. Fears of conspiracy in St Petersburg itself and among his Polish subjects also made it unlikely that he would permit investment in major railway systems, but his empire's mining and metallurgical industries had their own transport needs. Here, both state-owned plant and private undertakings could draw on trained mechanics and craftsmen, men who might legally be serfs but who were both skilled and talented.

Swifter, more effective overland communication conferred an evident commercial advantage, and might take the form not only of roads and improved navigation but also of iron railways. Pragmatic British industrialists and bankers saw them essentially as utilitarian improvements or as evidence of how money could generate yet more money, but many continental Europeans felt they might be the means of developing the strong, cohesive states for which they yearned. In this way, the railway could also become an apt symbol of Romantic nationalism.[4] The German-American economist Georg Friedrich List (1789–1846) argued strongly that they were an essential element of any national–commercial system. He corresponded with the Bavarian engineer Joseph von Baader (1763–1835), whom we met as an early proponent of monorails (chapter 2), another tireless advocate of railways in the German-speaking world, who explicitly identified patriotism with capital advantage.[5] So did the merchant Ludolf Camphausen (1803–90) of Cologne and Friedrich Harkort (1793–1860), 'the father of the Ruhr', for whom England was 'the high-school of European trade', and who constructed two short colliery railways (chapter 8).[6] In 1825 Goethe observed ruefully that 'Railways, express mails, steamboats and all possible means of communication are what the educated world seeks'. He did not approve, though he relented to the extent of giving his grandchildren a model of the *Rocket*, its tender and a passenger coach four years later.[7] Palmer's monorail patent inspired a resident of South Carolina to propose a suspension system to carry mails between New Orleans and Washington DC.[8] The economist Michel Chevalier (1806–79), a follower of the French utopian socialist Henri de Saint-Simon, took an international view: having heard of the Liverpool and Manchester's success, he advocated a network stretching from Astrakhan and Odesa to Spain and Portugal.[9] Chevalier had spent time in prison for his beliefs, and it is no surprise that those with other affiliations began to view railways very cautiously indeed, if they really were the harbingers of Enlightenment rationalism and of universal brotherhood, or even if they were simply another proof of British bumptiousness.[10]

Bumptious or not, it was clear that the British had achieved great things in the way of technology, and of railways in particular. Engineers from France, Austria, Prussia, Bavaria, Russia and the United States were sent to find out more and to report back.

This would not have happened had not the body of knowledge and practical understanding within the United Kingdom itself continued to grow significantly after 1815, when technical writers such as Thomas Tredgold and John Nicholson began to publish for a literate class of mechanics with some disposable income.[11] In particular, George Stephenson's experiments with the colliery viewer Nicholas Wood and with the iron founder William Losh (see chapters 2 and 4) meant that over the next 20 years railways began to assume the form they would take for the remainder of the nineteenth century.[12] Engineers were finding work all over the country, taking their skills with them.[13] Producing railway components had itself become a significant business for foundries and ironworks.[14] Established social networks, such as the long-standing links within the Quaker community, were proving important.[15] Quaker capital brought into being the engineering firm of Robert Stephenson & Co. in Newcastle in 1823 – named after its 19-year old managing partner, George's son – as well as the Stockton and Darlington Railway two years later, which successfully combined innovative forms of locomotive traction with wrought-iron rails. Events like the Rainhill locomotive trials on the Liverpool and Manchester in 1829 and the ceremonial opening of this double-track city-to-city railway the following year were stage-managed affairs in which engineering talent, capital and political influence were brought together on an international scale. British firms began to supply overseas from 1827, when locomotives and rolling stock were designed and built by Robert Stephenson & Co. for a railway from the collieries of Saint-Étienne to the Rhône and the Saône at Lyon, a trade made easier by British government policy to reduce export tariffs under legislation introduced by the President of the Board of Trade, William Huskisson (1770–1830).[16]

The appeal of British technology was such that study visits to see and assess these marvels were considered a worthwhile professional and commercial investment, however time-consuming and costly. An Atlantic voyage was a matter of a month or more, and even within the United Kingdom a long-distance stagecoach journey might still take the best part of a week.[17] A well-qualified specialist could not undertake productive work for a considerable period, and had to find bed and board as well as purchase books, plans and models. For their visit to England in 1828, three American engineers were allowed £2,000 expenses by the Liverpool office of the Baltimore and Ohio Railroad, which was then at the planning stage.[18] William Strickland of Pennsylvania was paid $2,500 for his 12-month study tour of

England, Wales and Scotland, of which £400 was placed on order in the United Kingdom.[19] Accountants at head office sometimes complained. They challenged the expenses incurred by a deputation from the Saint-Étienne–Lyon railway to Newcastle over the winter of 1826–7, in particular the purchase of two travelling cloaks, but conceded that those who were already familiar with cold, heretical England should decide.[20] Horatio Allen, 26 years old, representing the Delaware and Hudson, with letters of introduction to a Liverpool banking house, was made to feel very welcome when he docked on 15 February 1828, and was not only introduced to George Stephenson and Jesse Hartley, the dock engineer, but taken to dinners, church services and dances, noting in his diary which of the young women had made the deepest impression on him at all these social events ('much the same as in America, but rather inferior in point of beauty'). He was struck by the ceremonial surrounding the Lord Mayor, far in excess of what was afforded the President of the United States.[21]

For some, time and money were no object. In the flurry of diplomatic activity after Waterloo, Archduke John of Austria and his brother Louis were interested to observe a Wylam locomotive at work in 1815, and the following year the Grand Duke Nicholas Pavlovich of Russia (1796–1855) saw Blenkinsop's locomotives at Leeds, and was sent a model on his return home. Nicholas had studied engineering as a young man, and throughout his life remained interested in technical education and in public works.[22] Once this most conservative of individuals himself succeeded to the imperial throne after the death of his brother, Tsar Alexander, he began cautiously to consider railway building in his own territories – so cautiously indeed that the first mainline railway in Russia was not built until 1837. A trusted aide of the royal house came with him to England, Dr Joseph Christian Hamel (1788–1862), Collegiate Assistant at the Ministry of the Interior, whose professional role was to identify developments in science and technology and to assess how they might benefit Russia. Hamel, an ethnic German Protestant from Astrakhan, was

1. *(above)* The wooden way. A chaldron full of coal on the Heddon waggonway descends a 'run' to the River Tyne with a horse as brake, as an empty one is hauled back on a 'pull'.

2. *(right)* A recreated wooden waggonway at the open-air museum at Beamish, north-east England.

3. Granite from Haytor in Devon was carried to the Stover Canal on a railway built of the quarry's own output.

4. The 1805 Risca 'long bridge' on the Sirhowy plateway was one of the wonders of Wales. On the right, a young woman points out these improvements to an older man.

5. Iron railways were increasingly used in major construction works. This view shows St Katharine's Dock in London being built in January 1828.

6. The Croydon, Merstham and Godstone extension of the Surrey Iron Railway crosses Chipstead Valley Road in Coulsdon.

7. *(left)* Rope haulage: the inclined plane carrying railed boats to and from Ketley Ironworks forms the background to a portrait of the versatile and talented Shropshire ironmaster William Reynolds. He holds plans for the innovative iron aqueduct at Longdon-on-Tern on the Shrewsbury Canal.

8. *(below)* The Liverpool and Manchester's inclined plane from Edge Hill to the Mersey docks occupies the centre arch, the passenger route from Crown Street Station the right-hand arch. The 'pillars of Hercules' are the chimneys for the winding engines but also symbolise the way to the open sea.

9. Early-morning operations: two trains pass each other on the Philadelphia and Columbia's 'Belmont' inclined plane. At the foot of the plane is the building over the lower sheave, and beyond, the covered bridge over the Schuylkill, with the city of brotherly love on the far bank.

10. Rope haulage was used on industrial and constructional railways for many years. Edmund John Niemann's painting shows the fixed steam engine, boiler, winding drums and haulage cable on a railway from Baddesley Colliery in Warwickshire to the Coventry Canal.

FROM 1804 TO 1830 LOCOMOTIVE DESIGN WENT THROUGH
FUNDAMENTAL CHANGES. THESE SIX RECONSTRUCTION
DIAGRAMS SHOW THE KEY DEVELOPMENTS.

11. The 'tram engine' diagram of 1803 shows a Trevithick locomotive for a 3-foot gauge plateway. Long thought to depict the first Merthyr locomotive or William Reynolds's design for Coalbrookdale, it may have been a proposal for Tredegar Ironworks.

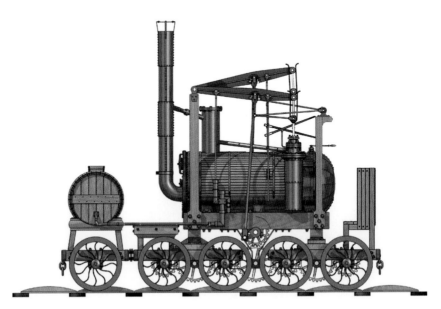

12. *Puffing Billy* and *Wylam Dilly* went through many alterations during their long working lives between Wylam Colliery and the Tyne. By 1825 they were tall, ungainly machines running on four axles to spread the weight on the cast-iron plate rails, coupled to a single-axle tender.

13. *Locomotion No. 1* hauled the first train on the Stockton and Darlington Railway on 27 September 1825. It reflected the 'Killingworth' design with two vertical cylinders set in line with the boiler top which George and Stephenson had initiated eleven years earlier.

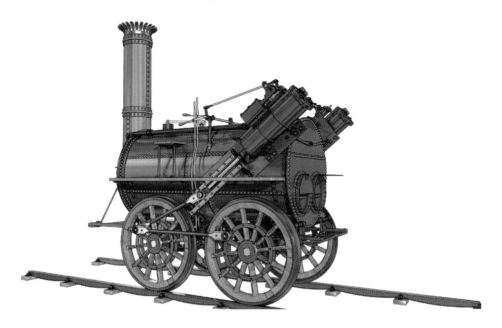

14. *Lancashire Witch* broke away from the 'Killingworth' design. The double flue in the boiler increased the heating surface, and the cylinders drove the axles directly.

15. *Rocket* was a breakthrough technology when it won the Rainhill 'ordeal' in 1829 but within a year the 'Planet' and 'Samson' designs had made it out of date.

16. The Camden and Amboy Railroad's *John Bull* is preserved in the Smithsonian, showing American changes to the basic British 'Samson' design, including a pilot axle and a headlamp, and an American tender.

then on an eight-year-long journey which took in Ireland, France, Belgium, Holland, Germany, Switzerland and Italy as well as England. This polymath intellectual dined with the Romanovs and knew everyone from Sir Humphry Davy to Alexander von Humboldt. Two other Russian visitors who, by contrast, did not frequent palaces were the engineer-serfs Yefim Alekseyevich Cherepanov (1774–1842) and his son Miron (1803–49). They were the property of Count Paul Demidov, one of the richest men in the world, who owned mines and ironworks at Nizhny Tagil. Yefim Alekseyevich had been sent on a study tour of industrial locations in the north of England in 1821, and observed one of the Middleton 'machines' pulling a train (chapter 5), but was unimpressed by it. Miron Yefimovich was sent to assess the Liverpool and Manchester in 1833, despite being unable to speak English, and the two of them constructed the first steam locomotive in Russia the following year.[23] They were a gift to historians in the Soviet period, as evidence that untutored proletarians could accomplish more than a decadent aristocracy or exploitative capitalists. In fact, they were drawn from a well-established elite of managerial and engineer serfs whose education and training might include an apprenticeship or even study at a foreign mining institution. Without Demidov's financial support the Cherepanovs could have done little, but it is true that they were exceptionally gifted men who built a functioning locomotive and a steam railway in Russia when the Tsar, his ministers and the state bureaucracy could only talk endlessly around the subject.[24]

Despite the social gulf between them, the Cherepanovs had in common with Hamel the fact that they were able individuals brought up in an innovative technical culture, and that they had powerful patrons. The same can be said of George Stephenson and Timothy Hackworth (chapter 4). One American visitor described Stephenson as 'a common digger in the Newcastle coal pits (who) took to engineering instinctively' and Hackworth as 'another engineer by instinct ... a blacksmith (whose) engineering consists in tinkering and refining locomotive engines', a judgement which recognised their skills while failing to appreciate the inventive and experimental milieu in which they had been brought up, and the patronage systems which supported them.[25] Hackworth had indeed been a smith as a young man, but he was a significant figure in engineering circles, and in the religious and social life of his area.[26] He was overshadowed by George Stephenson, but both men, and others like them, already belonged to an elite, one formed by the technical and managerial culture of the north-east of England coalfield, already centuries old yet one which had remained resilient, outward-looking and innovative.[27] In the comparatively open society of Britain they could be – had to be – advanced socially, a point grasped by Stephenson's bankers, Pease and Richardson: '... he is a clever man, but he must have leading straight ... he should always be a gentleman in his dress, his clothes real and new, and of the best quality, all his personal linen *clean every day* his hat and upper coat conspicuously good, *without dandyism*'.[28]

Although accounts stress Stephenson's welcoming personality and his evident charisma, there is nothing to indicate how he was initially approached by those who sought his guidance. Letters of introduction were sometimes furnished.[29] Perhaps distinguished visitors from foreign parts would write in advance to express the hope that he would be available on such-and-such a day, or once they had settled themselves in one of Newcastle's many comfortable inns, they simply walked up to Killingworth in the hope of finding the great man at home and with time on his hands to show them the colliery and the waggonway, and then to have a discussion around the table at Dial Cottage with them.[30] Horatio Allen met George Stephenson in Liverpool in the more refined circumstances of a dinner party, when Jesse Hartley teased Stephenson that a north-easterner sitting between a Yorkshireman and a Yankee must be in a dangerous predicament, a dig at their different accents.[31] Stephenson never lost his strong Northumbrian dialect, but was able to conduct conversations with visitors for whom English was not their native tongue. One French engineer admitted to him that he had learnt English from reading Shakespeare, and was complimented by Stephenson on sounding like Elizabeth I.[32] Britons were notorious for their unfamiliarity with languages other than their own, and visitors from abroad who ventured into Wales would find that even their laboriously acquired English was little use to them.

Visitors were not always welcome. In August 1829 Robert Stephenson was annoyed to find that the engineer Timothy Burstall of Edinburgh 'had walked into the manufactory this morning and examined the Engine with all the coolness imaginable before we discovered who he was'.[33] Burstall was also proposing to submit a locomotive for the Liverpool and Manchester Railways trials at Rainhill, and the machine he saw was the *Rocket*.

George Stephenson by and large stayed at home and let others come to him.[34] Other English experts travelled widely and took their knowledge with them, in some cases beyond the United Kingdom. Colliery viewers had always been peripatetic individuals but none travelled as far as John Henderson (1781/2–1835), who had worked in the Shropshire and north-east of England coalfields before being sent out in 1807 as assistant viewer at Höganäs Colliery in Sweden. On his return, he found employment in collieries in South Yorkshire, near Berwick, and at Dunfermline in Fife before making his way to New South Wales as manager of the Australian Agricultural Company's colliery at Newcastle in 1826, where he ended his days. He installed iron railways in Höganäs and in Australia, the first in Scandinavia and Oceania.[35] Other technical experts who might find themselves far from home were locomotive drivers. Once steam traction became established, it became common to send someone out with the machine, either to teach locals how to operate it or to be its long-term driver. This practice is recorded in Wales in 1819, as well as later in France and the USA.[36]

The spread of the iron railway across Europe, the United States and even to Australia between 1814 and 1834 was due both to men who wore fustian jackets and to those who wore frock coats. Waterloo and the Congress of Vienna had supposedly made the world safe for gentlemen once again, but gentlemen had begun to understand what the lordly coal owners of Tyne and Wear had known for a long time, that their fortunes, even their survival as a ruling class, depended on scientists and scholars from outside the traditional elite, and on skilled men who might be of very lowly origin indeed. Conversely, George Stephenson, Nicholas Wood and Timothy Hackworth all depended on the patronage of wealthy, landed men, and accepted the social order; Robert Stephenson became a Tory Member of Parliament.[37] Individuals from very different backgrounds – engineers, technicians, landowners, bankers and theoretical scientists, Catholics, Orthodox, Protestants and Jews – proved willing to talk to each other and to promote each others' causes, not only in the newer, more democratic society of the United States but also in the conservative monarchies of Europe. The 'spirit of association' (chapter 1) was fundamental to the spread of the iron road.

As well as personal visits, publications formed an essential element in this transfer of railway technology between and within different countries. Newspapers were by this stage widely published throughout Europe and the USA, and the 'penny press' made its appearance in 1830. Abraham Rees's massive *Cyclopædia*, with its emphasis on agriculture and scientific theory, appeared serially between January 1802 and August 1820 and ran to 39 volumes. The *Mechanics' Magazine* began publication in 1823, the first low-priced English-language scientific weekly, and soon became an enthusiastic supporter of the railway movement. American publications about British railway developments or homegrown proposals first appear in 1806, but the trickle became a flood in 1825.[38] That same year the St Petersburg *Gornyi Zhurnal* was established, and Russian newspapers began to take an interest in railway developments after the opening of the Stockton and Darlington.[39] France had had its *Annales des Mines* since 1794 and the *Annales des Arts et Manufactures* since 1800. This last journal published an article in 1804 explaining how many horses were required to haul slate from Penrhyn Quarry to the sea at Abercegin in Wales, even though Britain and France were then at war with each other.[40] The *Bulletin de la Société d'Encouragement pour l'Industrie Nationale*, founded in 1801, similarly brought the Blenkinsop steam locomotive (chapter 4) to French attention even while Napoleon was endeavouring to regain his throne.[41]

More, and more detailed, written descriptions became available after the wars came to an end. The long-standing continental European tradition of engineering study visits had always assumed publication of the results on return. Dutens's *Memoires sur les travaux publics de l'Angleterre* (1819) is one of the earlier postwar examples of this genre: though it refers only briefly to railways, it is an important source for the Hay inclined plane in the Ironbridge Gorge (chapter 3). Joseph von Baader had lived in Britain where he had been granted a patent for the monorail described in chapter 2.[42] The second and more detailed edition of his *Neues System der fortschaffenden Mechanik* in 1822 circulated widely; the author set out to show how British practice could be married with Teutonic theory. Tsar Alexander I (to whom it was dedicated, as was the first edition of 1817) ordered 100 subscription copies; others went to the Empress of Austria and to the kings of Prussia, Saxony, Wurtemburg and Bavaria, as well as to noblemen and civil servants.[43]

Karl von Oeynhausen and Heinrich von Dechen's *Ueber Schienenwege in England in den Jahren 1826 und 1827*, as its title suggests, was only concerned with railways, though it misleads in suggesting that these two Prussian mining engineers only visited England, as they took in Wales and Ireland as well. They quoted at length from Nicholas Wood's *A Practical Treatise on Rail-roads, and Interior Communication in General* (1825), which ran to 314 pages.[44] Wood's *Treatise* also informed *Reports on Canals, Railways, Roads, and Other Subjects*, published in Philadelphia in 1826 by the architect and civil engineer William Strickland (1788–1854) following a study visit to England. This was more succinct, at 51 pages, with 14 of the 71 engravings depicting aspects of railway construction and operation. Strickland was specifically instructed by his patrons, the Pennsylvania Society for the Promotion of Internal Improvements, that on his arrival in England '*Locomotive machinery* will command your attention and enquiry. This is almost unknown in the United States.'[45] The Society was made up of influential and well-connected Philadelphia subscribers but also received donations from sympathetic individuals and from the coal companies.[46]

Books published in German might quote publications in English and include in parentheses the English or French terms for unfamiliar innovations.[47] In some cases, translations were undertaken. The Bohemian engineer Franz Joseph von Gerstner's *Zwey Abhandlungen über Frachtwägen und Strassen und über die Frage*, published in Prague in 1813, appeared in French as *Mémoire sur les grandes routes, les chemins de fer et les canaux de navigation* in 1827.[48] His son's three-volume *Handbuch der Mechanik*, the first comprehensive German-language study of applied mechanics, circulated widely, and the first chapter was translated into English in 1834.[49] Part of Nicholson's *Operative Mechanic* was soon translated into French.[50] In 1829 von Baader published in French a volume on the advantages of railways which quotes Tredgold and Strickland. It was dedicated to the Vicomte de Martignac, a moderate royalist French politician, Charles X's Minister of the Interior and effective head of government.[51]

Many of these publications rehearse the same technical problems: how did roads, canals and railways compare with each other, and in what sort of environment would they be most effective? In the United Kingdom ambitious road and canal schemes were still being constructed 20 years after Waterloo.[52] In the USA and Canada, canals were favoured for longer routes. The Erie Canal connecting New York with the Great Lakes was built between 1817 and 1825, the Rideau Canal from Ottawa to Lake Ontario between 1826 and 1832. American mountains were higher, frosts more bitter and snows deeper than in Britain, which were arguments against canals. On the other hand, railway technology was unfamiliar, engineers few and far between and imported machinery likely to prove unsuitable. A pamphlet and newspaper war between canal partisans and railway promoters in the USA had begun in 1812 and went on for years.[53]

France, Belgium, the Netherlands and Sweden also completed ambitious canal projects in the 20 years after Waterloo. Central Europe was less favourable to inland waterways. A pet project of the Bavarian King Ludwig I was a canal to connect the Rhine and Danube watersheds, but in 1835 the merchants of Nuremburg opened a short test-railway along the metalled highway to nearby Furth in the hope that this might be extended to provide the link instead. Undeterred, Ludwig gave it his grudging patronage but pressed ahead with the canal anyway (chapter 13). In essence, it was established that canals were useful for bulk transport at slow speeds but were costly; roads were suitable for the movement of individuals and documents over long distances at some speed, as well as for the slow movement of carts, but railways required far fewer horses, and steam locomotives were evidently well on the way to outperforming them.

The case for steam and for railways generally could be demonstrated by models or small prototypes, which were also a useful way of carrying out assessing improvements.[54] Dr Hamel wrote to John Buddle in 1815 to ask for both exact drawings and a model of Stephenson's locomotive.[55] Von Baader began his experiments with large and half-size models.[56] The lawyer and engineer Col. John Stevens III (1749–1838) of Hoboken, New Jersey, built and operated a small-scale demonstration track on which he ran a wood-fired locomotive in 1825. This was the first in the United States, and was meant to show what he believed was possible on his proposed Pennsylvania Railroad. A trip on the railway was compulsory for all visitors to Castle Point at Hoboken. 'Gentlemen must cram their beavers a bit tighter on their heads; ladies must gather their wide skirts about them and clutch the handles of their tiny parasols. At the appalling speed of six miles an hour, one and all must try the first American steam railroad.'[57] Stevens joined the Pennsylvania Society and considered that it was at his prompting that Strickland had been sent to England.[58] When a model of a 'Killingworth' locomotive (chapter 8) built for Strickland in Leeds arrived in Philadelphia in 1825, the first issue of *The Journal of the Franklin Institute* recorded delightedly:

> ... it is now as much in the power of our state, or of any individual or corporation, to erect and adopt a railway of the best, or of any form of construction, as it would be, were *Wood*, and *Jessup* [sic], and *Rennie*, and *Telford*, and *Tredgold*, to become resident citizens of the Commonwealth of Pennsylvania. A working model of a loco-motive engine, on the most approved plan, and having a power equal to the strength of two men, was purchased by Mr. Strickland, and is in the possession of the Society. A machine so valuable, and of such astonishing competency for the purpose to which it may be applied, ought to be more generally known in our country.[59]

From 1825 onwards, the Franklin Institute and its *Journal* were an important force in the professionalisation of American science and technology, honouring and emulating the founding father Benjamin Franklin's commitment to rational inquiry and Enlightenment values. Franklin himself had been instrumental in reviving the American Philosophical Society, which maintained a standing committee on American improvements. Its home in 104 South Fifth Street in Philadelphia contained a museum of curiosities where on 25 April 1831 Strickland's model was joined by one based on Braithwaite and Ericsson's Rainhill *Novelty*, constructed with the aid only of 'imperfect published descriptions and sketches' but which was capable of hauling two small passenger cars on a circular track made of pine boards. Its builder was Matthias Baldwin (1795–1866), a man of

12. William Strickland's 1826 *Reports* and model locomotive inspired Americans to think of railway building.

little formal education but whose childhood had been spent making and dismantling mechanical toys. He went on to establish what ultimately became the world's largest locomotive manufactory.[60]

Philadelphia, with its access to capital, its revolutionary and free-thinking spirit and its well-established industries, could promote native wit and talent, and reward invention but its Institute was not a degree-awarding body. The budding American engineer who wished to write 'A.B.' after his name had to go to New York. Here, the Rensselaer Polytechnic Institute was established in 1824 for the 'application of science to the common purposes of life', and awarded degrees from 1835. From its foundation in 1754, Columbia College valued science, engineering and mathematics. James Renwick (Class of 1807; 1790–1863) held sway in its department of Natural Philosophy from 1820; he designed the railed inclined planes on the Morris Canal (see chapter 7) and published a *Treatise on the Steam Engine*.[61] Other alumni included John Stevens (Class of 1768;

1749–1838), and Horatio Allen (Class of 1823; 1802–89). It was from New York in 1832 that the weekly *Rail-Road Journal* came to be published.[62]

The increasingly scientific turn of military organisation, and the need for capable surveyors, led to the establishment of West Point, the United States Military Academy in New York State, which from its reorganisation in 1817 offered tuition in civil engineering. A spin-off was Alden Partridge's American Literary, Scientific, and Military Academy at Norwich, Vermont, founded two years later. Johan (John) Ericsson (1803–89), builder of the *Novelty* locomotive (see chapter 8), had been trained both by Sweden's Royal Navy and by its Army. Whereas George Stephenson had to teach himself to read and write as an adult, his son Robert was able to spend several terms at Edinburgh University, leaving, as most did, without graduating once he had gained the learning he felt he needed. Scottish universities had a long tradition of imparting practical information and teaching the useful arts. Durham University, founded in part to divert the attention of reformers from the embarrassingly lavish colliery-derived stipends enjoyed by the cathedral dean and chapter, only instituted teaching in engineering some years after it opened in 1833. It was a while before the longer established universities, Oxford, Cambridge and Trinity College Dublin caught up, though this last institution did produce the remarkable figure of Dr Dionysius Lardner (1793–1859), Professor of Natural Philosophy and Astronomy at the newly founded London University from 1828 to 1831. Lardner is now largely remembered for being bettered by Isambard Kingdom Brunel (chapter 13) in argument and as a compulsive adulterer, but he proved a remarkably effective popular writer, publishing a flood of textbooks aimed at the aspiring mechanic. James Renwick thought well enough of him to collaborate on a publication.[63]

In Russia, the Institute of Transport Engineers was founded in 1810 to stand alongside the existing military and mining departments. Teaching on railway building began as early as 1820, with the appointment of two French engineers, Benoît Paul Émile Clapeyron (1799–1864) and Gabriel Lamé (1795–1870).[64] They founded the first Russian journal for engineers, published in French as the *Journal des Voies de Communication*.[65] It only appeared in Russian in 1836. Lamé was sent to Britain to study 'noteworthy construction' in 1830. Reprimanded for returning late to St Petersburg, he drily replied that this was due to Russia's poor roads, and gave public lectures at the Institute on railway building and road building. Clapeyron and Lamé fell into disfavour as Tsar Nicholas became increasingly suspicious of foreign influence, and were forced to return home, but left behind a corps of young engineers whose work would eventually bear fruit.[66] Remarkably, formal instruction in railway engineering at France's École des Ponts et Chaussées seems to have begun as late as 1833, and then only as a part of a course on *navigation intérieure*, by Professor Charles-Joseph Minard (1781–1870).[67]

Chapters 6 to 12 trace developments from 1814 to 1834 as they came to be understood in the light of the breakthrough technologies of this period – the steam locomotive and wrought-iron rails – and chapter 13 shows how they in turn made possible the spread of the railway by mid-century. Though railways were indelibly and inevitably associated with the figure of George Stephenson and with the north-east of England, it is in this next period that they became more widespread, more technically varied and less dependent on their English origins. Strong commonalities remained but it was a mark of the times that by the early 1830s the Baltimore and Ohio, the 'railroad university of the United States' was exercising some of the same functions in the new world as the Stephenson railways in England were doing, as a proving ground and a point of learning for the iron road.[68]

6

'Geometrical precision'
Wrought-iron rails 1808–1834

... parallel lines of iron bars, laid with almost geometrical precision.[1]

The second stage of the adoption of iron rails was the substitution of rails made of wrought iron for those of wood or of cast iron. Wrought iron is a material which is 'malleable' – in other words, it does not easily fail when it deforms under compressive stress. For this reason, it could carry heavier loadings than its fibrous or brittle alternatives, making it possible to build and operate locomotives heavy enough to apply tractive force to the rails. It could be bent into curves and laid in longer lengths, both of which made for a smoother journey. A plateway or a railroad could only offer a slow and jolting ride, even if very well laid out by a capable engineer. Wrought-iron rails made possible the 'geometrical precision' of the first generation of main lines, the high capacity of the new coal carriers and the sinuous mountain formations of the Allegheny and the Ffestiniog.

Like the locomotive, it was a technology in which the hand of George Stephenson is very evident, and it evolved over a number of years. As with cast iron, it represented a very substantial part of the investment in any railway, and no competent engineer would spend significant sums on an untried technology, any more than a properly cautious accountant would allow an important component to be replaced when there was still life in the old materials.² It took 20 years for wrought-iron rails to prove themselves, almost exactly the time frame within which the major initial locomotive developments took place. Their first experimental use was in 1805. They became general practice in 1825, when the Stockton and Darlington showed how they could be used to build railways where support and guidance from the rails harmonised with mechanical traction to offer high capacity operations, and the prospect of speed.³

Wrought iron

Wrought iron is so called because it is hammered, rolled or otherwise worked while hot enough to expel molten slag in order to produce a semifused mass of iron with fibrous slab inclusions which gives it a 'grain' resembling wood. It is tough but also ductile, capable of being bent and also of being passed through a mill to produce a uniform cross-section. Blacksmiths would hammer it out on anvils to produce straps and ties, and this was how the thin strips laid on wooden rails were produced (chapter 1), but it was the puddling process which reduced its cost considerably and enabled it to become a major structural material. Perfected in 1791 at Cyfarthfa Ironworks in South Wales, this involved stirring the molten pig iron in a reverberatory furnace, in an oxidising atmosphere, thus decarburising it. Grooved rolls enabled bars or rods of this material to be produced in almost any profile.

Its adoption as a track component reflects the major growth of Britain's metallurgical industries in the late eighteenth and early nineteenth centuries, when the country first challenged, and then surpassed, Sweden and Russia as a major producer. Just as cast iron had earlier begun to supplant timber and masonry as the main structural materials, it was now being displaced by, or used in conjunction with, wrought iron, even in bridges and shipbuilding. Not only was wrought iron much in demand for rails, it had many other railway uses, such as plates for boilers and other mechanical components. Production demanded ever-higher levels of energy, since in addition to the needs of a blast furnace – coke, fluxing material and an air-blast – it needed fuel for reheating, and powerful waterwheels or steam engines to operate the forges and rolls through which the red hot iron would be passed and repassed.

The first dedicated rail-rolling mill was set up in 1820 at Bedlington Ironworks in Blyth Dene, Northumberland, where the proximity of iron ore, coal, abundant water and a convenient nearby river port had sustained furnaces and a nail-making business since at least 1750. It was established by Michael Longridge (1785–1858), an astute and well-connected entrepreneur who had a close business relationship with George Stephenson and later with the firm of Robert Stephenson & Co., and who benefitted from their advocacy of locomotive traction. Bedlington also rolled other railway components such as axles for the Stephensons. Other rolling mills were set up in South Wales and in the English West Midlands, and these between them supplied most of the world's railways until the 1860s.[4]

The first known use was by Charles Nixon of Walbottle Colliery in the north-east of England, who laid down 2-foot-long narrow wrought-iron rails with a half-lap joint in 1805, but they wore the wheel rims and were superseded by cast-iron rails.[5] More apparently successful was their introduction in a slate quarry in the Nantlle Valley in north-west Wales around 1811 by its manager, John Hughes, *John y gôf* ('John the blacksmith'), and better known is the relaying of part of the Tindale Fell railway to the Tyne near Brampton between 1808 and 1812 with wrought-iron rails 1.5 inches (38 mm) square, on stone blocks. These replaced cast-iron fish-bellied rails on stone blocks, which had themselves replaced a wooden 'way' some years earlier.[6] It was soon found that lengths of bar iron in cast-iron chairs or even simply slotted into wooden sleepers made excellent temporary railways, since they could be lifted up and relocated, and they were widely used in mines, quarries and construction sites by the mid-1820s. The Prussian engineers Von Oeynhausen and von Dechen commended the ones they saw in mines around Redruth and St Austell in Cornwall in 1826–7 for 'their simple and light construction, and therefore their cheap cost of laying out'.[7] In Welsh slate quarries, they were widely adopted, and were still being used as late as the 1960s.[8] The Scottish engineer Alexander Nimmo (1783–1832) favoured them for his construction projects in Ireland.[9]

Their success piqued the interest of the Scottish engineer Robert Stevenson (1772–1850, grandfather of the novelist Robert Louis Stevenson), who drew attention to them in his report on his proposed (but not built) Edinburgh Railway in 1818 'as likely to be attended with the most important advantage to the railway system'.[10] He sent a copy of this report to George Stephenson at Killingworth, who put it into the hands of Longridge at Bedlington. Longridge's agent, John Birkinshaw, proposed that an intended railway from a colliery about three miles away be laid with such rails, but rolled in a T-section so that the head of the rail should be as broad as needed and the depth of the web (the vertical section of the T) increased, without adding unnecessarily to its weight. On 23 October 1820, Birkinshaw took out a patent for this new type of rail, and shortly afterwards the first were installed on the line from the colliery, in which George Stephenson was a partner.[11] At the suggestion of the viewer John Buddle, a 'fish-belly' was added to the rail by running it through a concentric roll, so that the web was deepest, and hence strongest, at the weak point midway between the chairs. Bellies came in 3-foot lengths, and such rails might have between three and seven of them.

Stephenson's preference for Birkinshaw rails, and his insistence on them for the Stockton and Darlington, led to a rift with William Losh (chapter 2), with whom he had collaborated on producing cast rails, and some commercial disadvantage to Stephenson himself, but he clearly saw long-term benefit in the new technology.[12] Even so, a proper caution dictated that part of the Stockton and Darlington be laid in cast-iron rails and these were furnished by another Quaker enterprise, the Neath Abbey Ironworks in South Wales, which also provided the chairs.[13]

The railway promoter William James, whom we will meet again in chapter 9, waxed lyrical about them: 'Light has at length shone from the north and I pronounce as my decided opinion that the Malleable Iron Rail Road at Bleddlinton [sic] Works is by far the best I have ever seen both in respect of its material and its form'. This was in June 1821. Some engineers feared that the rails would laminate, while others stated that the evidence pointed otherwise.[14]

After the Stockton and Darlington had shown the way forward in 1825, Bedlington supplied similar T-section fish-belly rails to many of the new public railways mentioned in the pages that follow. These included the Canterbury and Whitstable (1825), the Monkland and Kirkintilloch (1825), the Nantlle (1827), the Garnkirk and Glasgow (1827) and the Dundee and Newtyle (1831).[15] The large order for the Liverpool and Manchester (1827) was carried out jointly by Bedlington and by John Bradley of Stourbridge, and rails 'of exactly the same sort' were laid on the Bolton and Leigh.[16] The Leicester and Swannington was supplied by a local firm, Cort of Belgrave Gate.[17] Dowlais Ironworks in South Wales and Jevons, the Liverpool iron merchants, between them supplied the Ffestiniog Railway, once its formation was ready for track, in December 1834.[18]

The years 1825 to 1834 were a time of considerable experimentation, when engineers often had to rethink and rebuild systems that had only just been put in place. Taking the most meticulous care to identify the optimal form and strength of these rails did not guarantee the right result. The engineer and the promoter of the Dundee and Newtyle, for instance, together carefully considered all their options and only resolved on iron edge rails rather than plate rails after lengthy discussion and correspondence with Timothy Hackworth, George Stephenson and Edward Pease. On a visit to Bedlington, they were told that the 25 lb rails used on the Stockton and Darlington were not

13. Wrought-iron fish-belly rails on stone blocks were widely used from 1825.

strong enough, and that they should buy 35 lb rails. They bought some 28 lb rail and some 35 lb, only to find within a few years that they were both quite insufficient.[19]

Although the fish-belly came with a recommendation from no less an authority than Buddle, as early as 1827 the Liverpool dock engineer Jesse Hartley had questioned the need for it, and the French engineer Marc Seguin accepted a bid from the ironworks at Le Creusot for a parallel profile rail for the Compagnie du chemin de fer de Saint-Étienne à Lyon, the first to be rolled in continental Europe.[20] Harford Davis's ironworks at Ebbw Vale in South Wales rolled a distinctive form of parallel cross-section rail which used iron wedges rather than wooden keys in the chairs, and these made their appearance on the St Helen's and Runcorn Gap in Lancashire and on the Clarence in the north-east of England (1831), then on the Bodmin and Wadebridge in Cornwall (1833).[21] Harford Davis seems also to have reintroduced the practice of scarfing, or half-lapping, the joints, as set out in Losh's patent for cast-iron rails.[22] Otherwise, rails were butt-jointed in chairs. Some of the parallel profile rails for the Dublin and Kingstown were produced by Bradley of Bilston, near Wolverhampton.[23]

Cast-iron components

The chairs into which rails were set were castings just as they had been with plate rails, and were generally made in a way which permitted a wooden or iron wedge (a 'key') to be driven on either or both sides of the rail web to keep them in place but also to allow them to be removed when necessary. Whereas rolled rails could only be produced by a few specialist firms, chairs could be cast locally. Though the Ffestiniog's rails came from a distance, for the chairs the company turned to a foundryman in nearby Caernarfon, who set up a branch at the lower terminus to deal with the business, which was continued by his son.[24]

Other cast-iron components were sometimes used, such as on the Leeds and Selby, where a tie held the blocks, and hence the rails, to gauge.[25]

Stone and wood

There was no unanimity on the best form of support for the chairs. They were set either into individual stone blocks, sleepers extending from one rail to the other or stringers running underneath them, and these might be made of dressed stone or wood, depending on what was available. For reasons of economy, part of the Stockton and Darlington was laid with wooden blocks purchased from shipbreakers on Portsea Island near Portsmouth at 6d (2.2p) each, whereas stone blocks cost between 5d and 8d (2–3.3p).[26] Whatever form they took, they would be drilled to receive the chairs, two holes for each one, with a wooden plug in each, into which the spike would be driven. Ballast might be broken stone, or sometimes small coal.

The use of stone and wooden rails in conjunction with thin wrought-iron strips under the wheels on two railways in the USA noted in chapter 2 reflects the comparative local cheapness of these materials and the cost of imported iron. The Baltimore and Ohio took its cue from the Quincy Granite Railway, and intended its use of wooden rails to be temporary, with stone becoming the permanent way.[27]

A typical technology and its variations

Though wrought iron became the 'typical technology' of the railway, very few systems in the USA made use of them in the way that the British, Irish and French did. There was very little rolling capacity on the North American continent, and insufficient iron, whereas timber and stone were easily obtained. For iron, ordering from Britain was the only option, but in 1828 Congress was debating an increase in the import duty. The Delaware and Hudson Canal Company, which incorporated a railway feeder, had persuaded the Treasury that rolled rails should only pay an *ad valorem* tax of 25 per cent, less than bar iron, and now joined forces with the Baltimore and Ohio and the Charleston and Hamburg railroads, which were then at the design stage, to urge that the duty be dropped altogether. A newspaper campaign supported the proposal, arguing that the choice lay between imported iron or no American railroads, to no avail. The Baltimore and Ohio requested a stock subscription, and the rails were ordered from Liverpool agents.

The Baltimore and Ohio ordered 10 tons of T-section fish-belly for evaluation purposes, tried out in sidings, but began life with wrought-iron straps on its main lines. It took 31 tons to lay a mile of Liverpool and Manchester fish-belly, whereas a mile of strap rail would only require 22 tons. These were laid on wooden or stone stringers, initially on their outside edges, anticipating the use of outside flanges, an intention which was very soon abandoned. The Baltimore and Ohio's engineers learnt as it was being built. Another change was the gauge, originally intended to be 4' 6" (1372 mm) but soon changed to the Stephenson standard of 4' 8" (1422 mm), which was itself enlarged to 4' 8½" (1435 mm) on the Liverpool and Manchester and thereafter generally adopted.[28]

The Quincy's and the Baltimore and Ohio's use of stone stringers was adopted on the Philadelphia and Columbus. More commonly, American railroads laid wrought-iron straps on timber beams, even on the new public carriers which were being used into the 1840s and beyond.[29] Those laid on the Delaware and Hudson and the Charleston and Hamburg came from Wolverhampton, after an initial offer from Dowlais Ironworks had been rejected.[30]

There were exceptions. The first section of the Boston and Lowell was laid with rails similar to those of the Liverpool and Manchester, laid on two parallel stone walls, but they were 'inadequate in both form and strength' and soon replaced.[31] The heavier Harford Davies track was found on two systems in New Jersey and Pennsylvania, the Allegheny Portage Railroad and the Philadelphia and Columbia, both of which were state-funded.[32] The link may be the long-standing friendship between their engineers, Major John Wilson (1789–1833), his son William Hasell Wilson and Charles Blacker Vignoles, who had used them on the St Helen's and Runcorn Gap.[33] These Ebbw Vale rails replaced strap rail on granite stringers which had already 'proved an entire failure' and on timbers which had rotted. Curiously, the Americans initially failed to appreciate that wrought iron could be bent, and ordered both 9- and 15-foot lengths, presumably in the hope of forming 'threepenny bit' curves.[34]

By the mid-1830s, American practice in terms of rails, chairs, support and ballast was very varied, depending on available capital, topography, traffic, traction – for instance whether locomotive, horse or fixed engine – or the conclusions of bitter experience. Not until 1844 were rails rolled in the United States, at Mount Savage Iron Works in Allegheny County, Maryland.[35]

'GEOMETRICAL PRECISION'

It was nevertheless an American engineer, Robert Livingston Stevens (1787–1856), who devised the form in which rails are most commonly used all over the world today, albeit made of steel, what we now know as 'flat bottom'. Stevens was President and Chief Engineer of the Camden and Amboy. Anxious to construct the railway to a high standard, he set off for Britain to discuss the purchase of rails and locomotives, and whiled away the time on the voyage by carving a block of wood into a variant of a Birkinshaw rail which incorporated a foot enabling it to be spiked directly to wooden sleepers without the need for a chair. He also seems to have advocated a 'fishplate' to join the rails, normally credited to William Bridges Adams on the Eastern Counties Railway in the 1840s.[36] A wider flare in the foot where the rail was to be laid on a cross-sleeper was rejected in favour of a parallel form and, as 'flat-bottom' (UK) or 'T-section' (USA), this type of rail is now near-universal. By the mid-1830s, railroads in the USA were using a variety of wrought-iron rail sections, including not only variations on the T-section and flat bottom but also 'bridge rail', an inverted U.[37]

A benefit that was initially unappreciated, or not entirely understood when wrought-iron rails were first widely adopted, was that wheels with a conical or bevelled tread profile as well as a flange, running on a railhead sloped in a similar way, could negotiate curves better than cylindrical wheels running on a flat-head rail.[38] Nicholas Wood explained the principle:

> All the wheels now used on railways, especially where curves occur on the line, are constructed, so that the outside rim is conical, or is enlarged in diameter next the flanch [flange]; when, therefore, the carriages are passing round a curve, the wheels being connected together by the axle, forms, as it were, a conical roller, running upon the rails with different radii; the larger radii being on the outside curve of the rail. This increase in the diameter of the wheel, running on the outside, compensates, to a

certain extent, for the increased length of the outer curve of the rail; and if the radius of the curve, is not less than the line which the two wheels of unequal radii would describe, the wheels will travel along the line of the curve without rubbing against the flanches.[39]

This meant a smoother ride since in practice the flanges are rarely called upon to do any guiding when a train is in motion.[40] This was not possible on flat straps, nor on short cast-iron Losh rails or on a plateway, since, in each case, the flange battered either the rail or the wheel. It was this which added the extra half-inch to the Stephenson gauge, to prevent the wheel flanges wearing against the inside edges of the rail, the now-standard British, continental European and North American gauge.[41]

These developments seem to have been the result of gradual experiment and observation – there was no single moment when their advantages came to be understood and then proclaimed to the world. The principle of coning wheel profiles was appreciated by the Baltimore and Ohio's engineers in 1829–30, perhaps as a consequence of discussion on study visits to England.[42] Certainly, George Stephenson had already been telling Horatio Allen how the Liverpool and Manchester would be built with coned rails, how the rails would also be 'canted' on a curve, by raising the outside one slightly to prevent sudden changes in lateral acceleration, and how revolving axles were to be preferred to fixed axles on which the wheels ran loose. These, he assured his young American visitor, had been used on the railways with which he had already been involved.[43] Coning and fixed axles became the common practice from the 1830s onwards on all newly built railways in Britain and those in the USA which could afford T-rail. The use of loose wheels on loose or fixed axles became confined to some short industrial railways, where they could still be seen in use in some locations in the late twentieth century.[44]

'GEOMETRICAL PRECISION'

These developments also led to the use of track transition curves, otherwise known as spiral easements, the mathematically calculated change in alignment at the point where a straight section changes into a curve, initially at infinite radius, then at the end of the transition, the same radius as the curve itself.[45] Not only did this enable locomotive-hauled trains to run at speed but it also reduced the effort required of horses when they were used to pull waggons and made it possible for the slate-carrying Ffestiniog in north-west Wales to run loaded trains by gravity to a tight timetable. Preparing the formation of a railway became a more complicated matter but the end result was a faster, smoother journey.

Rail technology as use

By the mid-1830s the long-term form of rail had emerged, and the only significant difference was a change in material that becomes evident in the 1860s, when wrought iron began to give way to steel, an alloy of iron in which carbon improves strength and fracture resistance. The 1860s coincidentally saw the effective end of strap rail on chartered railroads in the USA and the final abandonment or conversion of some of the major surviving plateway systems in Britain. Some of these, in Leicestershire and South Wales, went through an interim period in which cast-iron plates were replaced by wrought iron, which might additionally be rolled in such a way that they could accommodate both flanged and unflanged wheels; another method involved wheels designed for both plates and edge rail. These made for systems of extraordinary complexity, especially at points and crossings, with a tendency to derail.[46] It also saw the reappearance of the wooden railway, in the misguided decision of a provincial government in New Zealand to spend as little money as possible on its locomotive-worked line from Invercargill to the Otago goldfields and, more rationally, in the context of the lumber industry of the USA and Oceania.[47]

THE COMING OF THE RAILWAY

The reign of the wrought-iron rail was short-lived, but it was one of the developments which enabled the Stephenson railway to evolve from a colliery 'way' into a main line running from city to city. The other essential component was effective mechanical traction, and for some time it remained unclear whether this would be in the form of fixed or locomotive engines.

7

'Most suitable for hilly countries'
Rope and chain haulage 1815–1834

Fixed Engines with ropes are most suitable for hilly countries.[1]

Chapter 4 described the first applications of steam to railway use, and how by 1815 it made possible the operation of mineral trains by locomotives in a reliable and systematic manner. Over the next 20 years locomotives evolved into machines capable of travelling long distances and at significant speed, demonstrated by the success of Stephenson's *Rocket* at the trials held by the Liverpool and Manchester Railway at Rainhill in 1829, and the subsequent evolution of the 'Planet' and 'Samson' classes, as well as of American builders' distinctive and practical designs. These are discussed in the following chapter. Although improved locomotive design was undoubtedly a fundamental stage in the history of the 'iron road', it has obscured the way in which other forms of rail traction also evolved, both in conjunction with locomotives and as alternatives to them. Many engineers and investors were persuaded that fixed engines operating rope or chain systems offered more effective mechanical traction, especially since locomotives of this period struggled on anything other than the slightest gradients.[2] Rope or chain haulage formed a design element in most of the significant railway systems built in England and Wales between 1815 and 1834, including the ones which are better known for their innovative use of locomotives, such as the Stockton and Darlington, and the United Kingdom's first main line, between Liverpool and Manchester. It was an important element in many of the railroads built in the USA, and saw some use in France also.

As chapter 4 also described, the introduction of iron rails had led to the widespread adoption of the Shropshire technology of the steeply graded inclined plane, whether in its counterbalanced or mechanically operated form. It was not long before the possibilities of fixed steam engines for longer and more ambitious systems came to be discussed. As early as 1802 Richard Lovell Edgeworth had suggested their use to operate endless chains to haul his railed platform cars along major roads.[3] However, their most persistent and vociferous advocate was the ironmaster and coal owner Benjamin Thompson (1779–1867), who used fixed engines to operate level sections as well as adverse gradients on the railways he built or updated in the north-east of England, and who extolled their virtues in print.[4] He replaced horses with a stationary engine at Bewicke Main Colliery in 1821, using the longest rope then made – 1.5 miles. On the Fawdon waggonway to Scotswood staithe, he introduced his 'reciprocating system' around 1822, whereby a succession of fixed engines operated two ropes between each one – a head rope which hauled the train and a tail rope attached to the last vehicle, their functions alternating when they hauled a train in the opposite direction. An American visitor in 1826 noted that wherever it crossed a public road:

> ... the rope was bent down by friction rollers, and carried under a plank bridge, to the other side of the road, where it again rose above ground. When the wagon comes to the public road, the boy who rides on it, releases the rope from the vice; the velocity of the wagon carries it over the road, and the boy again hooks up the rope into the vice, while the wagon continues its motion.[5]

He considered that it worked very effectively, but his account does indicate some of the potential problems Thompson's system created wherever it encountered roads and where human error might be a factor: if the boy failed to release the vice in time, or if the waggon encountered an obstacle.

Successful and highly profitable though the Fawdon waggonway proved to be, it was soon to be made redundant by Thompson's next undertaking, the Brunton and Shields, in 1826, which rerouted Fawdon coals to staithes at Hawdon by a mix of horse haulage and fixed engines on five inclined planes, the longest rope-hauled railway at the time. The Rainton and Seaham ('Benny's bank') which he built for the Marquis of Londonderry, and which opened in 1831, used both self-acting and steam-powered inclines along its entire five miles.[6]

Thompson was an aggressive and ruthless individual whose commercial ambitions fed his capacity for both technical innovation and public controversy.[7] He clearly saw great merit in rope haulage. George and Robert Stephenson, by contrast, preferred to use locomotives wherever possible, but steep inclined planes elsewhere, on the canal feeder plan. Their Hetton Colliery railway, opened on 18 November 1822, was the first in the world to be built entirely for mechanical traction and involved two fixed engines to raise waggons over Warden Law, as well as five self-acting planes and near-level locomotive-worked sections in between. George Stephenson followed the same approach when the Grand Allies commissioned a railway from Springwell colliery, operational from 1826. Both he and Robert did the same on the Stockton and Darlington, which had two inclined planes, one at Etherley at its western end linking the Wear and Gaunless valleys, the other at Brusselton to overcome a ridge.[8] Most of the systems built in emulation of the Stockton and Darlington used them, including the Canterbury and Whitstable and the Bolton and Leigh in 1827, the Dundee and Newtyle and the Cromford and High Peak in 1831, the Leicester and Swannington in 1832 and the St Helen's and Runcorn Gap in 1833. An extension of the Brecon Forest plateway to reach the Swansea Canal, completed in 1834, included a substantial inclined plane at Ynysgedwyn. Above all, the Stanhope and Tyne, opened in 1834, had no fewer than 15 rope-hauled sections of one sort or another on its route from the limestone workings in Weardale and the collieries of Pontop and Medomsley to the mouth of the river at South Shields.

However, the best known such system at the time, not to say also the most ambitious (see chapter 10), was the two-mile-long double-track Wapping Tunnel below the streets of Liverpool, operational from 1830, which connected the Mersey docks with the main course of the railway to Manchester at Edge Hill. This was where visiting engineers from the United States landed when they came to Britain, and it is not surprising that this particular feat of engineering should have made a profound impression on them, the more so as many of them were already thinking how to drive their projected railroads over mountain ridges once they had returned home.

Some of these early American systems faced environments far more challenging than the lowlands of Lancashire and Cheshire, or even the moors of Stanhope and the slopes of the Tyne Valley, so rope-hauled sections were inevitable. On a small scale they had been a feature of systems in the USA since about 1805, and the Quincy Granite Railway (chapters 2 and 3) had a gravity inclined plane 315 foot (96m) long, at an angle of about 15 degrees on an endless chain, so they were far from being a novelty.[9] The best-recorded inclined planes were those on the Allegheny Portage Railroad, which connected two canals to offer a route across the barrier range in central Pennsylvania separating the Eastern Seaboard from the Gulf of Mexico watershed, and which began operations in 1834 (chapter 12). Their design echoes the Wapping Tunnel, which also influenced practice on the Philadelphia and Columbia (1831–4) as well as on the Charleston and Hamburg (1833).[10] Other public railways in the

'MOST SUITABLE FOR HILLY COUNTRIES'

14. The Morris Canal used wheeled cradles on railed inclined planes to transport water vessels.

USA which used inclined planes were the Mohawk and Hudson, the Baltimore and Ohio and the Ithaca and Owego.[11] Coal carriers also needed them. The Delaware and Hudson made use of six powered and three self-acting planes on its 16.5 mile route through Rix's Gap in the Moosic Mountains from the collieries at Carbondale to the canal.[12] However, their most intensive use in the USA came not on a railway at all but on the Morris Canal, which was built to move coal across the northern New Jersey hills to New York. Operational from 1831, it used no less than 23 railed inclined planes (as well as 23 locks) in the course of its 102-mile journey from Phillipsburg on the Delaware to Paulus Hook on the Hudson, raising boats on hinged 16-wheel carriages.[13]

Two coal-carrying railways built in France in this period also used them. The short Épinac railway to the Burgundy Canal involved two, one operated by a 20-horsepower steam engine, the other self-acting, between sections where horses and oxen hauled trains uphill and gravity powered them on the down gradient.[14] Two powered inclines were built on a railway from Andrézieux to a wharf on the Loire at Roanne, operational from 1833.[15]

Technology

Rope and chain haulage was complicated and took many different forms, depending on the weight, frequency and direction of the loads to be passed and on the terrain. Ropes and chains might be wound around drums or around sheaves or pulleys. They could be end to end or continuous. Gradients might be steep or formations might be entirely level.

If the loads ran consistently down a gradient, and only empties had to make the climb, a self-acting plane was an option. Where the traffic ran in both directions, a water tank on rails could be used to counterbalance loads. Otherwise, if loads needed to be hauled up a gradient, there were several possibilities. In these circumstances, an inclined plane required a prime mover, generally a fixed steam engine, though a waterwheel or other power source might serve, as chapter 3 describes, operating a revolving drum or wheel at its summit, which activated a rope or a chain. The plane might be single track, with the loads pulled uphill mechanically and empties descending by gravity on the rope, regulated by a brake. Where loads needed to pass each other, the plane could either be built with double track or with a shared common rail, opening out to a passing loop at mid-point, with the section below the passing loop either uniting to single track or reverting to a three-rail formation.

Some planes were designed to cross a hill or a ridge, where the load had to go both up and down. The least complicated arrangement in such circumstances was a single-track system with a prime mover at the summit operating a rope which hauled the load up the gradient and then regulated its descent on the other side. Alternatively, double drums or wheels could be installed at the summit, connected by gearing. Where traffic was heavier, two parallel lines of rails were laid, which enabled loads to pass each other.

Thompson's system involved dividing the entire railway into separate sections, each one with a fixed engine which drew the waggons towards itself and at the same time unwound a rope from the engine on the other side of the waggons. The consuming question by 1828 was whether or not the Liverpool and Manchester would use some such method on its whole length, not purely for the steeply graded sections at the Liverpool end.[16] The original proposals of 1824–25 had anticipated the use of locomotives, confirmed by a visit to Killingworth and Hetton to see George Stephenson's machines at work undertaken by an independent engineer, Charles Sylvester. He recommended short inclined planes on steeper sections on which fixed engines would operate continuous ropes to haul trains, with locomotives providing part of the motive power. Sylvester returned to Killingworth and Hetton to undertake further trials in the company of Nicholas Wood and William Brunton, both of whom we met in chapter 4, along with Stephenson himself, James Walker (1781–1862), John Urpeth Rastrick (1780–1856), of whom more shortly, Philip Taylor (1786–1870) and other civil engineers and representatives of London and Birmingham railway interests.[17]

The initial Liverpool and Manchester Bill failed in parliament, but, once it was revived, its directors challenged George Stephenson's assumption that the primary form of motive power would be locomotives provided by the firm of which he was partner and that fixed engines would be restricted to gradients. James Cropper, one of the directors, who favoured fixed engines, and the secretary and treasurer, Henry Booth, who favoured locomotives, were sent to Darlington, 'the great theatre of practical operations',[18] and reported that rope haulage would be suitable for the Liverpool and Manchester. Stephenson produced a document offering detailed comparative costs, and pointing out that rope haulage was problematic at level crossings and junctions.

That still did not settle the matter so far as the committee of management was concerned. Recognising that it had urgently to be resolved, they engaged James Walker and John Urpeth Rastrick to examine the Bolton and Leigh and the Middleton, as well as the principal railways of the north-eastern coalfield. Here they spoke at length with Thompson, were allowed to consult his private papers (not least because a heavy fall of snow impeded fieldwork for the best part of a week) and were impressed by the 'neatness of execution, despatch, and methodical system and arrangement' of his system.[19] Their hurriedly written report, received in January 1829, saw the advantage of locomotives where loads were light, but recommended that rope haulage be adopted for the Liverpool and Manchester.[20] The rate per ton per mile with fixed engines would be, they believed, 0.653 of a penny cheaper than with locomotives.[21]

They recommended that the fixed engines be located at intervals of between 1.25 and 2 miles on the double-track main line. Trains were to operate like a crack stagecoach service, with the same capacity for a quick changeover of haulage ropes at stations as with a nimble fleet of horses in an inn yard.[22] However, the Liverpool and Manchester directors resolved against their proposals and it is easy to see why. Walker and Rastrick's scheme was overcomplicated and inflexible, requiring stations to be constructed as the gradient and length of rope required, rather than where passengers or merchants might need them, and the system of crossovers from one running line to another which it required was potentially problematic. All of these were set out in an 83-page report drafted by Robert Stephenson and written up by Joseph Locke, an engineer on the Liverpool and Manchester and a Stephenson protégé who was now coming into his own (chapters 8, 11 and 13).[23] Nicholas Wood attempted to explain their operation in the third edition of his *Practical Treatise on Rail-Roads*, but as he wearily admitted:

'MOST SUITABLE FOR HILLY COUNTRIES'

> The inconvenience, delay, and risk, attending the carriages thus passing from one line to the other, and travelling, in both directions, upon different parts of the same line of road, confine the use of this mode of transit, exclusively, to private lines of road; upon public lines, it would produce inexplicable, and irremediable, confusion. Upon public lines, where fixed engines are obliged to be used, to surmount steep declivities, some mode of application, different from any of those previously described, therefore became necessary. On the Liverpool and Manchester railway, which terminates, at the Liverpool end, at a high level, it was necessary to apply a fixed engine, to drag the goods up from the low level, or station, near the docks, to the higher level; or where the locomotive engines could be used, the length of the plane being 2250 yards, and the inclination, 1 in 22.[24]

The inclined plane to which he refers here was the Wapping Tunnel, which was operated by a continuous rope running down one track and returning up the other, with horizontal sheaves at the summit and the foot and a tensioner mechanism to keep it taught. The summit at Edge Hill was also the foot of another plane, which ascended from this point to the passenger station at Crown Street.[25]

Visiting engineers considered the Wapping plane to be the state of the art. Moncure Robinson (1802–91) discussed its progress with George Stephenson when it was in its early stages before returning to the USA to survey and design the Allegheny Portage Railroad, the beginning of his long and illustrious career as an engineer and financier.[26] When the even more youthful Edward Miller (1811–72) made his way to England in 1831 to learn about railroad building, he also visited the Liverpool and Manchester, as well as travelling to Derbyshire to see the Cromford and High Peak (see chapter 10).[27] On his return, now principal assistant engineer of the Allegheny, he designed and superintended the construction of the machinery for its 10 inclined planes by five separate foundries in Pittsburgh.[28] They closely followed Liverpool and Manchester practice, down to the use of a tensioner carriage at the summit, and only significantly departed from it in the use of a double sheave arrangement at the summit rather than a single one as at Edge Hill, and in the use of brake cars permanently attached to the rope to which rolling stock could be coupled, to halt the vehicles should there be an accident.[29] These were found on some other American systems, but on only one British railway, the Dundee and Newtyle, on its inclined plane over the Law hill.[30]

The 'Belmont' inclined plane on the Philadelphia and Columbia was evidently similar to those on the Allegheny – both railroads were sponsored by the Commonwealth of Pennsylvania, and both formed part of a line of communications conceived to link the Atlantic Ocean and the Schuylkill River to the Ohio. It was powered by two stationary machines of 60 horsepower each, installed in a two-storey stone building, composed of two wings joined by a wooden platform crossing the double track, which wound what was by 1836 at least an endless iron rope.[31] A lithograph prepared a few years later shows two trains passing each other and, at the foot, a structure spanning the tracks which may have protected the lower sheave and the brake cars from the elements, as well as providing shelter for workmen.[32]

Much less is known about the inclined plane on the Charleston and Hamburg, though it is also said to have followed Liverpool and Manchester practice.[33] A native touch was the use of brake cars, but, unlike the Allegheny, a single sheave operated at the summit, powered by two steam engines.[34] On the Baltimore and Ohio, the four inclined planes over Parr's Ridge on the Frederick branch were short-lived. The railroad's engineers had discussed these schemes with Benjamin Thompson when they visited England, and its consulting engineer, Stephen Harriman Long (1784–1864), had worked on the Allegheny, but when it opened on 1 December 1831 the carriages had to be pulled up the grade by horses, since the machinery, from Mattawan of New York, did not arrive until the following August. In the event it was not installed until after the first successful locomotive-hauled trip over the branch in December 1834.[35]

Waterwheels were used in some locations. Of the six planes on the Delaware and Hudson which ascended against the load, five were steam-powered and one operated by a waterwheel.[36] However, the system which made the most use of them was the Morris Canal. Every one of its 23 planes was powered by water. Their basic design was the work of James Renwick, Professor of Natural and Experimental Philosophy and Chemistry at Columbia College, New York, but the work was carried out by different contractors and there were many variations between them – some were double track as Renwick had proposed, some single track, one triple-track.[37]

The hinged 16-wheel carriages on the Morris Canal were similar in principle to those on the Allegheny Portage Railroad, and revived the well-established system of moving boats by railed incline plane that had first been used at Ketley in 1798, which in turn influenced the design of lifts and canals over the world.[38] The use of railed platforms to carry other forms of rail vehicle up and down a gradient, first evident in the inclined plane at Cyfarthfa Ironworks discussed in chapters 3 and 4, also became more common. An American example was to be found at Carthage, New York State, on the Rochester railroad, powered by a winch to connect with the ship landing on the Genessee River.[39] The Stanhope and Tyne used them to negotiate a ravine called Hownes Gill, 160 feet down and 800 feet across. Tomlinson, the historian of the North Eastern Railway and of its predecessors, describes it thus:

> When the line was first planned, the projectors intended to throw a bridge over the ravine, but it was afterwards decided, presumably at the suggestion of Robert Stephenson, to lower the waggons down one side and to haul them up the other by the operation of a small engine of 20 horse-power situated at the bottom. The sides of the Gill being so precipitous, 1 in 2½ and 1 in 3 respectively, it

was considered necessary, in order to keep the waggons in a horizontal position, to provide cradles or trucks – one for each incline having front and hind wheels of unequal diameter: upon these the waggons travelled side foremost. Only one waggon could be taken across at a time.[40]

This system was echoed in the Ynysgedwyn incline built in 1832–4 by the engineer John Brunton (son of William, see chapter 4) as part of an extension of the Brecon Forest plateway to the Swansea Canal, where narrower gauge waggons and road vehicles could be rolled on and off the platforms at the foot and summit of the plane, without having to attach them to the rope. Joseph von Baader had recommended a similar system in 1822.[41]

Prime movers and their manufacturers

Different types of winding machine were to be found on these systems wherever a load needed to be raised. If the winch on the Rochester Railroad was hand-operated, it is unlikely to have been very powerful. Neither, probably, were the horse whims on the substantial double-track planes on the Ithaca and Owego and, more plausibly, on a short section of the Tuscumbia, Courtland and Decatur in Alabama.[42] Otherwise, fixed steam engines and waterwheels were employed. The demands on these prime movers were considerable, as they needed to achieve a high starting torque whenever operations began. Some hybrid systems balanced ascending loads as much as possible. The steam engines on the Mohawk and Hudson's inclined plane at Prospect Hill in Schenectady were assisted by a counterbalance car containing stone.[43] The Whitby and Pickering used a water balance on its inclined plane, whereby railed water tanks filled at the summit on each track drew up the loads.[44]

There was no shortage of manufacturers in Britain willing to tender for the construction of fixed steam engines to operate rope systems.

'MOST SUITABLE FOR HILLY COUNTRIES'

In 1818 Fenton, Murray & Woods of Leeds constructed a 20-horsepower condensing engine for Fawdon Colliery, Thompson's stamping ground.[45] The firm of Robert Stephenson & Co. in Newcastle built its first high-pressure fixed engine in 1826, which was installed at Mount Moor Colliery, south of Newcastle. Its first opportunity to apply this technology to railway winding came the following year when they contracted to supply three to the Canterbury and Whitstable, one for the Bolton and Leigh, two each for the Stockton and Darlington and for the Liverpool and Manchester, the winding machinery in this case being provided by William Fairbairn.[46] Other builders included the Butterley Company, for the Cromford and High Peak in 1829, R. & W. Hawthorn of Newcastle, who built a replacement for the troublesome Stephenson engine on the Brusselton incline in 1831, Mather & Dixon of Liverpool in 1834 for the tunnel to the new Lime Street terminal on the Liverpool and Manchester, Hawks of Gateshead for the Stanhope and Tyne, the Horsley Coal and Iron Company for the Leicester and Swannington, both 1833, and Neath Abbey Ironworks in South Wales for Ynysgedwyn in 1834.[47]

Most British winders were beam engines of one sort or another, built integrally with their housing, reflecting well-established colliery traditions. The two Stephenson engines at Edge Hill on the Liverpool and Manchester, supplied in 1829 and 1830, were each built with 24-inch diameter, 6-foot stroke cylinders with 20-foot flywheels, set in ornamental structures on either side of the railway's Moorish arch, and supplied by return-flue boilers cut into the rock on either side of the cutting. One wound goods trains up through the long Wapping Tunnel from the docks, the other passenger trains up the shorter tunnel to the terminus at Crown Street.[48]

Exceptions to the conventional beam engine rule were three side-lever types with the reciprocating beam placed below the vertical cylinders. One was installed at Etherley on the Stockton and Darlington, which was originally intended for a steam boat, and the other two were built by Mather & Dixon for a third plane at Edge

Hill in Liverpool to connect with a new terminal station at Lime Street, which was under construction from 1833 to 1836.[49] Only one British horizontal-cylinder winder from this period is known, built for the Leicester and Swannington in 1832.[50] By contrast, this type seems to have been the rule in the USA. On the Allegheny, two 2-cylinder horizontal engines powered each of the 10 double-track inclined planes, some 35 horsepower, some 30 horsepower, with 14-inch cylinders and a 5-foot piston stroke, each fed by three boilers, allowing ascending cars to move at about 4 mph. An initial plan to use single-cylinder engines with flywheels was abandoned as unsafe – better to have two engines which could be properly regulated. Arrangements on the Charleston and Hamburg were similar; here two fixed engines capable of operating independently turned the drive axle through gearing without a flywheel. The boiler was divided into two compartments, enabling one to be repaired while the other remained in use, and a stock of spare parts was retained for general maintenance.[51]

The fixed engines on inclined planes in the USA were all American-built. The United States had by now decisively entered the steam age, its foundries and manufactories turning out machines for southern sugar plantations, for saw and grist mills out west, textile mills, printing and navy yards on the east coast and, above all, for river vessels.[52] Railway engineers clearly had every confidence in them, even if they more cautiously looked to England when they purchased locomotives (chapter 8). The Allegheny Portage Railroad placed its orders with five separate Pittsburgh manufactories.[53] The firm of Rush & Muhlenberg, operating Oliver Evans's old Mars Works ('Iron foundry, and steam engine manufactory'), supplied the Philadelphia and Columbia, and the West Point Foundry of New York the Charleston and Hamburg. Peter Clute's foundry at Schenectady provided the engine for the Mohawk and Hudson, and Mattawan of New York those for the Frederick branch of the Baltimore and Ohio.[54]

Little is known about the engines used on the Andrézieux–Roanne line except that both were by the long-established Chaillot Foundry at Paris. The one at Biesse was 70-horse power; the surviving buildings suggest a beam engine operating two drums. The Neulize (Neulise, Neulisse) machine was a feeble affair, reused from a sunken river boat, which could only haul a load if there were empties to counterbalance it. After its boiler blew up on 21 October 1834, a chronic lack of funds prohibited the purchase of a replacement for many years, and traffic was hauled up the gradient by horses and oxen.[55]

Four stationary engines from this period survive, all in England.[56] One remains *in situ*, at Middleton Top on the Cromford and High Peak, built in 1829. Like the other seven on the railway, it is a twin engine, the two cylinders being low pressure condensing double-acting types, with separate controls, operating a common crankshaft to drive what was originally an endless chain from a wheel in the basement. All of these were built by the Butterley Company, which was based near the south-eastern terminus. Much of the Clowes Wood engine from the Canterbury and Whitstable also survives, though in dismantled state. This was a 25-horsepower machine by Robert Stephenson & Co.[57]

The other survivors are departures from the Watt-type beam engine model. The Stanhope and Tyne's Weatherill engine of 1833 in the National Railway Museum at York is a single vertical cylinder beam engine like a Durham colliery shaft-winder of the period. It was built to power a plane 1 mile 128 yards (1726m) in length on an average gradient of 1/13 and was equipped with Phineas Crowther's parallel motion and a large flywheel. It operated two drums, was rated at 50 horse power and worked at a boiler pressure of 60 lb per square inch. The massive ashlar-stone engine-house was an integral part of the design.

The Swannington engine is also preserved in the National Railway Museum. Built to a design by Robert Stephenson to haul coal trains up a 1 in 17 incline, it is fitted with an early form of piston valve gear later used extensively on locomotives. Stephenson evidently feared that gravity would cause the weight of the horizontal piston to wear the cylinder bore into an oval shape, so a substantial tail-rod arrangement with slide bars and slippers was fitted.[58]

Ropes and chains

Rope-hauled systems such as these depended on improvements not only in steam technology but also in the manufacture of hemp ropes, chains and ultimately wrought-iron wire ropes. The north-east of England was well placed in this respect since its industries had relied on hemp rope for a long time, not only for shaft haulage in collieries but also for rigging the ships on which the London trade depended. As a consequence, high-capacity mechanical production had already superseded the traditional ropewalk in this part of the world even before the end of the eighteenth century. John Grimshaw (1763–1840), who established the first steam-powered rope-making plant in Sunderland, became a partner with Benjamin Thompson at Fatfield Colliery, and carried out experiments with him on railway traction. It was he who produced the ropes used on inclines on the Stockton and Darlington and on the Canterbury and Whitstable.[59] His firm, as Webster's of Sunderland, made the 6 inch (15.24 cm) circumference rope for the Bolton and Leigh's longer Chequerbent inclined plane, which when worn was transferred to the shorter one at Daubhill.[60] In 1822 Grimshaw was granted a patent for flat ropes, a type used at Épinac.[61]

American railroad inclined planes also began their working lives with hemp ropes, a crop found in abundance in Kentucky, though the Allegheny used Russian or Italian hemp, which was reckoned to last no more than one season.[62] Ropes on the Philadelphia and Columbia lasted at most 18 months and cost $3000 to replace. A break in the rope was 'not an extremely rare occurrence'.[63] Hemp ropes were a significant cost on any system and were apt to stretch. However, experimental use of India rubber on the Crown Street plane on the

Liverpool and Manchester and on the Stanhope and Tyne offered no improvement.[64] Iron chains, on the other hand, were notoriously as strong as their weakest link, and were initially little favoured for industrial haulage.[65] However, the Royal Navy had been conducting trials since the early nineteenth century, and was increasingly replacing hemp with chains.[66] The development of suspension bridge technology and a growing understanding of metal fatigue made chains a safer option, and the Butterley Company was evidently confident enough to install them on the Cromford and High Peak in 1829.[67]

The French-American surveyor and army officer Guillaume Tell Poussin stated in 1836 that the 'Belmont' inclined plane on the Philadelphia and Columbia Railroad was operated by an endless iron rope.[68] If so, this was a very early instance. Ropes of twisted steel had been introduced at Clausthal mine in Saxony in 1834 by Wilhelm August Julius Albert (1787–1846), but the Allegheny and the Morris Canal only went over to wire rope haulage in the following decade, reflecting the work of the Prussian-American engineer John Augustus Roebling (1806–69). An iron rope was introduced on an incline on the Room Run railroad in 1839, 'an iron band, one-twelfth of an inch in thickness by three inches in width, as a substitute for rope or chain'.[69]

'Swift engines upon a double way, I am convinced, may be used to the utmost advantage'

The Liverpool and Manchester opened as a locomotive-worked system which made some use of rope haulage, but discussion continued as to their relative merits. Not only were the Wapping and Crown Street tunnels operated by fixed engines, but so was the tunnel to the new town terminus at Lime Street, where an order was placed in April 1834 for two side-level noncondensing engines.[70] The Rainhill trials made it clear that, on the main line at least, the future lay with locomotives, but once the railway started operating it became equally evident that they were still struggling on the steeper sections at Sutton and Whiston. The directors resolved against ordering powerful locomotives capable of hauling trains up these gradients by themselves, and decided on the use of 'help-up engines' ('pushers') at the back of trains as assistance. This proved difficult and, after nearly a year's operation, Robert Stephenson was asked to prepare estimates for fixed engines, winding houses, rope and pulleys for these two sections. Given the likely cost (over £9,000), the Board reconsidered yet again and continued with locomotive operation, but it is clear that for some time these two parts of the system were still difficult to operate, that the 'help-ups' would bang into the rear of the trains they were assisting after following them up the gradient, and that drivers and firemen would engage in some potentially very dangerous practices to make sure that their locomotives reached the summit.[71]

The Baltimore and Ohio's experience with inclined planes was more decisive. When the Mattawan Foundry was late in providing the machinery, the railway found that horses could do the job. When the engines and fixed ropes finally did arrive, locomotives had also proved capable of hauling the trains.[72] The truth was that not only were footplate crews becoming more adept at getting the best out of their existing locomotives by coaxing them up gradients once thought impossible, but new designs were becoming more powerful still.[73] This was a development to which Walker and Rastrick had given insufficient weight but which Timothy Hackworth had divined some time previously. Writing to reassure George Stephenson when it was rumoured that the Liverpool and Manchester would indeed invest in fixed engines, he remarked:

> As to my general opinion as to the locomotive system, I believe it is comparatively in a state of infancy. Swift engines upon a double way, I am convinced, may be used to the utmost advantage ... Do not discompose yourself, my dear Sir; if you express your manly, firm, decided opinion, you have done your part as their adviser. And if it happened to be read some day in the newspapers

'MOST SUITABLE FOR HILLY COUNTRIES'

– 'Whereas the Liverpool and Manchester Railway has been strangled by ropes,' – we shall not accuse you of guilt in being accessory either before or after the fact.[74]

Hackworth, Wood and Stephenson were all to be vindicated, as the remarkable design transformation which locomotives underwent made them capable of climbing progressively steeper gradients. By contrast, rope-hauled systems were found to be complex and inflexible. Had they been used on the main line of the Liverpool and Manchester in 1829, they would have set back the railway cause significantly.

This did not mean that the use of inclined planes and of rope haulage came to an end. The Mather & Dixon engine at Edge Hill hauled trains in and out of the Liverpool and Manchester's new passenger terminus in the centre of the town for many years after it was first put to work in August 1836.[75] The following year, Robert Stephenson installed an inclined plane on his London and Birmingham on its first stretch from Euston to Camden, an enlarged version of the Wapping system, and then on his suburban London and Blackwall in 1840 (chapter 13). The idea that railways could be powered by a fixed engine, though with pneumatic, rather than mechanical, connection to the trains, was soon to be revived in the form of the atmospheric railway, which met with little success, but electrical transmission from central steam plant by means of a third rail or overhead wires to a locomotive came into use from the late nineteenth century onwards and is now common. Fixed steam engines still power many railways across the world, but now do so by generating electricity in a power station.

After the 1830s, rope haulage became increasingly restricted to industrial and constructional railways over the globe, where they survived long and late.[76] The 'era of rope and steam' on colliery systems in the north-east of England, for instance, lasted into the late twentieth century.[77] Rope haulage is also still used on the San

Francisco cable-car system and at Llandudno in North Wales on the Great Orme Tramway. It was not, any more than the plateway, an ill-considered technology but it was fortunate that, just as the double-track mainline railway came into being, and when the claims of rope and chain haulage were being pressed, the steam locomotive underwent its remarkable and sudden phase of development.

8

'That truly astonishing machine'
Locomotives 1815–1834

I took my passage to Manchester by that truly astonishing Machine called Steam Coach.[1]

By 1815 perhaps as many as 30 locomotives had been put to work, all in the United Kingdom. They hauled trains carrying coal or iron at a walking pace. Within 20 years, there were several hundred, built in places as far apart as Pennsylvania, Lyon, Berlin and the Ural Mountains. Major events such as the openings of the Stockton and Darlington in 1825 and of the Liverpool and Manchester five years later focused public and professional interest in the possibilities they offered. Above all, the trials held at Rainhill near Liverpool on the as yet uncompleted railway in October 1829 to assess the possibilities and merits of locomotive operation, established a design path which would be followed for a hundred years and more. A world that had never known anything faster than a galloping horse now learnt of machines travelling at 35, even 60, mph.[2] Towns and cities which had relied on the carrier's cart, the canal barge and the stagecoach could now plan to move their goods and their citizens at speed and on a much greater scale by locomotive-hauled trains. Simultaneous developments in steam road haulage and in ocean navigation contributed to a sense that the world was drawing closer together.

The years from 1815 to 1834 divide neatly into two nearly equal periods of development in Britain. For the first 10 years, locomotives remained largely confined to colliery railways in the north of England, where the basic Murray–Blenkinsop configuration was developed by George Stephenson and his colleagues. Thereafter, designs changed radically in several different ways in the light of the Rainhill trials, and their use spread over the United Kingdom and to continental Europe, Russia and above all the USA.

The first phase

The Steam Elephant

Steam Elephant was a six-wheeled locomotive built in 1815 to operate a 1.5 mile section of the wooden-railed Wallsend waggonway to a design by John Buddle and William Chapman, with components supplied by Hawks of Gateshead. Like the Killingworths (see below), it had a centre-flue boiler with two vertical cylinders set into its top centreline, driving motion beams on slide bars which turned crankshafts operating the axles through 2:1 reduction gears. The lower part of its tall tapering chimney was surrounded by a feedwater heater. It did not work effectively on the wooden rails but was more successful after the waggonway's conversion to iron Losh rails, and it may have been sold on to the Hetton Colliery railroad (chapters 2 and 7).[3]

Stephenson locomotives

Developments began at Killingworth Colliery, on the north bank of the lower Tyne, in lease to the 'Grand Allies', whose forebears – Baron Ravensworth, the Earl of Strathmore and James Archibald Stuart-Wortley-Mackenzie, who became Baron Wharncliffe – had been working here since the late 1720s. Their viewer was Ralph Dodds (1792–1874), initially assisted, and in 1815 succeeded, by Nicholas Wood (chapter 4), a highly literate and scientifically minded man. They ran a technically advanced pit. George Stephenson, 33 years old in 1814, was their enginewright, responsible for the pumping and winding machinery. He had not so far built a locomotive, but it was only a short walk from Killingworth to Coxlodge, where the Murray-Blenkinsops were at work, while a trip home to his birthplace on the back of 'Squire', the Irish horse the Grand Allies allowed him, would have shown him the newly built *Puffing Billy* and *Wylam Dilly* making their way to and from the staithes (chapter 4).[4]

The Grand Allies had long experience of the coal industry, had invested heavily in their new operations and knew that long-term output would require effective mechanical haulage. As the Napoleonic wars drew to a close, Stephenson, Dodds and Wood began to experiment on the Killingworth Way and then on the first purpose-built steam railway, the Hetton. Before long, Wood and Stephenson were even recording the resistance of railway vehicles using an early form of dynamometer.[5]

Stephenson's first locomotive was constructed at Killingworth in 1814, a small trial machine known as *Blücher* (chapter 4), followed by one for Killingworth in 1815, two in 1816, two in 1818 and one in 1821. One of the 1816 batch survives, much altered, at the Stephenson Railway Museum, as *Billy*. Five were built for the Hetton in 1822–3, two for the Stockton and Darlington in 1825 and two for the Grand Allies' Springwell Colliery in 1826. Others went to the Kilmarnock and Troon plateway in Scotland in 1816, and to Hendreforgan Colliery at Llansamlet near Swansea in 1819, which probably also ran on a plateway.[6]

Though the first locomotives were built at Killingworth, others were perhaps constructed at William Losh's Tyneside engineering workshops, and the later ones at the manufactory established on 23 June 1823 off Forth Street in Newcastle by the Stephensons, the

Quaker banker Edward Pease and Michael Longridge of Bedlington Ironworks. It took its name – Robert Stephenson & Co – from the 19-year old managing partner. It was all the more strange that within 12 months Robert should have left on a three-year contract to work gold and silver mines in South America. His complicated relationship with his father and a desire to prove himself on his own terms prompted the move, but the firm was deprived of his counsels at a crucial time, and his relationship with Longridge was damaged.[7]

Perhaps for this reason, all the locomotives built up to 1826 followed the Killingworth design path. Each one included some incremental improvements but had in common a single-flue boiler with a projecting chimney and dished ends, with two inset vertical cylinders, driving motion beams which activated connecting rods. Those on *Blücher* drove gear wheels which engaged with the drive axles, synchronised by a central spur wheel, whereas on the later ones the rods directly drove axles connected by an endless chain, described in a patent which George Stephenson took out with Ralph Dodds in 1815.

He obtained another patent in 1816, with William Losh, for cast-iron rails (chapter 2) as well as for steam locomotive springs and an eccentric to drive the valves instead of square tumblers. The diagram accompanying the patent shows a six-wheeler, reflecting Stephenson's concern to spread the weight on the track, but the only one known to have been built to this wheel arrangement was the Kilmarnock and Troon's *The Duke*, running on cast-iron plates. Neither it nor the Llansamlet locomotive lasted long in service, a fact which underlines the success of the others, and their compatibility with edge rails. The achievement of George Stephenson and his colleagues was to enable motive power and track to operate harmoniously, by equipping their locomotives with springs and with wheel treads of a slightly conical cross-section, by making use of Losh cast-iron rails and then, when it became available, the wrought-iron Birkinshaw type. They were not simply building locomotives for customers but obliging them to accept a complete traction system.[8]

'THAT TRULY ASTONISHING MACHINE'

The first Stockton and Darlington locomotive inaugurated the service on 27 September 1825. It dispensed with the chain between the axles, and was the first locomotive to be fitted with coupling rods. Instead of the complicated steam springs, its rear axle was carried in a tube, which acted as an axle box, pivoted at the centre to allow it to rock, thereby giving a three-point suspension. It was built with cast-iron wheels, each with eight spokes. One of these broke only a few days after the opening of the railway; it was soon replaced but this resulted in everything having to be hauled by horses in the meantime. They were later changed for two-piece cast-iron plug types. It came to be known as *Locomotion*, and was followed by *Hope*, *Black Diamond* and *Diligence* later in the 1820s.[9]

Though the Killingworth types were soon eclipsed by the Stephensons' *Rocket* design and its successors, two were built for the Monkland and Kirkintilloch by a Glasgow firm in 1831, albeit incorporating some later design features, and the last ones were not constructed until around 1848.[10]

Robert Wilson's Chittaprat *and Timothy Hackworth's* Royal George

The French engineer Marc Seguin (1786–1875; see below and chapter 9), on a study tour of England, inspected the Stockton and Darlington Railway in December 1825 with his brother Paul, and was intrigued by a locomotive then on trial, known as *Chittaprat* because of the sound of its unsyncopated exhaust. It had been built not by Stephenson, but by Robert Wilson in Newcastle and it had four vertical external cylinders directly operating one axle in pairs, connected to the other by a coupling rod.[11] Some parts may have been used by Timothy Hackworth in October 1827 to build the *Royal George*, an 0-6-0 locomotive for the Stockton and Darlington. Its boiler had a return flue like the Wylam engines, and vertical cylinders working downwards to drive one axle which operated the others through coupling rods. It was considerably heavier than the

Stephenson locomotives, giving it greater adhesion. As a six-wheeler, it distributed its weight more effectively, and, being built for power rather than speed, there was little danger of oscillation or of it damaging the track.[12] It was the most powerful locomotive yet built.

Stephenson locomotives for France, and the remorquers

Seguin returned to England in December 1826 to assess traction options for his Saint-Étienne to Lyon railway, which was then at the planning stage.[13] He duly placed an order for two locomotives with Stephenson & Co., the first such purchase from outside the United Kingdom, and the beginnings of a long-term and highly successful British export industry.[14] They arrived in the late summer or early autumn of 1828. Their boilers incorporated water tubes crossing the flue, and the four-coupled wheels were driven by two vertical upward cylinders on either side of the boiler, driving the wheels by parallel motion connecting rods. One went initially to the engineer Alexis Hallette (1788–1846) at Arras for evaluation, the other directly to the railway's workshops at Perrache near Lyon. Here Seguin conducted trials on it to improve draughting, including a boiler of his own design, which instead of water tubes incorporated many small fire tubes passing through the whole water space. These offered a far greater heating surface area for the same overall volume, and the increased speed of gases through the tubes improved conduction. These led Seguin to build his *remorquer* ('tug') locomotives at Perrache, the first of which he completed in October 1829, in which cylinders and drive mechanism to Stephenson's design were mounted on each side of a multi-firetube boiler. The firebox lay beneath the boiler barrel, and both were water jacketed; the ashpan was closed and air blown in by two large rotary fans mounted on the tender, driven by cords and pulleys from the axles to ignite the bituminous coal it burnt.[15] This was soon abandoned in favour of a blastpipe, which meant that the locomotives could be rebuilt with a firebox at one end of the boiler and, at the other, a smokebox, the component which supports the chimney, assists combustion and enables ash and soot to be removed.

The firetube boiler was a crucial breakthrough, and is further discussed in the context of the *Rocket* (see below), which was completed in September 1829, after the experimental reconfiguration of the locomotive at Perrache and a few weeks before the first *remorquer*.[16] Though it has been claimed as both a French and an English invention, it is far more likely that the growing use of both water and fire tubes reflects regular meetings and correspondence between Seguin and the Stephensons – a client with a scientific and a practical grasp of steam production, and an experienced engineering company.[17] Invention was also claimed by Henry Booth, Treasurer and Secretary of the Liverpool and Manchester Railway.[18]

Rastrick's locomotives

Foster, Rastrick & Co. in Stourbridge in the English West Midlands produced only four locomotives, all in 1828 and 1829, including their best known the *Stourbridge Lion*, one of three exported to the Delaware and Hudson, of which components remain, and a fully surviving example, *The Agenoria*, for a 1.375-mile section of the Earl of Dudley's Shutt End Colliery Railway to the Staffordshire and Worcestershire Canal, where it first ran on 2 June 1829.[19] The firm's managing partner was the same John Urpeth Rastrick who had favoured cable haulage on the Liverpool and Manchester (chapter 7) and who then acted as a judge at Rainhill. His four locomotives were 0-4-0s and combined established components such as external vertical cylinders and grasshopper beams like the Wylam 'dillies', with innovative technologies such as a bifurcated fire tube in the boiler of the type that Robert Stephenson was developing for his own locomotives.[20] Whereas the American engines were barely used, *The Agenoria* operated until 1865 and is now preserved in the National Railway Museum.

The Delaware and Hudson order came about when the company's chief engineer, John B. Jervis, instructed their agent Horatio Allen while on his visit to Britain to contract for three locomotives suitable for hauling coal to the company's canal barges. In 1828, orders were placed for three from Foster, Rastrick and a fourth from Stephenson.[21]

In May 1829 the Rastrick locomotives were landed at the wharf of the West Point works in Beach Street, New York City, to be assembled for exhibition to 'gentlemen of science and particular intelligence', and to be put into steam.[22] This confirmed that they would work successfully on the Lackawaxen coal which the Delaware and Hudson was itself mining. They arrived by canal at Carbondale on 23 July, and the *Stourbridge Lion* was steamed on Saturday, 8 August. After running it up and down at the terminus, Allen drove it for a mile and a half 'light engine' as far as Seelyville, where a low bridge prevented it from going further.[23] The *Lion* confounded pessimists by not derailing where the rickety timbers on which the track was laid reached a height of 30 feet on a curve over the Lackawaxen Creek, but it did cause the trestles to vibrate alarmingly. Allen returned safely to the cheers of the spectators and the booming of cannon, but the locomotive and the trestle timber formation were an evident mismatch. Although sporadic trials continued that month and the next, the locomotives were never used in service. Jervis recommended that the track be upgraded, advice which was wisely ignored, as the traffic could be handled by horses.[24]

The Rainhill trials

When the Liverpool and Manchester Railway was nearly completed, the directors, still unpersuaded by the competing claims of different forms of traction, decided to hold what they called an 'ordeal' to assess the possibilities that locomotives offered, on the lines of the earlier trials their engineers had conducted at Killingworth and

Hetton, and in the spirit of experiment and research characteristic of the north-east of England coalfield. This initiative was welcomed by the Stephensons and by Joseph Locke (chapter 7).[25] The 'ordeal' was held at Rainhill only three months after it had been announced, giving little time for anyone to produce a machine worth entering for the prize of £500, but putting both Hackworth and the Stephensons at an advantage, since they had recently built several locomotives to a promising new configuration. The judges were John Urpeth Rastrick, despite his preference for cable-hauled systems, Nicholas Wood and the Manchester mill proprietor John Kennedy. The rules went through several revisions; the final set stated:

1. The weight of the Locomotive Engine, with its full complement of water in the boiler, shall be ascertained at the Weighing Machine, by eight o'clock in the morning, and the load assigned to it shall be three times the weight thereof. The water in the boiler shall be cold, and there shall be no fuel in the fireplace. As much fuel shall be weighed, and as much water shall be measured and delivered into the Tender Carriage, as the owner of the Engine may consider sufficient for the supply of the Engine for a journey of thirty-five miles. The fire in the boiler shall then be lighted, and the quantity of fuel consumed for getting up the steam shall be determined, and the time noted.
2. The Tender Carriage, with the fuel and water, shall be considered to be, and taken as a part of the load assigned to the Engine.
3. Those Engines that carry their own fuel and water, shall be allowed a proportionate deduction from their load, according to the weight of the Engine.
4. The Engine, with the Carriages attached to it, shall be run by hand up to the Starting Post, and as soon as the steam is got up to fifty pounds per square inch, the engine shall set out upon its journey.
5. The distance the Engine shall perform each trip shall be one mile and three quarters each way, including one-eighth of a mile at

each end for getting up the speed and for stopping the train; by this means the Engine, with its load, will travel one and a-half mile each way at full speed.

6. The Engines shall make ten trips, which will be equal to a journey of 35 miles; thirty miles whereof shall be performed at full speed, and the average rate of travelling shall not be less than ten miles per hour.
7. As soon as the Engine has performed this task, (which will be equal to the travelling from Liverpool to Manchester,) there shall be a fresh supply of fuel and water delivered to her; and, as soon as she can be got ready to set out again, she shall go up to the Starting Post, and make ten trips more, which will be equal to the journey from Manchester back again to Liverpool.
8. The time of performing every trip shall be accurately noted, as well as the time occupied in getting ready to set out on the second journey.
9. Should the Engine not be enabled to take along with it sufficient fuel and water for the journey of ten trips, the time occupied in taking in a fresh supply of fuel and water, shall be considered and taken as part of the time in performing the journey.[26]

Over 10,000 people turned up to watch the eight-day proceedings begin on 6 October 1829.[27] Ten locomotives were originally entered but only four appeared – five if one counts Thomas Shaw Brandreth's *Cycloped*, powered by two horses mounted side by side on a carriage, which with their hooves moved boards on an endless chain working on one axle by gearing. It reached speeds of 5 mph but was withdrawn after a horse fell through the floor. Each of the four steam locomotives departed significantly from the Killingworth model. These were *Perseverance* by Timothy Burstall of Leith in Scotland, Hackworth's *Sans Pareil*, Braithwaite and Ericsson's *Novelty* and the Stephensons' *Rocket*.

Perseverance was damaged on the way to Rainhill when the waggon carrying it overturned. Burstall spent the first five days of the trials trying to repair it before it could take part but after only reaching 6 mph it was withdrawn and he was awarded a prize of £25. Its design reflected Burstall's earlier road locomotives, and its vertical boiler may, like them, have had a circular plan firebox tapering to the chimney with no flues. Their common weakness was steam raising.[28]

Hackworth's *Sans Pareil* is a smaller version of the *Royal George*. At first there were doubts whether it could compete, as the judges claimed that it was still over the weight limit, until it was eventually agreed to let Hackworth show what his new locomotive could do. It carried out eight trips and reached a top speed of just over 16 mph. Though it made a promising start, it suffered a cracked cylinder – ironically (or suspiciously), this had been cast by Robert Stephenson & Co. It was heavy on coke and like *Royal George* it had no springs, which had mattered little in a six-wheeler slowly pulling coal trains but which caused this short wheel-base 0-4-0 running at speed to move 'like an Empty Beer Butt on a rough Pavement ...'. After the trial it went to the Liverpool and Manchester but was soon sold on to the Bolton and Leigh, ending its working life as a fixed engine in a colliery.[29]

The *Novelty* was the work of John Ericsson (1803–89), a Swedish-born inventor and mechanical engineer, and of John Braithwaite (1797–1870), an Englishman who built the first steam fire-fighting engine – it was in essence a hasty adaptation of this design. It had two vertical cylinders, which operated on the drive axle by means of a bellcrank, and a two-part boiler which ran under the full length of the running plate. Through it ran a tube carrying the hot gases in an 'S' shape, which made cleaning and repair practically impossible. A mechanical blower provided a forced draught. Weighing only 2 tons 3 cwt, the *Novelty* was much smaller than the other entries. It was also the quickest, reaching speeds of 28 mph on the first day. It was extremely popular with the crowd, and was the favourite to win the competition. However, on the second day the boiler pipe became overheated and was damaged, and Braithwaite and Ericsson had partially to dismantle the boiler. The steamtight joints were made with a cement which normally took a week to harden. *Novelty* had to perform the following day and when it reached 15 mph the joints started to blow. The damage was considerable and it was withdrawn from the competition.

Although *Novelty* was unsuccessful, Ericsson and Braithwaite were permitted to carry out further trials on the Liverpool and Manchester, and Charles Blacker Vignoles (1793–1875) was impressed enough with it (or disliked the Stephensons sufficiently: see chapter 11) to put it to work on the nearby St Helen's and Runcorn Gap Railway, which he was then constructing. Here it earned its keep along with two improved versions, *William IV* and *Queen Adelaide*, aesthetically designed with recessed panels and fluted decoration on the cylinders.[30] The design was not perpetuated for long, though it inspired Matthias Baldwin (1795–1866, see below), a Philadelphian already known as a builder of stationary engines, to construct a model for the American Philosophical Society (chapter 5).[31]

The evident winner at Rainhill was *Rocket*, proof that once again the Stephensons had identified the winning technology and its optimal design path.

The *Rocket* – predecessors, rivals and successors; the 'Planet' and 'Samson' types

The *Rocket* is the most famous locomotive ever built. It made a brave sight at Rainhill in its stagecoach yellow and black livery, its white-washed chimney demonstrating how clean it would be in operation. It is an 0-2-2 with bar-iron frames and inclined cylinders and a separate copper firebox attached to the backplate of a multitube boiler, like Seguin's. This was particularly suitable for burning coke, a hard grey porous substance with a high carbon content made by heating coal in the absence of air, and which does not produce smoke. This became the locomotive fuel of choice in Britain until the 1850s.[32]

Draughting was accomplished by the steam exhaust. Initially, the gases from the tubes were collected in a smoke chamber at the base of the tall chimney, soon replaced by a smokebox. Another alteration was to *Rocket*'s cylinders, lowered from 38° to 8° in 1831. Revolutionary though *Rocket* was, it became design-expired so quickly that after 1831 it was only used on engineers' trains and other secondary duties.[33]

Much was at stake at the trials, so no part of the design was left to chance. Each element had been tried and tested as part of Robert Stephenson's thoroughgoing component review once he returned from South America in November 1827. This he carried out with the assistance of the draughtsman George Phipps (1807–188) and the works manager William Hutchinson (1792–1853). Within only 33 months, locomotive design was completely reconfigured, stimulating the use of materials that had yet been barely developed, such as copper plate for fireboxes, brass for robust flue tubes and higher specification iron for wheels and crank axles.[34]

The first tentative departure from the Killingworth model had taken place while Robert Stephenson was away, in the form of a locomotive appropriately named *Experiment*, which had two cylinders incorporated horizontally within the boiler operating on the levers on the backplate, which gave it the nickname 'old elbows'. Like *Chittaprat*, the drive ran to one axle, and another innovation was the use of water tubes crossing the grate.[35] Back in post, Robert rebuilt it as an 0-6-0, as its initial four-wheel configuration damaged the track, and set to work designing further locomotives which similarly incorporated an increased evaporative surface and a direct drive from the cylinders to the axles.[36]

The immediate predecessor of the *Rocket* was *Lancashire Witch*, constructed in 1828 for the building of the Liverpool and Manchester but diverted by agreement to the Bolton and Leigh.[37] It was seen by the French engineers Coste and Perdonnet, who published an elevation in the *Annales des Mines*, and by Horatio Allen, who ordered a similar machine as the fourth locomotive for the Delaware and Hudson, the *Pride of Newcastle*. This saw little or no use, and may have blown up at Honesdale in July 1829.[38] Both were fitted with twin-flue boilers. Four six-wheelers were also built, one for the Stockton and Darlington in the autumn of 1829, one for the Liverpool and Manchester and two for service in south-east Wales.[39] The following year, Tredegar Ironworks built the first of what proved to be a long series of locomotives to this design.[40]

Invicta, designed for the Canterbury and Whitstable Railway in February 1830 and dispatched in April, was similar to *Rocket*, but was built as an 0-4-0 to increase tractive effort on the two-mile stretch for which it was intended, which was partly on a gradient of 1 in 50/57. *Invicta* was nevertheless defeated by this climb and was relegated to working a mile-long level section. It is the first purpose-built passenger locomotive – *Rocket* had been designed primarily to win at Rainhill.[41]

This basic design was again transformed within a year or so, continuing on a design-path that led into the mid-twentieth century. The next locomotives for the Liverpool and Manchester from Newcastle incorporated lower cylinders, to minimise swaying, and a smokebox.[42] It may have been on one of these, rather than the *Rocket* itself, that the playwright and acting sensation Frances (Fanny) Kemble (1809–93) rode on 25 August 1830. She was as charmed by George Stephenson as she was delighted with 'the little engine which was to drag us along the rails'. She likened it to a dragon out of the *Arabian Nights*, but clearly understood its mechanical properties.[43]

The *Northumbrian* is the best recorded of this series. It had a firebox incorporated in the boiler and a smokebox the full diameter of the boiler, after an initial square smokebox had been replaced. *Northumbrian* was dispatched in July 1830, but already new designs were emerging.

The first of these was *Liverpool*, built by Edward Bury (1794–1858), whose Clarence Foundry by the Mersey docks turned out industrial and marine engines, and an earlier, unsuccessful locomotive of which little is known.[44] Work began on *Liverpool* after the Rainhill trials, but it had been reconfigured before it was put through its paces on the Liverpool and Manchester in July 1830; the multi-tube boiler and smokebox may not have been part of its original conception, but it was the first locomotive to be equipped with the prominently raised firebox which became a Bury hallmark. Like the *Invicta*, it was an 0-4-0. The cylinders were set horizontally under the smokebox, operating the rear drivers, again to obviate the tendency to sway which was manifest in *Rocket*, but which did also introduce a potential design weakness in that crank axles require a very high standard of forging and have a tendency to break. Its 6-foot diameter wheels made it capable of speed, but George Stephenson considered it unsafe; sure enough, it derailed on the Kenyon and Leigh Junction Railway on 23 July 1831, killing its crew.[45]

Another was Hackworth's *Globe* for the Stockton and Darlington, designed in March 1830 for passenger service; it was built by Robert Stephenson & Co. and delivered in December for the opening to Middlesbrough. Its name derives from its spherical copper dome. It was clearly meant to be aesthetically pleasing, like the *Rocket* and the Braithwaite–Ericsson designs, with its ornamental curlicues on the splashers over the large (5-foot diameter) wheels and a frame guard over the footplate. Hackworth thought well enough of it to show it on his business card.[46] It was also an 0-4-0. Its horizontal cylinders drove the cranks on the leading axle from behind the rear drivers.[47]

Globe took nine months to build because it had to be put to one side for Robert Stephenson to construct a locomotive named *Planet* for the Liverpool and Manchester, which also incorporated some of the changes evident in *Liverpool* and *Globe* but in a more evolved way. *Planet*'s cylinders were near-horizontal and were enclosed within the smokebox to prevent loss of heat by radiation, driving a wheelset placed in advance of the firebox rather than immediately behind the smokebox as on *Rocket*. It was also the first to use 'sandwich' frames

From Timothy Hackworth's business card.
THE "GLOBE" ENGINE.

15. Timothy Hackworth depicted the *Globe* locomotive on his business card.

formed of ash or oak, strengthened by iron plates inside and out. These gave some flexibility and great strength.

Robert Stephenson & Co. went on to build about 40 2-2-0 'Planets' and a lesser number of 0-4-0 versions known as 'Samsons', not only for the Liverpool and Manchester but for other railways in the United Kingdom as well as in the USA and France. Newcastle did not have it all its own way, as Bury types following on from *Liverpool* were also built in their numbers. Their bar frames were well suited to winding tracks, which made them popular in the USA, where the round-topped fireboxes also became a standard design feature. However, they were seriously underpowered. Bury's stubborn adherence to these small locomotives led him to inflict them on the Leeds and Selby in 1834 and on the London and Birmingham a few years later.[48]

The Stephenson types were better suited to European railways and were soon at work in other countries. One was built in Russia in 1833–4 by Count Demidov's two well-travelled engineer-serfs Yefim Alekseyevich Cherepanov and his son Miron Yefimovich (chapter 5). Demidov had attended the opening of the Liverpool and Manchester and subsequently sent Miron Yefimovich to inspect the railway and assess its motive power. He could not speak English, carried no letters of introduction and was unable to obtain plans, but on his return he and his father built a locomotive at one of their owner's factories at Nizhny Tagil in Perm Province in the Urals. It was tested in September 1834. It had 80 boiler tubes and ran on a cast-iron 5' 5.75" (1670 mm) gauge demonstration railway 924 yards long. Its reversing mechanism was claimed to be the Cherepanovs' own invention. A second and larger locomotive was built in 1835 and sent to St Petersburg for evaluation. A half-size working replica of uncertain date, preserved in the Central Museum of Railway Transport in St Petersburg, is believed to represent it and is certainly 'Planet'-inspired. The inside cylinders, cranked axle and valve gear follow Stephenson practice closely.[49]

Attempts were made in the United Kingdom to adapt and improve 'Planets' by employing vertical cylinders and a bellcrank drive, which did away with the potentially troublesome crank axle. Sharp, Roberts & Co. of Manchester constructed several such locomotives. The first was another *Experiment*, a 2-2-0 in which the cylinders were placed either side of the boiler, driving bellcranks placed over the carrying wheels, with the chimney situated above the dome. *Experiment* was trialled on the Liverpool and Manchester in October 1833, and sold to them the following February despite its high coke consumption. It was not a success. The locomotives Sharp, Roberts supplied to the Dublin and Kingstown in 1834, *Britannia*, *Hibernia* and *Manchester*, were similar, though the boilers were more conventional and the bellcranks were placed behind the cylinders and piston rods, making the connecting rods shorter. This made them even more unstable, and their front ends were said to oscillate violently up and down when they were running at speed, damaging the track. Charles Vignoles ran *Hibernia* at the unprecedented speed of 60 mph on 1 November that year, but vertical cylinder engines were simply unsuited to fast mainline operation.[50] The railway fared no better with three locomotives built by Forrester of Liverpool in 1834, *Kingstown*, *Dublin* and *Vauxhall*, the first to be built with horizontal outside cylinders, which were attached to outside plate frames. The distance between the cylinders caused them to sway so much that they came to be known as 'boxers'.[51]

James & Charles Carmichael of Dundee tried a similar design for three locomotives built for the Dundee and Newtyle Railway in 1833 and 1834, the *Earl of Airlie*, *Lord Wharncliffe* and *Trotter*, 0-2-4s with vertical cylinders driving the front wheels through bellcranks; the trailing wheels were mounted in a bogie to mitigate the pitching effect.[52]

The 'Patentee' type

In September 1833 Robert Stephenson & Co. built the first 'Patentee' locomotive for the Liverpool and Manchester, a development of the 'Planet' and 'Samson' types incorporating a trailing axle, with flangeless wheels on the middle axle to protect the frames from lateral force on curves. This design responded to calls from railways for more powerful and faster locomotives, and permitted a larger boiler and firebox grate area. 'Patentees' were built into the 1840s, and many were exported to continental Europe. Most were 2-2-2s, but some were 0-4-2, 2-4-0 or 0-6-0 wheel arrangement.[53]

The Hackworth series

Hackworth was not discouraged by *Sans Pareil*'s difficulties at Rainhill and continued building locomotives to this basic plan. His *Victory* of 1829 closely resembled *Royal George*, and the overall design led to 12 further locomotives of two different types between 1830 and 1832. Both retained the vertical cylinders and the six-coupled cast iron wheels, but the motion and boilers were different. A curious feature of these locomotives to modern eyes was that they each had two tenders, one for water at the footplate end, the other for coal at the fire door/smokebox end of the return-flue boiler. The six engines of the 'Majestic' class had the cylinders placed on an overhanging platform in front of the smokebox, which operated not on a drive axle but on a jackshaft with connecting rods to the wheels. The six 'Wilberforces', of which three were built by Robert Stephenson & Co. and three by R.W. Hawthorn, all had return multitubular boilers and inverted cylinders which again operated a jackshaft but at the driver's end.[54] Double-tender types with a return-flue boiler were later supplied to a colliery railway in Nova Scotia and to the Llanelly Railway in South Wales, and variations on these basic designs were still being built as late as 1846.[55]

Neath Abbey and other South Wales builders

Another builder which like Hackworth specialised in slow-moving goods locomotives was Neath Abbey Ironworks near Swansea in South Wales. A foundry in the shadow of the Cistercian ruins at the mouth of the Neath River had been producing pig iron from Cornish ores since the end of the eighteenth century but from 1818 it also manufactured machine parts and complete steam engines under the management of the Quaker Joseph Tregelles Price, supplying not only the surrounding industrial region but also commercial operations in England, France, Spain and South America.[56] In 1829–30 it built *Speedwell* for Thomas Prothero to haul coal from his Blaencyffin Isha Colliery along the Monmouthshire Canal plateway to Newport. This resembled *Novelty* in having vertical cylinders, a bellcrank drive with offset connecting rod and a feed pump driven off the crank, though *Speedwell* had a better designed boiler and is an altogether much more convincing machine. Price had been at Rainhill and evidently noted what he saw, since no published account of the *Novelty* was available by the time *Speedwell* was on the drawing board, only two months later.[57] He had offered the Liverpool and Manchester directors an improved version of Braithwaite and Ericsson's design, and they had agreed to give it a trial and purchase it if satisfactory.[58] *Speedwell*, however, was better suited to pulling coal trains than passengers, and nothing came of the suggestion. It was followed by *Hercules* the following year.[59] In 1830 and 1831 Thomas Brown of the Blaina Ironworks also put two locomotives to work on the Monmouthshire system, of his own design and build.[60] Neath Abbey's *Perseverance*, built for Dowlais Ironworks in 1832, was the world's first rack-and-adhesion locomotive.[61]

Neath Abbey constructed at least eight locomotives by 1834 (and more thereafter), including one for the Gloucester and Cheltenham Railway and one for the Bodmin and Wadebridge. Each one was individually designed and built at a time when locomotives were becoming standardised. Most, for instance, did not have a direct drive from the cylinder to the drive axles, instead using bellcranks, rocking beams or a jackshaft and gearing. Some had single-flue boilers, others increased the evaporative area by incorporating a return flue, water tubes across a single flue or multifire tubes. In 1831 they designed, and in 1838 constructed, a bogie locomotive with coupled wheels, the inboard axles driven from a central crankshaft.[62]

These variations reflected the fact that they were being built for very different types of railway, in terms not only of length, gradient and loading but also of gauge and rail type. Neath Abbey was the first builder to offer locomotives not only for operation over significant distances but also for shunting, reflecting the much denser networks of internal tracks in ironworks in South Wales than in, for instance, a colliery. One such was *Yn Barod Etto* ('ever ready') for Dowlais Ironworks in 1832, unusual in being designed to run on both edge rails and a plateway. It was otherwise a straightforward machine with inclined rear cylinders driving the front axle.[63] Neath Abbey's locomotives were better suited to their tasks than some of the more apparently sophisticated options which were becoming available after Rainhill, and they were reasonably priced.[64] Although they differed from those built in the north-east of England, Price was on friendly terms with his fellow-Quaker Edward Pease, and was well informed about developments in Darlington and Newcastle.[65]

'THAT TRULY ASTONISHING MACHINE'

Vertical boiler locomotives in Britain and the USA

Although Burstall's *Perseverance* had been unsuccessful, from the 1820s onwards other locomotives were built with vertical boilers, which can raise steam very quickly.[66]

In 1829 Robert Stephenson & Co. built what they called the 'Liverpool Coke Engine' as a construction locomotive for the Liverpool and Manchester. It anticipated the *Rocket* in having inclined cylinders operating directly on one of the three drive axles, but had two vertical boilers, which was why it was named *Twin Sisters*. After some initial modification, it seems to have worked well.[67]

Benjamin Hick of Bolton in Lancashire built two such locomotives, the first for the Bolton and Leigh, named *Union*, with a spiral boiler flue and horizontal cylinders operating the drive axle through a rocking beam, and another for a local lawyer and entrepreneur in 1833 which had three cylinders powering a jackshaft and gearing.[68]

It was in the United States that vertical boilers saw most use. The first two were for demonstration purposes. Col. John Stevens' wood-fired locomotive of 1825 at his Hoboken estate was the first multitubular boiler locomotive ever built. Though no illustrations exist, the surviving boiler–tube cluster consists of 20 wrought-iron water tubes, arranged in a circle, connecting a water chamber at the bottom with a steam chamber at the top. A cast-iron rack between the rails meshed with a driving cog wheel.[69]

A little more is known about *Tom Thumb*, built by Peter Cooper (1791–1883), a New Yorker who became convinced that the Baltimore and Ohio Railroad would drive up the value of his Maryland property. He built it in 1830 from various old parts, including musket barrels, which served as fire tubes. It is famously said to have lost a race with a horse-drawn road carriage, but prompted the railway to commit to steam trials the following year.[70]

The first locomotive to enter revenue service in the United States was another 0-4-0 vertical boiler type, built by the West Point

Foundry of New York and delivered to the Charleston and Hamburg in October 1830, where it was given the name *Best Friend*. The boiler was mounted at one end of the frame, counterweighted by the two inclined cylinders at the other end. It did not last long, as the boiler blew up on 17 June 1831. Some parts were incorporated in a new locomotive, the *Phoenix*, where the boiler was placed between the drive axles.[71]

The Baltimore and Ohio made the most consistent use of vertical boiler designs. The locomotives were built at its semi-independent Mont Clare workshops to burn Pennsylvania anthracite, which was easily available but difficult to ignite, particularly in the small fireboxes of the time. Steam trials, perhaps prompted by their engineer Ross Winans, who had been at Rainhill, identified a more substantially engineered vertical boiler type as the clear victor. This was the *York*, built by Phineas Davis, originally a watchmaker, but who had built the first American iron-hulled vessel in 1825, in partnership with one Davis A. Gartner.[72] As originally built, *York* was an 0-4-0 with a vertical water-jacket boiler to which the two vertical cylinders were bracketed, operating trussed coupling rods. It was soon rebuilt by Winans with a tubular boiler, and the cylinders were moved to the leading face of the boiler, driving the front axle through gearing. It became the prototype for the 'grasshoppers' which followed from 1832, *Atlantic*, *Indian Chief*, *Arabian*, *American* and *Antelope*, with their vertical cylinders and boiler-mounted motion beams operating connecting rods to a jackshaft coupled to the drive axle or axles.[73]

Atlantic, with its drive to only one axle, was withdrawn after only three years, but the others lasted in mainline service until 1850, and then put in many more years' work shunting at Mont Clare. They long outlasted their builder, as Phineas Davis, who had become a Baltimore and Ohio engineer, was killed by one of his own locomotives on the Washington branch in 1835.[74]

'THAT TRULY ASTONISHING MACHINE'

British designs in the United States

The Baltimore and Ohio was unusual among early American steam railroads in operating no British imports, though a six-wheeler ordered from Newcastle was lost in a shipwreck.[75] Others did turn to England for their locomotives, importing a significant number of 'Planets' and 'Samsons' between 1831 and 1837. They were copied in the United States, beginning with Matthias Baldwin's first full-size locomotive, *Old Ironsides*, for the Philadelphia and Reading Railway in 1831.[76] Some put in many years' service but they were not always suited to American conditions. Their workmanship was good, but they damaged flimsy American strap rails and were not heavy enough to gain adhesion on steep gradients. With their rigid wheel bases, they were unsuitable for curves. The solution lay in an adaptation of the basic design.[77]

The evolution of the American-type locomotive

This British two-axle design was adapted in several different ways in the United States, evolving in 1837 into a distinctive form of locomotive which became the mainstay of the US railway system.[78]

The Mohawk and Hudson began its working life on 24 September 1831 with two 0-4-0 locomotives. *DeWitt Clinton*, named after the politician and naturalist (1769–1828) who served as a Senator, Mayor of New York City and Governor of New York State, was designed by John Bloomfield Jervis, the chief engineer, and assembled under the supervision of a young mechanic, David Matthew, at the West Point Foundry.[79] It resembled, and was perhaps influenced by, Stephenson–Hackworth practice in its inclined cylinders mounted on either side of the firebox, driving a front crank axle. Like *Globe*, it had a narrow-diameter boiler and a prominent dome. It also resembled Bury's *Liverpool* in its large wheels and in its use of bar frames. These might have made it better able to cope with strap rail but it proved not to be so, and after only a few months' service, it was dismantled.[80]

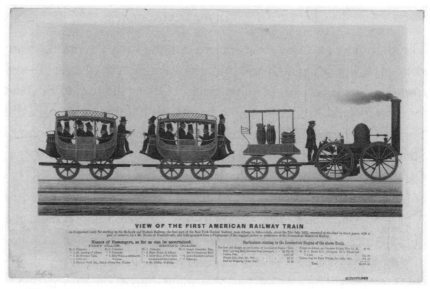

16. The Mohawk and Hudson's *DeWitt Clinton* had a short working life.

Its English stable-mate, a 'Samson' rejoicing in the name of *Robert Fulton*, did eventually give useful service but only after rebuilding in 1833 with a two-axle subassembly, known as a 'truck' in the United States and a 'bogie' in Britain, instead of the front drive axle to lead it into curves, while the driving wheels were moved behind the firebox, making it a 4-2-0. Thus altered, it became known as *John Bull*.[81] Just as *DeWitt Clinton* might have been Newcastle-inspired, so might *John Bull*'s bogie; in December 1828 Robert Stephenson recommended the principle to George Whistler, Jonathan Knight and William McNeill from the Baltimore and Ohio, all of whom went on to work on other American railroads and may have passed the idea on to Jervis.[82]

Another *John Bull* operated on the Camden and Amboy, and survives in the Smithsonian to show how these evolutionary changes could be made in a different way. A journey on the railroad's behalf to England by Robert Livingston Stevens (1787–1856, the son of Col. Stevens of Hoboken), in December 1830 resulted in an order to

Newcastle, which duly arrived in parts at Bordentown on 4 September the following year. There was as yet no operating railway for it – the Camden and Amboy did not open until 1 October of the following year. It was another 'Samson', though it had a Bury-style boiler, and was assembled by chief mechanic Isaac Dripps who had never seen a locomotive before and had no drawings or plans, but was able to work out what to do on the basis of his familiarity with steamships.[83] Matthias Baldwin inspected the *John Bull* and assembled another 'Samson', the *Delaware* for the Newcastle and Frenchtown, before building the 'Planet' copy *Old Ironsides*.[84]

John Bull derailed frequently on the Camden and Amboy, so it was given a pilot axle to guide it into curves, which required the removal of the coupling rods between the two drive axles, turning the 0-4-0 into a 2-2-2-0. Other additions were a cow-catcher on the leading axle, a headlight and a bell (reflecting the absence of fences on American systems and the dangers posed by livestock), as well as a balloon-stack spark arrestor (a metal cone fitted over the chimney to collect the embers from the wood fuel) and a cab. In this guise it worked into the 1860s.[85]

Despite their name, the two *John Bulls* are an apt metaphor for American ingenuity and 'on the spot' adaptation, but it was the leading bogie or 'swivel truck' rather than the pilot axle which took hold.[86] Jervis stated that he was not the inventor of the leading bogie, but that it had been derived from Hedley's Wylam locomotives in their eight-wheel phase (chapter 4).[87]

The 4-2-0 locomotive *Experiment*, later renamed *Brother Jonathan*, built by the West Point Foundry to the design of John B. Jervis for the Mohawk and Hudson in 1831, was effectively the prototype of the distinctively American locomotives which followed.[88] American builders soon followed his example. Matthias Baldwin built no more *Novelty*-inspired miniatures or 'Planets' after 1832, but concentrated solely on designs of this type, including one for the Charleston and Hamburg in 1834.[89] Word went to England that specifications now required this innovation, and manufacturers, eager to hold on to the US market, were happy to oblige. In 1833 Newcastle built a very similar 4-2-0 to Jervis's plans, the *Davy Crockett* for the Saratoga and Schenectady, with which the Mohawk and Hudson connected. A similar Stephenson locomotive followed for the Mohawk and Hudson in 1834. The Vulcan Foundry at Newton-le-Willows and Rothwell & Hicks of Bolton also produced these locomotives for the US market.[90]

One other fundamental difference between British and American practice was that most of the first generation locomotives in the USA burnt wood. The eastern states were well forested, and supplying the railways became a useful source of income for farmers anxious to clear their fields.[91]

Bogie locomotives on the Charleston and Hamburg

The potential of swivelling subassemblies on weak track was reflected in the unique design of four eight-wheel articulated locomotives built by West Point for the Charleston and Hamburg in 1832–3. The boilers had a common firebox with four barrels connected to smokeboxes at each end. The two cylinders were set into the smokeboxes, ball joints at the cross-head enabling freedom of movement as the pistons communicated with the cranks on the driving axle of each of the swivelling wooden subframes. The driver stood on top of the firebox next to the offset dome. They were not a success.[92]

Abortive proposals

Attempts were made to develop other forms of motive power, including horse-powered, compressed air and gas vacuum locomotives. The Liverpool and Manchester's directors were pestered with various fanciful schemes from the start.[93]

Nicholas Wood clearly regarded Thomas Shaw Brandreth's horse-powered *Cycloped* as an absurdity, without saying so directly. Brandreth (1788–1873) was a Fellow of the Royal Society (a distinction which eluded Wood until 1864), with several inventions to his credit.[94] Another horse locomotive was operated experimentally on the Charleston and Hamburg.[95]

Compressed air traction was advocated by the Bavarian engineer Joseph von Baader, who considered steam locomotives unnecessary, clumsy and apt to explode. In 1817 he erroneously believed that their use in England had been largely abandoned.[96] By 1822 he was proposing that fixed engines be placed by the side of tracks to compress air which could then be fed into a reciprocating-cylinder locomotive. Von Baader sought to address a perennial problem of the steam locomotive, and steam engines generally, which is that most of the energy is used inefficiently in relatively short bursts of work and is otherwise lost through the chimney. Fixed steam plant regularly powering a compressor would be less wasteful, though he suggested that wind and water could also be applied. He was still thinking that these might work with his monorail system, but also depicted a proposed nine-tank locomotive operated by two vertical cylinders powering a jackshaft on a central rack. This, he suggested, could recharge itself going downhill.[97] Experimental compressed air locomotives were constructed in the 1840s, but were only widely adopted in the late nineteenth century, for tunnelling contracts and in mining.

Von Baader also recommended the use of compressed air incline winders.[98] He had long experience of pressure vessels, and his proposals were addressed to the crowned heads and senior ministers of countries with little in the way of steam coal and who might be reluctant to buy from Britain, so they were not without their logic.

Another form of traction which only came into its own much later was internal combustion, a form of heat engine in which the ignition of a fuel in a chamber operates a piston. On 4 December 1823 and 22 April 1826 a London-based engineer, Samuel Brown, patented gas vacuum engines, in which the chamber could be filled with a gas flame, expelling the air, before the flame itself was condensed by injecting water. Experimental road vehicles and a marine engine were constructed, but Brown's attempt to interest the Liverpool and Manchester's directors led nowhere.[99] Internal combustion did not become a practical possibility until the 1860s.[100]

Several proposals were also made for forms of steam rail traction which proved impractical. William Henry James (1796–1873), son of William James the railway promoter (chapter 10), patented a 'Mode of Propelling Railway Carriages' in 1823, which involved cylinders on the locomotive powering a series of revolving drive shafts coupled by universal joints, running along an entire train, operating each axle by gearing. He pointed out that such a system would do away with costly earthworks by enabling trains to climb steeper gradients. The idea of powering each axle in a train became a practical possibility with the advent of electric traction and of multiple units in the late nineteenth century, but the mechanical transmission on which James's idea relied would have been costly and prone to failure.[101] Another proposal which would have minimised earthworks, though not strictly speaking a locomotive design, was the 'undulating railway' of Richard Badnall (1797–1838), on the same principle as roller-coasters or *montagnes russes*, which he claimed would allow steeper gradients.[102] Trials took place on the Liverpool and Manchester, and a particularly bad-tempered correspondence was conducted in *The Mechanic's Magazine* for 1834.[103]

Matthew Murray recommended to George Stephenson for use on the Stockton and Darlington a type of locomotive in which the cylinders were placed on one bogie and the boiler on another, which 'would reduce the great evil, viz., the weight of the engine, one-half, and would be a great saving of the rails'. The two would be connected by a jointed steam pipe. Such a design was put into effect on the Great Western Railway some years later, in the form of Thomas Harrison's 1838 *Hurricane* and *Thunderer*, which were as unsuccessful as Murray's design was likely to have been.[104]

American inventors were particularly ready to experiment. Jacob Perkins (1766–1849) developed ideas for water tube and high pressure boilers and uniflow engines, whereby steam flows in one direction only in each half of the cylinder, and predicted the expansive views of steam. He tried out his 'patent circulator' with some success on a Liverpool and Manchester locomotive but fell out with the directors. The American philosopher Ralph Waldo Emerson, on his European trip in 1833, bumped into him in a Liverpool hotel and was treated to his views on the subject.[105] Perkins' ideas were to be developed in the course of the nineteenth and twentieth centuries and influenced locomotive design, but they relied on material strength beyond the capacity of the 1820s and 1830s, and Emerson suggests that his obsessive enthusiasm was unhelpful.

Trials on the Baltimore and Ohio in 1831 also led to some impractical designs. William T. James of New York, who had experience of building road locomotives and who ran a steam carriage on the New York and Harlem Railroad a few years later, built one with a vertical conical boiler and two vertical cylinders.[106] Another was by George Welsh of Gettysburg 'upon an entirely new principle', of which nothing is known. It seems to have been rebuilt subsequently by William T. James to allow for the expansive working of steam – remarkably advanced if so – but exploded soon after arriving in Baltimore.[107] Young James Milholland (1812–75), who became a well-respected master mechanic on the Philadelphia and Reading Railroad, and whose career spanned nearly half a century, built a locomotive which had vertical cylinders, walking beams and a divided firebox, with chimneys at both ends of the boiler.[108] Another had oscillating cylinders.[109]

The 'drags' designed by Sir Goldsworthy Gurney (1793–1875) were light tractors with water tube boilers mainly designed for road use, though several rail versions operated at Cyfarthfa Ironworks in South Wales from March 1830. Here, some would 'rear up like restive horses instead of keeping to the rails and going ahead. With the next the wheels revolved violently, and that was all.'[110]

The *Rocket* was temporarily fitted with a positive displacement engine on the Earl of Dundonald's principle in 1834, taking advantage of its straight drive axle, the first of several unsuccessful attempts to operate a locomotive with a rotary engine.[111]

From enginewright to locomotive manufacturer: British and American builders

George Stephenson and his small group of trusted colleagues had worked wonders at Killingworth, but the successful operation of locomotives turned out in earlier years by Fenton, Murray and Wood at Holbeck and the Haigh Foundry near Wigan (chapter 4) showed that only a specialist firm with proper forging and machining capacity could work to a high standard and maintain quality control. By 1826, Robert Stephenson & Co. employed over 55 foundry men, millwrights, smiths, fitters and carpenters, though it was surprisingly slow to invest in machine tools.[112] Robert also opened the Vulcan Foundry with Charles Tayleur at Newton-le-Willows in Lancashire, with the intention of supplying the Liverpool and Manchester. In 1833 they completed two 'Samsons', *Tayleur* and *Stephenson*, for John Hargreaves (1780–1860), the carrier operating the Bolton and Leigh.[113]

In both Britain and the United States, a similar pattern emerged, in which small foundries and machine shops with experience of fixed and marine engines speculatively built a few locomotives, often to demonstrate a perceived or claimed improvement, but found themselves crowded out by larger firms.[114] Robert Stephenson & Co. dominated the market, but competed with, and also subcontracted to, other manufacturers mostly located along the Liverpool–Manchester axis: as well as the Vulcan Foundry, these included George Forrester, Bury & Kennedy, Rothwell & Hicks and Sharp, Roberts. In South Wales, Neath Abbey Ironworks produced its distinctive plateway locomotives for some years to come.

In the USA, the decision of Congress to retain a tariff worked to the advantage of native builders, and by 1835 there were 175 American-built locomotives in service.[115] Initially, Dunham of New York City, West Point and the Mill Dam Foundry at Boston and other small concerns contributed to the market, but the centre of American manufacture came to be Philadelphia, already well known for producing textile machinery. William Norris's American Steam Carriage Company began life in a converted stable and became for a while the largest locomotive builder in the world before it closed in 1866.[116] The Baldwin Locomotive Works outlasted it by 90 years, building steam locomotives until 1949, including the most powerful ones in the world, before going over to electric, gas turbine-electric and diesel-electric traction until it finally shut its doors seven years later.

Locomotive names

The custom of giving locomotives names in the United Kingdom and in the USA was formalised in this period. Instances in English, French, Latin and Welsh are recorded. A few may have been nicknames which achieved some grudging official acknowledgment, like *Wylam Dilly* and *John Bull*. *Tom Thumb* and *Globe* were named after a defining characteristic. *Lancashire Witch* was a tribute to all the beguiling women of the county, and not a reference to the notorious Pendle trials of 1612. Some names indicated civic ambition or commercial optimism, like *Dublin, Liverpool, Pride of Newcastle* and *Best Friend*, understood to mean 'of Charleston' for diverting traffic to the town. Others acknowledged a patron, like *DeWitt Clinton* or *The Duke*, or a royal personage or military or naval hero: the king and queen gave permission for the naming of *William IV* and *Queen Adelaide* 'in marking the pleasure they take in the success of the Liverpool and Manchester Rail-road'.[117] On the Leeds and Selby, a long way from the sea, the first four locomotives were named after admirals. *Blücher* commemorated a military ally with a reputation as an unstoppable force, whose victories inaugurated a period of peace and commercial optimism.[118] Patriotism is implied by *Dreadnought*, the name of a Royal Navy ship since the sixteenth century, by *Sans Pareil*, a captured French vessel, and by *Rocket*, with its echoes of Sir William Congreve's military weapons.[119] An American example is *Brother Jonathan*, the personification of New England. Perhaps Neath Abbey's *Yn Barod Etto* ('ever ready') was named to draw attention to its steam-raising capacity. The Liverpool and Manchester's directors bestowed their own names on locomotives rather than the ones given them by the builders, so *Wildfire*, with its suggestion of blazing hayricks, was prudently renamed *Meteor*.[120] *Northumbrian* commemorated George Stephenson's origins.

Lion, *Goliath* and *Hercules* were obvious references, whereas *Novelty*, *Locomotion*, *Experiment*, *Invicta* ('undefeated') and *Perseverance* tell their own story. It was presumably the Earl of Dudley, a well-read and cultured man, who named *The Agenoria* after the Roman goddess of activity and childhood.[121]

Ships and horses had always had names and they were often given to stagecoaches as well, so there was nothing surprising in the notion that locomotives should also have them, but they do suggest that sentimental fondness for these machines is nothing new. *Rocket* acquired nameplates at an early stage, and the Vulcan Foundry consistently attached them to its products.[122]

'A wonderfully practical and dependable unit of power'[123]

Locomotives were not easy machines to manage, as footplate crews and mechanics found out. David Matthew's first attempt to drive *DeWitt Clinton* on 9 August 1831 was long remembered for the way his sudden stops and starts snatched the couplings and knocked passengers off their seats, to the detriment of the gentlemen's beaver hats.[124] He nevertheless fared better than the driver who tried to make Neath Abbey's *Royal William* run on the Gloucester and Cheltenham plateway in 1831. Sixty-six years after the event, locals still recalled how 'he' (the locomotive) would not 'bile' but instead groaned, squealed and grunted until 'a terrible noise was yurd ... the blessed thing had busted', and how it was then dumped at the wharf 'till it got rusty'.[125] Even so, between 1815 and 1834, the steam locomotive became a mature technology, a remarkable collective achievement, to which England, Wales, France and the USA all contributed. It reflects not only growing familiarity with individual machines and better understanding of high-pressure steam but also the capacity to produce finely machined components, a readiness to experiment and re-evaluate and to share knowledge. Compressed air and internal combustion locomotives could be developed as experimental technologies for the future, and cyclopeds mercifully cast into oblivion.

The locomotives which captured the public imagination, then as now, were the *Rocket*, the 'Planets', 'Samsons' and the American variations – the ones which could haul passenger trains at something like speed, and which confirmed that main lines would soon connect towns and cities. However, Stephenson and Hackworth's six-wheelers showed how the heavy goods locomotive had also undergone a design revolution between 1825 and 1834. This was needed because coal-carrying railways were now becoming longer and were handling heavier traffic. They form the subject of the following chapter.

9

Coal carriers 1815–34

The years after Waterloo saw new and innovative systems being built to carry coal not only in places where railways had existed for centuries but also in parts of the world where they had never been seen before, or not in this form – in Prussia, Holland–Belgium, France, in the United States and even on the Australian continent, in the colony of New South Wales. Iron rails, whether cast or rolled, and fixed or locomotive engines also meant that the use of coal was becoming increasingly self-reinforcing in the context of railways, to say nothing of its increasing and general adoption throughout Europe and North America as an industrial fuel.

In some cases, very little is known about these systems. Chapter 4 discusses the locomotive-operated railway from Horloz colliery near Liège to a quay on the River Meuse about half a mile away, and the unsuccessful use of steam traction between Chorzow ironworks and a neighbouring colliery in Upper Silesia. This short system may have inspired the Prussian mineralogist Dr Karl Johann Bernhard Karsten (1782–1853) to propose a 128-mile rail link between the coalfields of Zabrze and the River Oder at Breslau (Wrocław).[1] He may have been seeking a peaceful domestic use for coal now that the need for armaments had lessened, but nothing transpired, despite the development of coal mining and metallurgical industries in continental Europe after the Napoleonic wars.

Another enthusiast was the Scottish engineer Robert Stevenson (1772–1850; see chapter 6), whose name is chiefly associated with lighthouse construction but who also actively promoted transport projects to connect collieries in the Lothians and Fyfe with centres of use and with the sea. He became increasingly convinced of the superiority of railways over roads and canals, as well as of wrought iron over other forms of rail. His works were many and his practical talents of the highest order, but he only ever built one railway, a short one at Newton Colliery, near Edinburgh, in 1818.[2]

In the event, both Prussia and Scotland were quick to adopt new coal-carrying technology after 1815, as were France, the USA and New South Wales, but it was England which showed the way, with major innovations taking place in the coalfields of the north-east and of south Lancashire.

The north-east of England coalfield

The wooden waggonways with which England's biggest coalfield had been endowed since the seventeenth century by no means disappeared with the coming of the iron road. Some were only converted after the Napoleonic wars came to an end, the Willington in 1819 and most of the Heaton in 1821, and a few lasted unchanged even into the 1840s. Even so, it was clear that this was a redundant technology, that the future lay with the innovative systems being constructed by the likes of George and Robert Stephenson, by John Buddle and Benjamin Thompson, and that finance was available to build them.[3] The Killingworth Way, the scene of George Stephenson's experiments with Ralph Dodds and Nicholas Wood (Chapter 8) which confirmed the practicality of locomotive adhesion, began replacing its Losh rails with wrought-iron Birkinshaw rails in 1820.[4]

Their experiments continued on the Hetton Colliery system, which has been overshadowed by the later Stockton and Darlington but which was the first railway in the world to be built entirely for mechanical traction – not only locomotive haulage but also two fixed engines and five self-acting incline planes. George Stephenson now had the chance to try out his ideas about how a completely new coal-carrying railway might be designed and operated, and his brother Robert and son Robert could put them into effect.

Hetton Colliery was a new venture, one which involved sinking a shaft through permeable limestone to reach the coal seams beneath, as well as laying out direct rail access to staithes at the mouth of the River Wear. It was a little over 7.75 miles long and used Losh cast-iron rails. Finance came from the Darlington and Durham Bank of Arthur Mowbray (1755–1840), an adventurer who until 1819 was also manager of Sir Henry Vane Tempest's collieries around Penshaw. The capital he raised from his family, partners and the City of London ran out before the colliery was in full production, and the Stephensons were removed under a cloud, though not before the railway had passed its first load of coal, on 18 November 1822. Its initial three 'travellers', Stephenson-built four-wheelers, *Dart*, *Tallyho* and *Star*, were joined by two others the following year.[5]

Between the collieries and the staithes the land rose considerably to a summit at Warden Law. Three powered inclined planes raised chaldrons on one side, and self-acting planes let them down the other; locomotive working was initially confined to two lightly graded sections, one at each end. The system underwent revision almost from the start, as new pits were sunk to the north and south of Hetton, and tonnages increased considerably. Locomotives struggled on the lower section and were replaced with rope haulage in 1827. It was described in some detail by Ernst Heinrich Karl von Dechen and Karl von Oeynhausen.[6]

This approach was also reflected in the Springwell Colliery railroad, built by George Stephenson for the Grand Allies, operational

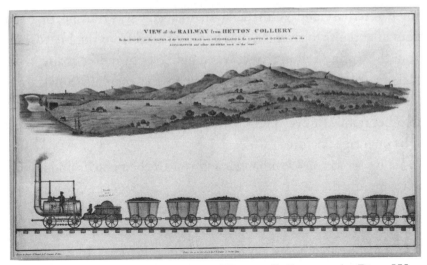

17. The first all-steam railway, from Hetton Colliery to the River Wear, across Warden Law, was completed in 1822.

from 1826, designed for fixed and locomotive steam in the Hetton manner.[7] The alternative system using fixed engines, as favoured by Benjamin Thompson (chapter 7), was reflected in the Fawdon waggonway around 1822, the Brunton and Shields in 1826 and the Rainton and Seaham in 1831.[8]

Stockton and Darlington

The Stockton and Darlington Railway, which opened on 27 September 1825, is one of the best-known railways in the world. It was the second railway of which George Stephenson was the chief engineer and surveyor, after the Hetton, and is the one with which he is particularly associated.

It was not the first all-steam railway, a distinction which as we have seen belongs to the Hetton. Nor was it the first railway to operate a steam passenger service, since other than on the opening day carriages were not hauled by locomotive until 1830, after operations had begun on the Liverpool and Manchester and even on the tiny Canterbury and Whitstable. Its mainstay was transporting coal from inland pits to river staithes but it differed from the Hetton and from the Springwell in that it was a public railway, like many of the ones described in chapter 3, the first in the coalfield, and carried general goods as well as people. Another 'first' it could claim was that no other public railway had owned and operated a steam locomotive in its own right before.

Yet it was, as von Oeynhausen and von Dechen noted, 'unquestionably currently the best [*gegenwartig unfreitig der beste*] in Great Britain', even though they pointed out it would shortly be eclipsed by the Liverpool and Manchester. The first 64 pages of their study are taken up with it, before they move on to describe the Hetton.[9] In essence, the Stockton and Darlington's claim on history is that it combined technical elements which had already justified themselves in their experimental phase into a successful and long-lived traction system. Its wrought-iron rails and locomotives in combination, on a well-engineered railway, proved very effective. It provided a test bed for the improved locomotive designs that Timothy Hackworth and Robert Stephenson soon started putting into effect, and these also operated successfully.

Its origins were subtly different from those of the railways of Northumberland. South Durham was a rural area, dominated by the prince-bishop and by Durham Cathedral's dean and chapter rather than by great lay proprietors like its neighbouring county.[10] The main pits around Bishop Auckland were a long way from navigable water; canal schemes, though technically feasible, had failed for lack of capital in 1767 and again in 1815. Yet, after Waterloo, it was clear that Darlington, on the Great North Road, now had money to invest. Joseph Backhouse's bank had survived the postwar crisis, one of the few that had, and Edward Pease (1767–1858) had amassed a fortune from the town's woollen trade. Both men were Quakers and had access to London capital through the Society of Friends.[11]

A plateway was surveyed by the engineer George Overton, who was based in South Wales but related by marriage to the steward of

Thomas Meynell (1763–1854), the squire of Yarm, one of the leading figures in the move to connect these collieries with the Tees. Overton presented evidence in favour of a bill to construct it which was presented to Parliament in 1819, but Pease favoured selecting a different engineer. The one he had in mind was Robert Stevenson, but George Stephenson was appointed, with the result that it was built as an edge railway to the Killingworth gauge of 4' 8" ins (1422 mm), and with locomotive traction in mind. Stephenson also insisted on fish-belly wrought-iron rails on stone blocks despite his commercial interest in the Losh patent, as he realised that this was the more promising technology.

On 23 May 1822, the first lengths of rail were laid by Thomas Meynell, accompanied by his neighbour and fellow-director Benjamin Flounders and by two Catholic priests, Thomas Storey of Yarm and John Bradley of Stockton – for Meynell, now chairman of the 'Quaker line', was from an old recusant family. The church bells were rung, ships in Stockton harbour were decorated with bunting, two or three hundred navvies marched in procession behind the gentlemen, a band played 'God Save the King' and a salute was fired.[12] As built, it was 26 miles long, and took a roundabout route between the collieries near Bishop Auckland and the staithes on the River Tees at Stockton in order to pass through Darlington. It crossed the Skerne at Darlington on a single sandstone arch flanked by smaller pedestrian arches, 'the chief architectural feature on the line', to the design of Ignatius Bonomi (1787–1870), a Durham-based architect.[13] The other river to be crossed was the Gaunless, by a horse-drawn branch. George Stephenson used a design whereby two curved girders in a lens shape, one above and one below, formed a balanced truss. This was not the first iron railway bridge, as is sometimes said, but it was the first lenticular truss bridge.[14] Stephenson had wanted a more direct route but was rebuked by Pease: 'George, thou must think of Darlington: thou must remember it was Darlington sent for thee'.[15]

THE COMING OF THE RAILWAY

The opening day, 27 September 1825, was a grander affair than the initial ceremony. The previous evening, Pease, Stephenson and other members of the committee had made a journey to Darlington on the newly arrived coach the 'Experiment', before taking it behind *Locomotion* to Shildon, with James Stephenson, George's elder brother, at the regulator. The 'Experiment' could accommodate up to 18 people sitting facing each other; a sketch made on the opening day shows a box-like body on four axles, suggesting it may well have been an articulated bogie.[16]

The following morning, 12 full coal waggons passed the Etherley Incline, where a waggon loaded with flour bags was coupled on. These were horse-hauled over the Gaunless Bridge and passed over the Brusselton Incline to Shildon Lane End, where *Locomotion*, 'Experiment' and 21 further coal waggons fitted with seats were waiting. This train left carrying between 450 and 600 people. Brakesmen were placed between the waggons, and it was led by a man on horseback with a flag. Despite a derailment and some mechanical problems, it reached Darlington, where six waggonloads of coal were distributed to the poor. Some passengers alighted, others got on, 2 waggons for the Yarm Band were attached and at 12:30 pm the locomotive started for Stockton, now hauling 31 vehicles with 550 passengers, including the musicians. A man clinging to the outside of a waggon fell off and his foot was crushed by the following vehicle. The train halted at a temporary terminus at St John's Well in Stockton, where Meynell had laid the first rails. The ceremony was considered a success, and that evening 102 people sat down to a celebratory dinner at the Town Hall.[17]

The railway settled down to regular operation. Horses continued to provide much of the motive power, aided by the introduction of 'dandy waggons' on which they rode on downward runs, which were believed to economise their power by a third and which enabled coal trains to move faster by gravity.[18] Branches to Yarm and to Croft were opened in 1825 and 1829, and in 1830 the railway was extended westwards to collieries at Haggar Leases and eastwards to deepwater staithes lower down the river at Middlesbrough, once it had become clear that the facilities at Stockton were unable to cope with the growing trade.[19] Passenger services were very much a secondary part of the business and were farmed out to contractors. 'Experiment', emblazoned with the company's motto, *Periculum privatum utilitas publica* (private risk for public good), began running in regular service on 10 October 1825, though the agreement with its operator contained a clause that it would become null and void 'the first time he is seen intoxicated'.[20] Later coaches were the 'Express', the 'Defiance', the 'Defence' and the 'Union' in 1826, and the 'Perseverance' the following year, which were all horse-drawn and built like stagecoaches but with back and front constructed similarly to enable them to operate in either direction.[21]

The eastward branch involved a suspension bridge over the Tees, designed by Captain Sir Samuel Brown (1774–1852), a former naval officer, and was the first such structure to carry railway traffic. This was the only major element of the railway that was not successful. It had lattice timber parapets but the deck was unstiffened and was unable to take the weight of a locomotive without undue deflection. During its short working life, it had to be propped at mid-span and the load was restricted to waggons only. The Menai Bridge, completed in 1826, had established suspension bridges as the most economical means of enabling wheeled vehicles to cross a wide span, and provided a basis for improvement in design, but Brown failed to consider the effect of a long, single moving load. The contract (for £2,200) was signed on 18 July 1829 and, despite disagreement with the Tees Navigation Company, work went ahead. An experiment carried out before the extension opened confirmed that the weight of 16 waggons, a locomotive and a tender caused the masonry on both towers to be affected and the cast-iron retaining plates on the Yorkshire side to split. It was concluded that waggons could be run in rakes of four, but only if coupled nine yards apart. It was observed that 'The failure of the bridge was a great disappointment to the Company'.[22]

The Clarence Railway

The complicated politics of the coal trade meant that some railways were beginning to form interconnected but competing regional networks, as distinct from the earlier systems which converged on a river outlet. In south Durham, the 11.5-mile Clarence Railway connected Coxhoe Colliery with staithes on the River Tees but also offered an alternative outlet for coal from Bishop Auckland by means of a junction at Simpasture with the Stockton and Darlington. A more direct northerly route from Auckland to the Tees had been considered since 1819, and unsuccessful applications had been made to Parliament for permission for such a line in 1823, 1824 and 1825.[23]

The south Lancashire coalfield

Two major railways were constructed to serve the south Lancashire coalfield. These were the Bolton and Leigh, which received its act in 1824 and opened four years later, and the St Helen's and Runcorn Gap, authorised in 1829 and completed in 1833. George Stephenson was the engineer of the first of these, and Charles Blacker Vignoles of the second, but both reflected the technology of the Stockton and Darlington in their use of wrought-iron rail, and of both fixed and locomotive-engine haulage.

The Bolton and Leigh Railway was the first public railway in Lancashire, opening on 1 August 1828. It actually connected two canals, the Manchester, Bolton and Bury at Haulgh, near its Bolton terminus, and a branch of the Leeds and Liverpool Canal which ran to the Bridgewater Canal at the town of Leigh, a distance of 7.75 miles. It was built for general merchandise but much of its anticipated revenue was to be derived from coal, by serving collieries at

Hulton, as well as by enabling coal delivered by barge to be transported to Bolton.[24] George Stephenson was familiar with the area, as he was working on the Liverpool and Manchester; he commissioned one of his assistants, Hugh Steel, to survey a route. Steel was in turn helped by Robert Daglish, who had earlier been involved in the construction of the Blenkinsop locomotives at Orrell, and who went on to supervise construction. The railway was to be a single track with two rope-worked inclined planes using fixed steam engines.

The promoters were locally based men, of distinctly Tory hue. William Hulton (1787–1864) was a scourge of Luddites, notorious for having issued the arrest warrant which led to the Peterloo Massacre at Manchester in 1819. Since the twelfth century his family had held sway over the Hulton estates, on which the productive Chequerbent pit was sunk. Benjamin Hick (1790–1842) was an engineer of humble origin and dissenting stock who had made his peace with both church and state. With his fellow-director Peter Rothwell, he ran the Union Foundry at Bolton, producing steam engines, boilers, hydraulic machines and locomotives; their first was the *Union*, a vertical boiler 2-2-0 for the Bolton and Leigh.[25]

This was not the railway's first locomotive. An 0-4-0 with inclined cylinders, the first to emerge from the Stephenson works at Newcastle and the immediate design predecessor of the *Rocket*, had arrived in July 1828. It carried out the ceremonial inauguration on 1 August, when the first section, from Bolton to Chequerbent, was put into service. Before 'an immense concourse of people' it hauled waggons and a carriage borrowed from the Liverpool and Manchester Railway, decorated with flags and streamers. Mrs Hulton named the locomotive *Lancashire Witch* and presented the driver with a garland. This he attached to the chimney, where, in the words of the *Manchester Mercury*, 'the complexion of the lilies and roses soon underwent a lamentable change'.[26]

The line was completed through to the Leeds and Liverpool Canal at Leigh by the end of March 1830, when a link to the

Liverpool and Manchester, the Kenyon and Leigh Junction Railway, was well advanced. Trains were horse-worked the short distance between Bolton and the foot of the two inclines, at Daubhill and Chequerbent, each of which had its own fixed engine, and were hauled by locomotive between this point and Leigh.

Passengers were carried from 1831, and by 1834 the railway was leased to John Hargreaves (1780–1860) of Bolton, who with his father, also called John, operated a long-established road and canal carrier which extended from north-west England into Scotland. He was required to provide his own locomotives (see chapter 8) and rolling stock, except ballast waggons, which belonged to the Company. Hargreaves acquired running rights to Liverpool and Manchester.[27]

Charles Blacker Vignoles, engineer of the neighbouring St Helen's and Runcorn Gap Railway, had had a 'sound and liberal education' at the hands of his maternal grandfather, a professor at the Woolwich Royal Military Academy, had studied at Sandhurst, served as an army officer and travelled extensively. His background was, in other words, very different indeed to George Stephenson's, and the two men did not get on. Stephenson, who could be vindictive, made life difficult for him when they were both working on the Liverpool and Manchester, and prompted him to resign. Even so, Vignoles did not depart from the Stephensonian model of railway building, except that he favoured Braithwaite and Ericsson's locomotive designs over those that emanated from Newcastle.

The railway Vignoles designed ran from Cowley Hill Colliery past the town of St Helen's, whose motto is *ex terra lucem* (from the ground, light), to a wet dock at Runcorn Gap on the River Mersey, and was opened formally on 21 February 1833, though some trains had been running before this date. Fixed engines were used at Widnes and near St Helen's. The total length of its main line was a little less than nine miles, and by 1831 its branches, into collieries and glass works and a connecting link to the Liverpool and Manchester, added nearly a further seven. Investors included local colliery owners and

James Muspratt, the soap manufacturer; the chairman was Peter Greenall, who had brewery, coal and glass interests.[28]

Scotland

Some short colliery lines in Fife running to the Firth of Forth were reconstructed with iron rails and inclined planes in this period, but it was the Lanarkshire field, south and east of Glasgow, which saw the most significant developments.[29] The Monkland and Kirkintilloch Railway, from Palacecraig and Cairnhill collieries to the Forth and Clyde Canal, was inaugurated on 17 May 1826, when a horse pulled 16 tons of coal to the canal. Remarkable though this was thought to be, it was nothing compared to the equine Rainhill which took place nearly two years later as the result of a bet, when a Clydesdale pulled a 14-waggon, 50-ton load from Gargill Colliery the whole 7-mile length of the line, partly on the level and partly on a slight downgradient.[30] Locomotives arrived in May 1831, Glasgow-built to the already old-fashioned 'Killingworth' design.[31] The following year the Monkland and Kirkintilloch provided early evidence of the environmental and public health challenges that railways brought in their wake when cholera arrived at its upper terminus.[32]

Feeder railways were the Ballochney, completed in 1828, and the Wishaw and Coltness, of which the first section opened in 1834. From 1831 it was also connected with the Garnkirk and Glasgow.

The railways of Saint-Étienne

The railways of Saint-Étienne in the French Massif Central were built under concessions of 1823, 1826, 1828 and 1833 to connect a major bituminous coalfield and its associated iron industry with the rivers Loire and Rhône and with the town of Lyons, following French loss of the Ruhr in 1815.[33] Initial proposals came from two engineers who had experience of mine railways, Louis de Gallois and

Louis-Antoine Beaunier. Beaunier had taken part in drafting the 1810 mine law, which imposed government concession and administrative control over extractive industries.[34] De Gallois had visited England and Wales to inspect their railway systems in 1816.[35]

The proposal underwent examination by the Conseil Général des Ponts et Chaussées, as was required for major transport projects, in order to establish its public utility and the choice of route, before the Prefects could be advised to approve the scheme and the appropriate royal ordinances be issued.[36]

The first part of the system ran westwards from the Saint-Étienne pits to Andrézieux on the River Loire, a single track contour formation, making use of Losh-type cast-iron fish-belly rails on stone blocks.[37] Designed and built by Beaunier, it was opened for horse traction and goods traffic in 1827, operated by the socialist engineers François-Noël Mellet and Charles-Joseph Henry. They were both disciples of the political and economic theorist Claude Henri de Rouvroy, comte de Saint-Simon, whose teachings were to be influential in the later development of mainline networks (chapters 12 and 13).[38] Passenger services were introduced in 1831 and locomotive haulage in 1832. An extension to Roanne, lower down the Loire to avoid a difficult passage of the river, was opened in 1832, inexpensively built using self-acting and powered inclined planes, horses and locomotives. It was slow and inefficient, and its operators were bankrupt by 1836.

The main route ran eastwards, to Lyon and to the confluence of the Saône and the Rhône. This was built on a substantial alignment, double-track, except in the tunnels, using wrought-iron rails. It was designed and built by Marc Seguin (1786–1875) and his brother Camille (1793–1852) in stages between 1830 and 1832. They had agreed with Baunier as early as 1826 to standardise on the north-east of England colliery gauge of 4' 8" (1422 mm).[39]

Marc Séguin was the designer of the *remorquer* on the Saint-Étienne to Lyon railway (chapter 8). An outstanding engineer, he is particularly associated with the development of the wire-cable suspension bridge and more particularly, from a railway point of view, with the multitubular boiler. His brother's contribution to their projects was financial acumen. They were nephews of Joseph Montgolfier, the pioneer balloonist, and developed an early interest in machinery. They were also both pioneers of steam haulage on the Rhône; Séguin's first multitubular boiler was installed not in a locomotive but in a tug, at a time when the relative merits of animal power and of fixed or self-propelling steam on both the river and the railway were much debated.[40] It is not surprising that the terminology used on the railway consistently reflects their role as accessories to water transport – not only were locomotives *remorqueurs* (tugs) but termini were *débarcadères* (landing stages), intermediate stations were *ports secs* (dry ports), the track a *chemin de halage* (towing path).

The Seguins considered English precedent carefully but critically, rejecting for instance the use of fish-belly rails such as on the Stockton and Darlington in favour of wrought-iron straight-section rails. They also refused to adopt fixed engines, instead making use of gravity for the loaded waggons where the gradient was sufficient, and of locomotives on level sections and to return the empties. Following visits to England by Louis-Antoine Beaunier and Marc Seguin, two locomotives were ordered from Stephenson and Co., as well as the first of several to Seguin's multitubular design (see chapter 8). Locomotives were used to haul coal trains; passenger and goods trains were operated by horses and by gravity. Oxen were also used. Seguin was determined to give the railway a steadily falling gradient of no more than 1 in 70 over its 35-mile length, which involved much tunnelling, bridge work and large-radius curves. He recruited a particularly able survey team, though he made emendations to the planned route after visiting the Liverpool and Manchester and after taking advice from John Rennie the younger.[41]

Construction was capably organised from the Perrache workshops at Lyon, with Marc Seguin taking a hands-on role, and was able to withstand both labour difficulties and political unrest in the summer of 1830.[42] Directors held regular working meetings at the Paris headquarters, appointing commissions of investigation whenever problems arose, and worked closely with their political and financial contacts as well as with their men on the spot. Calls on shares were honoured promptly.

The decision was taken to complete the central section first, from Rive-de-Gier to the Rhône at Givors, partly to ensure some traffic and income, and partly also to test the potential of the locomotives. By 1831 the stadium-plan wharves at the '*gare d'eau*' on the spit of land between the two rivers at Lyon had been built and a channel dug to the Saône, though they were not initially rail-connected.[43] The line opened in stages and was completed throughout in February 1833.

Passenger services were offered from 1832. On 25 February the *Mercure ségusien* announced that, from the first of March, passengers would be carried in a vehicle running on a road chassis from the contractor's office to la Terrase, where its body would be craned on to a railway underframe 'in less than five minutes and without a jolt, without shaking the travellers', before setting off for the Loire, making the last part of the journey to Montbrison on another road chassis. A well-known lithograph shows first class *financières* and second class *cadres*. Both types were side-entry three-compartment vehicles, the *financières* made up of two full stagecoach bodies and a coupé, with luggage space on the roof, the *cadres* having curtains instead of windows, benches on the roof and a platform for a guard or attendant at one end. All of them ran on a short wheelbase chassis, which cannot have made for an easy ride, and were horse-drawn or descended by gravity, as locomotives were reserved for coal trains.[44]

The Seguin brothers and their colleagues not only effectively married the best emerging English practice with the well-established traditions of French civil engineering (and furthermore did so in the challenging environment of the Massif Central and at a turbulent time in French politics) but also contributed several innovations of their own. These included the tubular locomotive boiler, the introduction of wrought-iron rails rolled in France, locomotive traction and a careful calculation of economic options. Part of this network is still used by the TGV.[45]

A shorter, purely coal-carrying system within the region connected collieries at Épinac with the Burgundy Canal at Pont d'Ouche, financed by the industrialist Samuel Blum with the encouragement of other followers of Saint-Simon. It was approved in 1830 and opened in 1835.[46] A Stockton and Darlington touch was the provision of dandy waggons (*wagons-écuries*) for the horses and oxen.[47]

Prussia

Two short coal carriers were constructed in Prussia. One was opened in 1828–9 from the Schlebusch coalfield to industries on the Ennepe River, the other a little to the north between 1828 and 1831 from Essen to serve the textile and metallurgical industries of Wuppertahl. Both were the brainchild of Friedrich Harkort, the industrialist and railway advocate (chapter 5); they were built to a gauge of 2' 8 9/32" (820 mm) and were a Prussian *meil* (a little over 4.5 miles) in length. The Essen railway was financed by the first German joint-stock railway company, and after its ceremonial opening by the Governor-General of the province, the King's brother, it was dignified as the *Prinz-Wilhelm-Eisenbahn*.[48]

Harkort had travelled to England and had been sufficiently impressed with Henry Robinson Palmer's monorail (chapter 2) to construct an experimental version of his own but was persuaded by the mathematician Peter Nikolaus Caspar Egen to construct the two railways as strap rail systems running on longitudinal baulks which were themselves supported on cross-sleepers.[49] Trains on the *Prinz-Wilhelm-Eisenbahn* were operated by relays of horses, one or two on the near-level near the Ruhr, three or four on the steeply graded sections.[50]

The USA – Pennsylvania

The war between the United Kingdom and the USA between 1812 and 1814 had brought to an end the Atlantic trade in British coal, and the Royal Navy's blockade had led to a shortage of supplies from the established coalfields of Virginia and Nova Scotia. The result was a new interest in US mineral reserves, in particular Pennsylvania anthracite, which is difficult to ignite but releases considerable heat and is practically smokeless. This was increasingly used as a domestic fuel and in foundries, but transport eastwards from its source to the main river navigations and the biggest cities meant cutting across the grain of the country.

The Schuylkill, which flowed into the Delaware just below Philadelphia, was navigable to the coalfield, and several short lines were built from the drifts to wharves on its upper reaches. Port Carbon, the site of the first lock on the navigation, was the focus of three such tributary systems. The earlier was built in 1826–27 by Abraham Pott to 4' 4" (1331 mm) gauge, on a constant downward gradient on its half-mile length from his mines at Broad Mountain. One horse could apparently pull 13 drop-bottom waggons each carrying 1.5 tons on the rough wooden track.[51] Longer but less well-engineered was a strap road down Mill Creek, of which the first section opened in 1829 on an undulating profile followed the natural land contour. By 1831, 200 one-ton cars, costing $50 each, were carrying 30,300 tons of coal annually, on an average haul of three miles. Also in partial operation by 1829 was a double-track 4-foot gauge strap road built under the auspices of the Schuylkill Valley Navigation and Railroad Co., connecting with mines to the northeast. By the following year, it extended 10 miles, had 15 branches and was also carrying passengers. Another double-track strap road, but this time to Stephenson gauge, was the Mount Carbon Railroad, operational from 1831.[52] This connected with the Danville and Pottsville, chartered in 1826 to link the Schuylkill with the Susquehanna using inclined planes and, by 1839, locomotives.[53]

A coal-carrying tributary of the Delaware and Hudson Canal made experimental use of locomotives. This waterway, 108 miles long and with more than 100 locks, had reached Honesdale in the Lackawaxen valley in 1828, but the most abundant deposits were found at a place which came to be known as Carbondale on the Susquehenna River, further to the west. Between them, the Moosic Mountains formed a formidable barrier more easily crossed by a railway than by a canal, though it took three counterbalanced inclined planes on the east side and five powered inclines, each with a fixed steam engine, on the west side, to complete the link. Each plane contained a single track, with an automatically operated loop at the halfway point where the loaded and empty waggons could pass. The sections in between were graded slightly to allow the loaded waggons to roll by gravity to the foot of the next plane, and to be hauled back empty in the other direction. One was six miles long, another four, and it was these that prompted the company's chief engineer, John Bloomfield Jervis (1795–1885), to carry out trials with locomotives. It was Jervis who appointed Horatio Allen (1802–89), only a few years out of Columbia College, as agent with expenses paid, to travel to England to buy locomotives and rails.

Much of the railway was built on trestles constructed from native hemlocks, which had already cracked and warped in the sun by the time it was due to open. The rails themselves were 6- by 12-foot hemlock timbers, approximately 20 to 30 foot long, placed on edge. The wrought-iron straps, from Sparrow of Wolverhampton, were screwed to the top. It was along this formation that on 8 August 1829, Horatio Allen, manifesting a courage he did not feel, drove the *Stourbridge Lion*, the first and last time he operated a locomotive (chapter 8). The lesson that locomotives and a timber formation do not go well together was ignored, as Allen used both on the Charleston and Hamburg shortly afterwards.[54]

A near-contemporary of the Delaware and Hudson railroad was the Mauch Chunk, another coal carrier, which ran from the Lehigh Coal

& Navigation Company's pits at Summit not to a canal but to the Lehigh River, a tributary of the Delaware. Construction began in 1827. Its engineer, Erskine Hazard (1790–1865), had been sent on a tour of UK railroads the previous year, and conceived a preference for strap rails for American use. They were, he argued, cheaper and easier to repair.[55] He built the Mauch Chunk to 3' 6" (1067 mm) gauge, the wooden sleepers 4 foot apart on a stone foundation, supporting wooden timbers carrying wrought-iron bars imported from England. Two self-acting inclined planes ('chutes') dropped waggons down to the river. Otherwise, full waggons ran by gravity and the empties were hauled back by horses and mules along a formation laid on an existing trackway.[56]

It soon became a tourist draw. Visitors who wanted to see the collieries could travel in one of as many as 14 or 15 'pleasure carriages' coupled together. On the downward journey they followed two trains each of 14 coal waggons and preceded the vehicles for the animals, which were equipped with feeding troughs. Speed was normally about 5 mph, though a driver who felt like showing off to passengers might let the run travel much faster, applying and releasing brakes by means of a continuous rope from one waggon to the next. The 'intelligent and liberal' among the US population were urged to enjoy this experience.[57]

The Australian Agricultural Company's railway at Newcastle, NSW

The first iron railway to operate beyond the Eurasian landmass and the North American continent was a short coal-carrying system in New South Wales, opened on 10 December 1831. A seam of coal evident in the sea-cliffs of Malubimba overlooking the Pacific Ocean had attracted the attention of settlers before the end of the eighteenth century. Here a penal colony, soon to be called 'Newcastle', was established for the most hardened and refractory of convicts, who were set to work as miners. Some sort of wooden railway seems to have been used underground by the late 1820s, but within a few years the workings were effectively privatised by being handed over to the Australian Agricultural Company. One of Australia's oldest still-operating companies, it was founded in 1824 by an Act of (the British) Parliament, with the right to select 1,000,000 acres in New South Wales for farming development.

The purchasing agent was Benjamin Thompson in Newcastle upon Tyne, whom we encountered earlier (see above), who was Sheffield-bred but had experience of iron working in South Wales and Tyneside. It was he who arranged for the supply of Losh-type cast-iron fish-belly rails, enough for half a mile of railway, from a foundry near Gateshead, one of which survived, with the letters 'A^{co}A' cast into it, to be discovered in 2004. In 1832 platerails were also ordered, probably for underground use. Overall design, however, was

18. The AACo railway included an inclined plane and a lifting bridge over what is now Hunter Street, Newcastle, NSW.

the responsibility of the man on the spot, John Henderson (1781/2–1835), with his wide experience in England, Scotland and Sweden (chapter 5). The railway, the counterbalance incline, the timber viaduct on which it ran and the staithes all have their parallels in the places where he had worked previously, though one feature which had no known precedent was a drawbridge which crossed what is now Hunter Street.[58] Though it was by then the furthest-flung descendants of the Newcastle way, the ancestry of the Australian Agricultural Company's railway was particularly evident, more so than its French and American cousins.

The railway of 1831 was the first of many in the New South Wales coalfield, which ultimately enabled Newcastle in Australia to take over from Newcastle upon Tyne as the largest coal-exporting harbour in the world.[59] The New South Wales coalfield also remained a stronghold of steam traction into the 1980s, just as the wooden way could still be seen in operation on Tyneside many years after the iron road first appeared. Coal-carrying technologies die hard.

10

Internal communications 1815–1834

The growing scale and ambition of iron railways conceived in the 1820s led to the construction of the first systems which effectively formed means of 'internal communications' with a national reach, which turnpikes and canals had already offered for some time. In particular the first railway connecting two great continental rivers was operational by 1832, the Budweis–Linz horse railway in the Austrian Empire. Its opening was narrowly preceded by a shorter railway in England, the Cromford and High Peak, which it did not greatly resemble technically, but which endeavoured to connect the watersheds to the west and to the east of the country, and was the first in northern Europe to be built to carry traffic from the opposite ends of the one jurisdiction. These followed on from earlier but abortive schemes to construct long-distance railways.

Long-distance railways and their proponents

After 1815 railways were being built within the United Kingdom which did more than simply connect mines, quarries and manufactories with navigable water. They were now serving rural areas and market towns, and offered a variety of services, including passenger transport. As chapter 3 points out, the English–Welsh border was home to two lengthy horse-drawn plateways whose principal goods were coal, iron, limestone, timber, corn and cider. One ran the 36 miles from the canal at Brecon to the Wye Valley and to Kington from 1820, the other from Abergavenny to the cathedral city of Hereford and to the River Wye, and was completed in 1829. These were the initiative of local gentlemen, clergy and magistrates, and were operated like the county turnpikes. They were undoubtedly of regional significance but did not in this sense have a national reach.

As we have seen, ideas about a network of railways that might service not a locality or a region but much wider areas of the country had been circulating for many years. In 1810 Thomas Telford surveyed, and William Jessop approved, a proposal for a 'cast-iron railway' from Glasgow to Berwick-on-Tweed, over 125 miles in length, the first credible proposal for a railway connecting the east and west coasts of Britain. Its promoters, landed gentlemen from the Scottish borders, needed coal and lime for their farms and hoped to supply Glasgow with grain.[1] In 1814, the French engineer Pierre-Michel Moisson-Desroches (1785–1865) urged Napoleon to build seven national railways from Paris.[2] In 1817 the radical English schoolteacher, author and publisher Sir Richard Phillips (1767–1840) anticipated double-track railways connecting London with Edinburgh, Glasgow, Holyhead, Milford, Falmouth, Yarmouth, Dover and Portsmouth, drawn either by horses at 10 mph or by Murray–Blenkinsop locomotives at 15.[3] By the 1820s these were becoming a serious possibility. Edward Pease in Darlington (chapter 6) wrote to his cousin Thomas Richardson, the London banker: '... don't be surprised if I should tell thee, there seems to us after careful examination no difficulty of laying a rail road from London & to Edinburgh on which waggons would travel & take the mail at the rate of 20 miles per hour, when this is accomplished Steam vessels may be laid aside!'[4]

Thomas Gray's *Observations on a General Iron Rail-way*, which went through five editions between 1820 and 1825, advocated a system connecting London and Edinburgh via Leicester, Nottingham and Sheffield, with branches to Birmingham, Bristol and beyond, to move coal and agricultural produce, though the 1822 edition added an engraving of a Murray–Blenkinsop locomotive hauling passenger coaches for good measure.[5] Gray's obsessive enthusiasm for the subject led him nowhere and set back the cause, though we will meet him again in chapter 12 in connection with main lines in Belgium. Another prophet was Thomas G. Cumming, a surveyor living in Denbigh, North Wales, who published *Illustrations of the Origin and Progress of Rail and Tram Roads* in November 1824, confidently proclaiming that steam railways 'are likely soon to be brought into full and effectual use for the purpose of conveying passengers, as well as merchandize, over very extended lines of country'.[6]

During the heady years of 1824 and 1825, 624 schemes were presented to the British Parliament, of which 30 were for railways; there was also an evident 'steamship mania' as well as considerable investment in overseas mining at this time.[7] The financial crash of 1825 put paid to nearly all of them. The most ambitious would have connected London, Liverpool, Manchester, Birmingham, South Wales and even Edinburgh.[8] British investors also submitted a proposal to Simón Bolívar and the government of Gran Colombia for a railway from the Changres River to Panama City, providing a link from the Atlantic to the Pacific, which was considered but declined.[9] Equally unsuccessful in the immediate term were proposals for cross-country 'land-bridge' railways connecting the east and west coasts of Britain. One, from Newcastle to Carlisle, was a revival of earlier canal and plateway schemes. Another would have connected Manchester to Hull, superseding existing canal and river links and providing a more expeditious outlet for Yorkshire woollens. An Irish proposal for a railway between Limerick and Waterford would have connected the busy Shannon navigation with a natural harbour on the southern coast of Ireland and anticipated locomotives or fixed engines fired on peat or Killenaule coal.[10] It took the second 'railway mania' of the mid-1840s (chapter 13) to bring these into being. What did emerge from these speculative years was Britain's first mainline railway, the Liverpool and Manchester, discussed in the following chapter, and the first railway in Britain built to connect different ends of the country, the subject of the latter part of this chapter.

William James and the Stratford and Moreton Railway

One who did do more than put his name to a prospectus or write a book was William James (1771–1837), a solicitor and land agent, whose very limited success as a railway promoter illustrates the challenges which these ambitious plans faced during what was essentially a speculative bubble.[11]

As early as 1802 he had become convinced that railways offered more than roads or canals, and in 1818 he prepared a detailed survey of a 'Central Junction Railway' to run from the canal at Stratford on Avon southwards to Oxford and ultimately to London – a railway served by a canal, providing an entirely inland route from his collieries and limestone quarries in Staffordshire and Warwickshire to their principal markets. His ideas coalesced over the next few years. The railway was, for instance, to be locomotive-hauled; when he paid a visit to Killingworth Colliery in 1821, Stephenson and Losh agreed to assign him one-fourth of the interest in their locomotive patents on the condition that James should recommend and give his 'best assistance for the using and employing the locomotive engines' on railways south of an imaginary line drawn from Liverpool to Hull. James called these 'Patent Land-Agent Engines'.[12]

As the economy expanded, his ambitions ceased to be confined to the English West Midlands. In 1823 he published his *Report, or Essay to Illustrate the Advantages of Direct Inland Communication*, advocating railways operated by Stephenson locomotives to connect London with Brighton, Rochester and Portsmouth, carrying goods, civilian passengers and troops.[13] This was written during a brief period of imprisonment as a consequence of a law suit, shortly before he was declared bankrupt. It was to have been the first of a series of 12 reports upon railway communication in various parts of England, but it was the only one to appear. He nevertheless became drawn into the Liverpool and Manchester, Liverpool and Birmingham and London and Birmingham projects.

Reality fell far short of ambition. Not only were the three mainline proposals rejected by Parliament, but his Central Junction scheme reached neither Oxford nor the metropolis, running a mere 16 miles to the market town of Moreton-in-Marsh, which had barely more than a thousand inhabitants. Horses were employed, following a report from the Quaker banker John Greaves, who had 'been introduced to some of the most eminent engineers and proprietors of railroads', and who advised against locomotives.[14] Greaves was neither an engineer nor a visionary like James but it was the right decision: George Stephenson's slow-moving locomotives would have been superfluous on a short line carrying light traffic, and totally impractical on a longer route. Had James had his way, the Central Junction would have been the greatest and most ambitious of the coal railways discussed in chapter 8, but also potentially a financial and engineering disaster: it is hard to imagine Killingworth locomotives slowly making their way the whole hundred miles to London on cast-iron rails, and in the process burning huge quantities of Staffordshire coal, never noted for its steam-raising qualities. In its incarnation as the Stratford and Moreton, it justified its existence as a horse-drawn railway supplying the parishes of rural Warwickshire with fuel, and carrying Cotswold building stone in the other direction.

The only other railway with which James was involved that was ever to see the light of day was the even shorter Canterbury and Whitstable, six miles long, opened in 1830, for which he surveyed the route and produced plans for improving the harbour. This was the world's first steam-operated general service railway (chapter 8), though its long-standing nickname, 'the crab and winkle line', conveys the nature of much of its traffic. It was in financial trouble right from the start.[15]

James is difficult to assess. The one patent he took out, a system of hollow cast-iron rails which could also be used as pipelines for gases or liquids, does not suggest a man with an eye for technology.[16] He seems to have been the sort of individual who, though unlucky, tends to make his own misfortunes, and he certainly lacked the ability to work with other people. He was also simply ahead of his time in projecting long-distance routes at a time when the technology was neither understood nor available, and when the British economy was in crisis.

The Cromford and High Peak Railway and the Budweis–Linz horse railway

Just as it was becoming clear that the Stratford and Moreton would never be anything more than of purely local importance, two other ambitious schemes were being evolved, both of which were actually constructed. One of these was built in the vigorous and recovering economy of the United Kingdom and, like the Central Junction Railway, also anticipated a heavy traffic in coal and limestone. It was less to be expected that the other was in the sprawling, slow-moving and impoverished dominions of the house of Habsburg in central Europe, but so it was. It ran from Budweis (České Budějovice) in the kingdom of Bohemia to Linz in Upper Austria, and connected the Moldau (Ultava), which flows into the Elbe and the North Sea, with the Danube, flowing east into the Black Sea.[17]

The Cromford and High Peak was one of the longer railway proposals in the United Kingdom, at 33 miles – a bold venture, and one of the few ultimately built of those incorporated in the speculative years of 1824–5.[18] It was to connect two canals, the Peak Forest and the Cromford, by crossing the 'limestone dome' of England. It was anticipated that some traffic would be generated by quarries en route but that most of the revenue would derive either from through traffic between the manufacturing districts of Lancashire at one end and Derbyshire at the other, or more particularly by providing a link

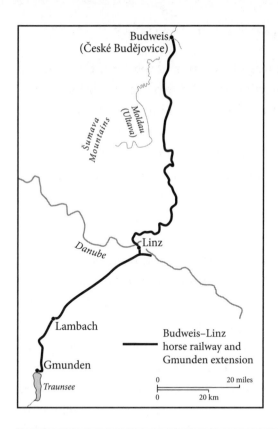
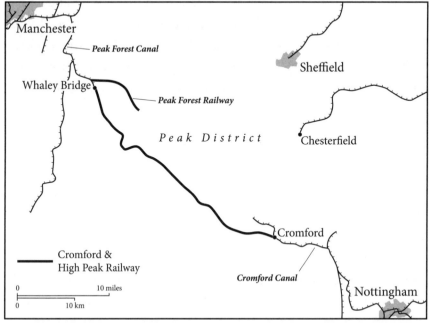

Map 3. The Cromford and High Peak Railway and the Budweis–Linz–Gmunden horse railway.

between the food-surplus areas of eastern and southern England and the hungry mouths of the north-west. The two Prussian mining engineers von Oeynhausen and von Dechen who reported on it (chapter 5) stated that 'it may have far-reaching consequences for the whole internal traffic of England ... [it] cannot fail to be profitable'.[19]

The Budweis–Linz, by contrast connected two continental rivers, the Danube and the Moldau. Improvements to the roads through the heavily forested Šumava mountains, which formed the watershed between them, had been under discussion since medieval times. This was the traditional route by which salt had been exported from its source south of the Danube into Bohemia and beyond, so there was every prospect that a railway would secure some of this lucrative trade. There was also a strategic imperative: it might unite the Habsburg dominions not only by drawing Bohemia further into Austria's orbit but more generally by linking the North Sea with the Mediterranean and the Black Sea. Central Europe had been thrown into turmoil by Napoleon's military victories, out of which the multiethnic Austrian Empire was established in 1804 (chapter 5). It emerged as the main beneficiary of the Congress of Vienna 11 years later due to the skill of the diplomat Klemens von Metternich (1773–1859), who nevertheless remained keenly aware that it comprised a formidable mixture of diverse nationalities and different interests.

Promoters and capital

Both railways were costly. The Cromford's £164,000 capital was the largest amount which had been contemplated for a railway project in the United Kingdom (though it was soon to be eclipsed by the Liverpool and Manchester), and it came from private funds, not state coffers. Earlier proposals for a canal had come to nothing but there was clearly an appetite for a long-distance through route, and a readiness to put up the money for it. Landed proprietors subscribed, as did some Manchester investors; so did the Butterley Company near its eastern terminus, which had extensive interests in collieries, limestone quarrying and iron production in the English East Midlands. The bill had an easy ride through Parliament and passed into law in May 1825.

The Austrian project had a longer genesis. As early as 1807, Professor Franz Josef von Gerstner (1756–1832), the Director of the Prague Polytechnic School, where he also held the chair of mechanics and hydraulics, had advocated a railway instead of a canal over the Šumava on the grounds that it would be cheaper, speedier, easier to maintain and not liable to stoppage during droughts or frost. He had recommended iron-strapped wooden rails and trains of horse-drawn waggons – an engineering vision, but, in an era when the Empire was struggling to discharge its debts, a financial impossibility. Better times came with the defeat of Napoleon and the arrival of French indemnity payments, further helped when the financier Salomon Mayer Rothschild (1774–1855) established a bank in Vienna alongside the family's existing houses in Frankfurt, London and Paris. In March 1820, Rothschild links with Austria were further strengthened when his brother Nathan Mayer Rothschild was appointed Austrian Consul in London; his other brother James was created Austrian Consul-General in Paris the following year. The Budweis–Linz was their first joint foray into railway investment, undertaken on the advice of the metallurgical expert Professor Franz Xaver Riepl (1790–1857) of the Polytechnic Institute of Vienna.[20] With capital of 1,000,000 *gulden* from the Rothschilds and two other banking houses (£100,000 at the current rate of exchange), and a grant from Vienna of the salt transport monopoly, the project boded well.[21]

Engineers and construction

The engineers who built these two systems came from very different backgrounds – one from a very pragmatic British tradition, the other from a more consciously intellectual and university-based central European approach to construction work.

The design of the Cromford, and the first phases of its construction, were carried out by the experienced and capable Josias Jessop (1781–1826), the son (and pupil) of William Jessop (1745–1814), one of the great canal builders of the late eighteenth and early nineteenth centuries (chapter 3).[22] At the age of 20, Josias was working for his father as an assistant engineer on the West India Docks. He had gained further practical experience on the Surrey Iron Railway and on an earlier line which served the Cromford Canal, the Mansfield and Pinxton. He had managed a large workforce and dealt capably with clients and shareholders, shrewdly representing their interests in Parliament. His home was Butterley Hall, not far from the projected eastern terminus of the railway, with his younger brother William (*c.* 1783–1852), who had become manager of the Butterley Company. So well organised was the project that even Jessop's death, on 30 September 1826, supposedly from overexertion, did not hold matters up, and the work was continued by the resident engineer, Thomas Jackson Woodhouse (1793–1855). Contracts were let in 24 lots. John Hodgkinson, with his wide experience on plateways, particularly in South Wales, contracted for the Cromford and Sheep Pasture inclined planes.[23]

It was a classic, if late, example of a 'dry canal', with the inclined planes concentrated at each end, eight of them steam-powered, one operated by a horse capstan, and level or lightly graded sections in between, the longest slightly over 25 miles. Curves were as sharp as 55 yards (50 m) radius through 80 degrees at one point – a late Victorian writer called it 'The skyscraping High Peak Railway with its corkscrew curves that seem to have been laid out like a mad Archimedes trying to square the circle'.[24] Jessop underestimated the cost of construction but overestimated the cost of the already archaic cast-iron fish-belly rails, a problem solved when the Butterley Company, which had contracted to supply them and the stationary engines, permitted its debts to be consolidated with interest payable at 5 per cent. Jessop's preference for cast-iron rails over wrought iron

19. The Sheep Pasture inclined plane on the Cromford and High Peak in 1952, showing the engine house on the left.

reflects the fact that Butterley had a foundry but no rolling mill, and so there were sound commercial reasons for the decision to use them even if no technical advantage.[25] The engines for the inclined planes were built to a standard, vertical-cylindered condensing pattern and, on surviving evidence, were housed in similar buildings of Derbyshire gritstone with large round-topped windows in three angled facets, like tall turnpike tollhouses.

On 29 May 1830 the first section, from Cromford Wharf to Hurdlow, was opened. The remaining half to Whaley Bridge and the

Peak Forest Canal opened 'to the public for general trade' on 6 July 1831. It was not a common carrier, though it had some waggons for hire.

The builder of the Budweis–Linz was Franz Anton von Gerstner (1795–1840), an associate professor at the newly founded Polytechnic Institute of Vienna (chapter 5), son of the Franz Josef who initially advocated a railway in 1807. The Gerstners were beneficiaries of the enlightened reforms of Joseph II in the eighteenth century which had enabled talented men of lowly origins to be promoted in the service of the state, and to earn themselves a title of nobility and a coat of arms in the process. As well as his Jesuit education and degree from the Charles University, the elder Gerstner had taught himself a wide range of artisanal skills, and moved at ease among practically minded engineers and technicians in the long-established industrial centres of the Kingdom of Bohemia and elsewhere in the Empire. He had been consulted on engineering and mining projects from Silesia to Hungary, and had built a stationary steam engine at his institution in 1806 to instruct the growing number of students who flocked to his lectures. He advised his son on the railway's construction, though he was not directly responsible for it.

They might have made a father and son team to match the Jessops, or indeed George and Robert Stephenson, and they certainly had a far stronger academic background than any of the English engineers, but Franz Anton lacked practical experience. This was implicitly acknowledged by a commission from Philip Ritter von Stahl, President of the Imperial and Royal Trade Commission, for him to travel to the northeast of England to see railway building and operation on the ground, and to assess how it might meet the needs of Bohemia and Upper Austria. The first of his visits to England was made in 1822.[26] In the event, he rejected emergent English practice in most important respects.

The railway that Franz Anton designed was true to his father's original vision of a wooden strap way, a *Holz - und Eisenbahn*, on which horse-drawn trains would run, rather than an iron railway on the English model, and he chose a narrower gauge – 3 ½ Austrian feet,

110.6 cm, or 3 foot 7.5 inches. In a densely forested region, using timber as the main rail component made sense. After 1829, cast-iron straps were used instead of wrought-iron, which had been found to give a rougher ride and to damage the sleepers and the wheels.[27] A further factor might have been that, although Bohemia's metallurgical industries were well-established, there were as yet no rolling mills to produce wrought iron.[28] Initially the railway was built without sleepers. On streets and at road crossings, cast-iron fish-belly rails were laid in U-shaped stone blocks.

The younger von Gerstner departed from English practice in one other important respect. The engineers he met on his visit advised him to build it as the Cromford and High Peak was to be built, using powered incline planes to surmount gradients, but von Gerstner grasped the potential of locomotive traction and chose instead to build a heavy formation with slight gradients but on a very long route to allow for the eventual adoption of steam.[29] This was forward-looking but in the circumstances self-defeating, as it meant that he had to construct the railway to an unnecessarily costly standard, for a technology that was not yet available. Locomotive traction on so narrow a gauge, and more particularly over a distance of 17 *meilen* (129 km), was not then a practical proposition.

Work began under his direction in 1825. In the name of Francis I, Emperor of Austria and King of Jerusalem, Bohemia and Hungary, von Gerstner was granted a concession to build and to operate for 50 years a railway between Budweis and Mauthausen, a small riverside market and a ferry on the Danube, the then intended terminus. The railway was to be governed by existing legislation for the building of public roads; any branches were to require additional permission, and landowners were to be compensated for any loss. The railway could run its own rolling stock. One *meile* (7.586 km) had to be operational within a year, the whole within six.[30]

He soon ran into trouble. Local interests offended at the loss of salt transportation revenues opposed him; peasants and landowners worried about a railway cutting their lands in two were not supportive; teamsters and workmen, of whom there were about 6,000, had no experience of this type of construction, and Gerstner did not always handle his subordinates well. Core stone walls in the embankments were found to be weak and had to be rebuilt, at additional expense.

He was not the last engineer to find that scientific understanding was no help in managing a project in terms of keeping a truculent workforce in line and shareholders supportive, but it nearly proved fatal to the railway. It may be that he lost his nerve in the freezing winter of 1826–7, when he left for his second study tour of the United Kingdom, rather than press on with his work and face down his detractors, though he did persuade George Stephenson's backer, Edward Pease, to write to Francis I praising his abilities as an engineer. Whether his Imperial and Royal Apostolic Majesty was impressed to receive a commendation from a Quaker in Darlington is unrecorded. This second visit seems only to have confirmed Franz Anton's enthusiasm for steam traction; on his return, he continued to build the railway to standards that could take the pounding and stresses of a locomotive. By July 1827 the entire project hung in the balance.

Thereafter the situation eased, if only temporarily. On 7 September a test train was operated, consisting of carriages built by a colleague of Gerstner's father, Josef Božek (1782–1835), the Czech technician, watchmaker and inventor who, inspired by English precedents, had built the first steam road locomotive in the Austrian Empire in 1815.[31]

A few weeks later the northern part of the line opened, but traffic was scant. The last act for von Gerstner came on 28 June 1828, when the directors ordered him to surrender the superintendency of the railway to his two subordinates, Eduard Schmidl (1794–1880) and Matthias Schönerer (1807–81).

Gerstner, true to form, then embarked on yet another study tour of Britain, returning in November 1829. It was at this stage that he wrote up his detailed *Handbuch der Mechanik* (chapter 5), published at Prague in 1831, based on his father's lectures and his own research, though it was published primarily under his father's name. This volume, as much for what it does not say as well as for what it does, suggests that he found it difficult to come to terms with his removal from the project. He references many railways in the United Kingdom, from the Newcastle waggonways to the Penrhyn Quarry railroad, but does not make it clear what he had personally seen or to whom he had spoken. He refers to the Cromford as 'not finished in 1829', suggesting he had seen it then but knew nothing more of its progress, includes circumstantial detail of the Brunton and Shields, and refers to Chapman, Rennie, Telford and George Stephenson.[32] He refers to, but probably did not witness, an experiment with horse haulage on the Monkland and Kirkintilloch on 7 February 1828.[33]

Schönerer prudently constructed the southern part of the route as a lighter, cheaper line, suitable only for horse traction, leading to the Danube not at Mauthausen but at the important town and river harbour of Linz. Bridges on this section were mostly built of timber, gradients were steeper and curves sharper. Something of this engineering approach is evident in the watercolour of the Emperor and the Empress dignifying the official opening of the completed railway on 21 July 1832 with their presence, riding in a railed barouche from the north bank of the Danube along the sinuous alignment as far as the medieval parish church of St Magdalena, the limit of the short imperial journey. It was Matthias Schönerer, not von Gerstner, who rode with them, on a back-facing dickey-seat, but twisting his neck to avoid the solecism of turning his back on the imperial and royal couple as he explained the works as they went along. Closed stagecoach carriages also formed part of this initial procession.[34] When regular services were introduced, intending passengers could choose between a first-class *Separatwagen*, like the surviving 'Hannibal' carriage of 1841 in the Vienna Technical Museum, an enclosed second-class *Stellwagen* or an open third-class waggon.[35]

Freight services ran the whole distance from Budweis to Linz from 1 August 1832. High, short-wheelbase waggons were built for carrying salt. Gerstner considered that the dead weight of British waggons, with their iron wheels and axles, limited the load capacity, so he only used iron at the hub, the fitting of the tread and the flanges – though they were still called *englischen Bahnwägen*. To cope with the sharper curves on the rerouted southern section in 1829, Schmidl devised a six-wheel waggon, Schönerer a single-axle vehicle.[36]

Operations

Very little is known about the early operation of the Cromford. Goods took about two days to make the journey from one end to another, a saving of time compared to the Trent and Mersey Canal.[37] Some sources suggest that a steam locomotive was put to work in 1833, presumably on the long near-level section between the topmost of the western inclines at Bunsall and the eastern inclines at Hopton. The carrier German Wheatcroft of Cromford, who first came to notice on the Peak Forest Railway from 1794 to 1809 (chapter 3), operated a passenger service from 1833, connecting with his road coach service from Manchester, and with barges on the Cromford Canal. There is no record that he built any dedicated facilities for travellers. Wheatcroft also acted as a goods carrier on the railway, possibly the only one. His business served London, Bristol, Bath and Nottingham.[38]

The Budweis–Linz is much better recorded. It was operated by the Bohemian industrialist Karl Adalbert von Lanna (1805–66), who had already taken over his father's shipping business on the Ultava, and who later developed railways in the Czech lands and in Silesia.[39] As well as the traffic in salt and timber, passenger transport had been anticipated from the start and was initiated in 1834. Main stations were established in the manner of a well-appointed inn on a *chaussée*, located at about 20 km intervals, with catering facilities. They were located so that every horse could return to stay overnight at its own stable, which would have mangers for anything between 25 and 100 animals, and be served by a farrier. The depots also functioned as crossing, transhipment, storage and loading points. Other stations were established to enable horses hauling passenger trains to be changed, were exclusively for freight or were purely boarding points for passengers with no facilities. There were also 51 guard houses at regular intervals for the men who inspected and maintained the track.[40]

Employees were given lanterns and tools for shoeing the horses and for carrying out any running repairs. It was one of the first railways to make a gesture towards putting its staff into uniform – for special occasions they wore suede gloves and what they called a *shako*, though, rather than the tall, cylindrical cap which was becoming common as military headgear, here it was a high felt hat with silver ribbon trimmings and a horse-tail. One such *bahnknecht* (railway worker), Leopold Viertbauer, became famous for entertaining his passengers by accompanying himself on the zither as he sang, and for always having a bottle of schnapps on hand. He died a wealthy man from the tips he received.

Traffic – hope and reality

Of the two, the Budweis–Linz represented the better investment, at least in the short and medium term, even though Gerstner's projected revenue proved to be unrealistic. The long-standing 'salt war' between Austria and the kingdom of Bavaria was resolved by a treaty of 1 November 1829, when the Emperor was compelled to relinquish the trade to Bavaria, thereby diminishing traffic on the railway. It nevertheless enjoyed some success. Between 1834 and 1836 it was extended south into Linz itself, spanning the Danube on the old wooden bridge authorised by Maximilian I in 1497, and on to Gmunden through Lambach, giving direct access to the salt-producing areas. This had also been earlier suggested as a canal, and the plan had been revived in 1818 as a horse-drawn railway, to no immediate avail. In 1829, one of von Gerstner's assistants, Francesco Zolla (1795–1847), had carried out a survey at his own expense but could not find financial support.[41] Finally, the backers of the Budweis–Linz received the contract in 1833 and the work was duly carried out by Schönerer.[42] The railway also promoted river traffic: in 1837 the Danube Steamboat Shipping Company (Donaudampfschiffahrtsgesellschaft), formed by two Englishmen, John Andrews and Joseph Prichard, in 1829 for the Austrian government, extended its services as far as Linz to connect with the railway, employing a wooden paddle-steamer, the *Maria Anna*.[43] The Budweis–Linz section lasted until the 1870s when it was succeeded by a standard-gauge locomotive railway, the Gmunden section until 1903.

The Cromford and High Peak never paid a dividend. Jessop's predictions about through traffic proved to be very overoptimistic. Coal and coke tonnages were approximately a quarter of what he had anticipated and, instead of the estimated 40,000 tons of corn, malt and flour making their way across the Derbyshire uplands to feed Lancashire mill workers and colliers, after five years of operation it carried a mere 5,200. Only limestone, quarried along the railway's route, brought it any significant revenue, and it was for this reason that much of it was still operational into the 1960s, effectively as a long mineral siding from the Manchester to Derby main line.[44]

Aftermath

These two systems, built to accomplish comparable tasks in different parts of Europe, have been criticised as a dead end in railway history, using technologies that were already time-expired – rope-worked inclined planes on a through route in one case, and outdated forms of track in the other.

Given the topography the Cromford had to face, there was perhaps little alternative to the use of inclined planes. As earlier chapters have suggested, these were not particularly problematic where compact minerals formed the main traffic. Even the cast-iron rails on the Cromford seem to have given good service; the northern section as far as Burbage was replaced with wrought iron in 1857, and contracts were invited to replace the remaining ones only in 1863.[45] Whether or not the Cromford used archaic features scarcely mattered from an investment point of view; it attracted the funding it needed, and by the time of its construction investors and industrialists in Britain were largely persuaded about the merits of this mode of transport. It was contemporary with the first recognisable mainline railways, discussed in the following chapter. Few new stretches of canal were being built by this stage. Roads remained busy, but parliamentary funding for new roads ceased in 1836.[46]

The real weakness in the Budweis–Linz railway lay not in its old-fashioned track but in its ambition for the future, in von Gerstner's insistence that it be built to a high and costly standard for a technology that was yet to come – narrow-gauge steam locomotives. It was another 30 years before they made their appearance.

Of the engineers involved in their construction, though Jessop did not live to see the Cromford's completion, Woodhouse went on to have a successful career on the Dublin and Kingstown Railway and the Midland Counties Railway, and like many other able men of his generation later found work on major railway and canal contracts in France and Italy.[47] The difficulties which von Gerstner encountered, or perhaps caused, did not inhibit him from developing further innovative railway projects within extremely conservative countries. He went on to build the first main line in Russia, from St Petersburg to Tsarskoe Selo, opened in 1837, having persuaded the conscientious but reactionary Tsar Nicholas I and his no less cautious ministers to approve the scheme (chapter 13).[48] Restless and inquisitive to the last, his final study tour was to the more open society of the United States, where he wrote the book for which he is best known, *Die innern Communicationen*, a description of the country's railroads and navigable waterways, published after his death in Philadelphia in 1840. A memorial erected in Linz, ironically a town he never intended for his railway's terminus, in 1955 described him as the builder of 'the oldest overland railway in the European mainland'. Matthias Schönerer distinguished himself in the Austrian railway service, becoming famous for dispatching some of the first ever troop trains during the uprising of 1848–9, and was raised to the nobility in 1860.

Although Emperor Francis I attended the opening day of the Budweis–Linz railway, to the day of his death in 1835 he could not bring himself to think that steam railways boded well for the lands he ruled. Neither did Metternich, for whom they were 'the attempts of a false and conceited civilisation ... and all the other extravagances of a morbid age'.[49] Yet in Austria, as ultimately also in Russia, the railway became unstoppable. Professor Riepl, who had first advised Salomon Mayer Rothschild to invest in the Budweis–Linz, was already developing plans for a railway network across the Empire from Krakow in Galicia to Trieste on the Adriatic, and was himself sent by Rothschild on a study tour of English railways in 1830 in order to inform both him and his banking associate Leopold Wertheimstein about their potential.[50] Salomon was granted the concession to construct a railway linking Bochnia and Vienna in 1832; this ultimately became the *Kaiser-Ferdinands-Nordbahn* after the new emperor agreed to give it his patronage.

The failure of William James's plans to develop beyond the purely local was probably fortunate; an ambitious but impractical early scheme might have set back the cause of railway building by deterring investment. The Cromford and High Peak, and even more the Budweis–Linz, demonstrated that a considered scheme did have the potential to attract capital, as well as state support (or to do without it), and to bring together an engineering team capable of creating an iron road to unite distant parts of the country, even though one was not profitable for years and the other struggled to be completed. By connecting navigations, they anticipate some of the American railroads that were completed a few years later, but they are coeval with the first generation of main lines in the USA and in Britain. These form the subject of the next chapter.

11

The first main lines 1824–1834

By the 1820s and early 1830s railways were for the first time being built to meet a need in regional economic life, rather than purely serving a locality by connecting a mineral region with navigable water. These developments took place within the same time frame as, and were closely related to, an associated step change, the construction of the first three railways designed to promote the fortunes of an ocean seaport, by ensuring faster and higher capacity access to its hinterland than canals and turnpikes could offer, and by carrying passengers as well as goods – the first main lines.

Crucially, because of the distances they had to cover, the loads they were expected to move and the tasks they were expected to perform, these railways also became the test beds for different types of haulage, confirming that innovative locomotive designs offered more than horses or fixed engines over long distances, and that wrought-iron rails offered the most practical track system. They were successfully completed and put into service between 1830 and 1834, at a time when other shorter but equally innovative public railways were also coming into being. These are described in the following chapter.

Of the three main lines, two were in the USA and one in the UK. These were the Baltimore and Ohio, the Charleston and Hamburg

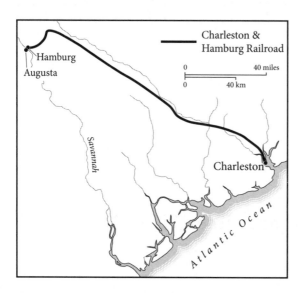
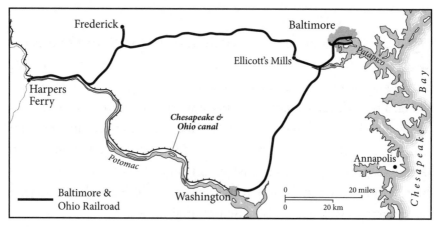
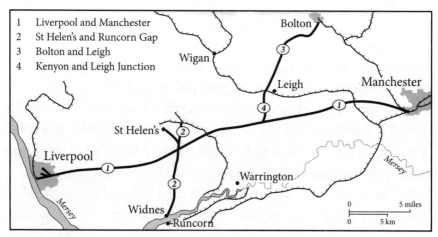

Map 4. The first main lines.

and the Liverpool and Manchester. If Hanoverian Britain alone possessed the conditions to initiate such developments, the Adams-era United States evidently also had the intellectual, material, financial and legislative preconditions that could make them a reality.

All three reflected current English thinking, yet the two American railroads reinterpreted it to different degrees, to suit their own circumstances. Given the Stephensons' role in designing and building the Liverpool and Manchester, it is no surprise that it followed evolving English practice in its track type – wrought-iron fish-belly rails on stone blocks – and 4' 8" (1422 mm) gauge, but it took the technology to a different level by sponsoring and adopting innovative locomotive design, to say nothing of the sheer scale and ambition of the undertaking. Its owning company had been one of the few put forward during the speculative years of 1824–5 that had actually received its act. It was completed and opened for traffic in September 1830. Meanwhile, the Baltimore and Ohio had been chartered and had opened its first 16 miles. The Charleston and Hamburg had operated its first demonstration train, locomotive-hauled, on Christmas Day 1830 on the one mile of track that had been completed – it was not opened throughout its length until 1833, but by then it was the longest railway in the world.

All three illustrate the way in which railways were now carrying out different functions. The Liverpool and Manchester carried bales of raw cotton imported from the southern USA and from India, to be spun and woven in Manchester and the western Pennines, before being sent back to be exported all over the world. The British textile industry had not hitherto had a significant impact on railway development but its rapid growth around the town of Manchester had already led to major technical changes as well as to new forms of organisation, mass production and marketing.[1] It confirmed the 'great divergence', when Western Europe and North America shook off premodern growth constraints to emerge as the most powerful and wealthy world civilisation, eclipsing India and China until the late twentieth century.

Global trade required swift and well-managed transport by sea and land. The town of Liverpool had invested considerably in its system of docks (with their own internal plateways) from 1811, and a highly significant development was the institution of packet ships to and from New York in 1817, which left on stated days whether they had a cargo or not. The assurance of regular Atlantic sailings made these the vessels of choice, and the practice soon became widespread.[2] However, existing means of inland transport, the Mersey and Irwell Navigation and the Bridgewater Canal, were slow and monopolistic. Liverpool had adapted successfully to the abolition of the slave trade in 1807 and was not about to contemplate any decline in its fortunes because links with its hinterland were becoming inadequate.

The idea of a railway had been first mooted in 1797–8, to be revived in 1822. The company was established two years later by Henry Booth (1788–1869), who wrote the initial prospectus, along with other merchants from Liverpool and Manchester. The first bill presented to Parliament failed because of stiff opposition from landowners and canal interests, as well as basic errors in the initial surveys carried out by William James (chapter 10) and George Stephenson, whose evasions and ignorance were humiliatingly exposed by Counsel in Committee. Persuasive negotiations with vested interests, the support of William Huskisson, one of Liverpool's two Members of Parliament, and a better survey on a different alignment prepared by George Rennie (1791–1866), his brother John (1794–1874) and Vignoles ensured that it was passed in 1826 and work could begin.[3] It ran directly from the southern Liverpool dock system to a terminus in Manchester which had good road and canal links, and was built to carry passengers as well as goods.[4]

The Charleston and Hamburg was in some ways the Liverpool and Manchester's transatlantic mirror image. Charleston had grown rich as the major port of entry for Africans to the southern USA, and was still the centre of South Carolina's brisk trade as a net exporter of enslaved people.⁵ Its principal commerce, however, was cotton, a crop increasingly cultivated in its Piedmont area between the Appalachians and the Atlantic coastal plain. Overland transport by waggon was slow and costly, and was in any case making its way not overland to Charleston, on the Atlantic coast, but to Augusta at the limit of navigation on the Savannah River and in the state of Georgia. Other exports were in sharp decline, and Charleston had barely recovered from the financial crisis of 1819. Canal projects and turnpikes had not addressed the problem, any more than had establishing the town of Hamburg on the Carolina side of the river, opposite to Augusta.⁶

Similarly, the fast-growing port of Baltimore, the state capital of Maryland, faced economic stagnation unless it opened routes to the western states. It was the largest port on Chesapeake Bay, exporting flour to the home market, to the West Indies and to South America, as well as trading regularly with Liverpool, but its river was the Patapsco, which appropriately means 'backwater' in Algonquian, and which gave no effective access to the hinterland. Baltimore was endeavouring to compete with Washington, which lay on the Potomac, and with Philadelphia on the Delaware, as well as with New York and the Hudson, which had been linked to the Erie Canal and to the Great Lakes since 1825.⁷

It was in these circumstances that, in 1826, Baltimore bankers Philip E. Thomas and George Brown sent Philip's merchant brother Evan to England to see the Stockton and Darlington. William Brown in Liverpool also sent his Baltimore brother details of the proposed construction of the Liverpool and Manchester, which suggested that a railway could be operated between Baltimore and – ultimately – the Ohio River and the Mississippi watershed.⁸

On 12 February 1827, 25 merchants of Baltimore held a meeting to consider this proposition. It was clear that, even in Britain, long-distance railways were an untried technology but also that the American climate was less conducive to canals. Copies of Gray's

Observation on a General Iron Rail Way and Wood's *Treatise* were eagerly consulted and other study tours organised, this time of engineers. Jonathan Knight, William G. McNeill and George Washington Whistler set off on 22 October 1828, not returning until 22 May of the following year; they were followed by George Brown and Ross Winans, who from 6 to 14 October 1829 observed the Rainhill locomotive trials on the Liverpool and Manchester.

Construction of the Baltimore and Ohio began that month. It opened over 16 miles to Ellicott's Mills, the Quaker enterprise which had developed into one of the largest centres of flour production in the United States, on 22 May 1830 and to Harpers Ferry in West Virginia, at the confluence of the Shenandoah and the Potomac, on 1 December 1834, a total of 82 miles. A single-track branch was built to the town of Frederick in 1831; in 1833 construction on a single-track branch to Washington, the federal capital, began and it opened on 25 August 1835. In the first years of operation, the line handled considerable freight and passenger traffic in horse-drawn vehicles on rails of wood or granite, topped by wrought-iron straps imported through Liverpool. The long granite sills, or stringers, inspired by the Quincy (chapter 3), were set directly into the ground, and proved expensive to purchase and extremely difficult to lay.[9] However, as chapter 6 notes, a consignment of fish-belly wrought-iron rails and cast-iron chairs, keys and spikes on the Liverpool and Manchester pattern, including 12 sets of points, purchased for evaluation purposes, was not adopted. Ten tons of iron would, it became clear, lay about a third of a mile of track on this basis, but 22 tons of strap rail would suffice for a mile of permanent way.[10]

Finance

Of the three railways, the Liverpool and Manchester was by far the most costly, at £600,000, £19,355 for each of its 31 miles, almost twice the per-mile cost of the Baltimore and Ohio by 1834, and about 12 times the cost per mile of the rickety Charleston and Hamburg.[11]

Ports had access to capital, but it was as well that Liverpool's merchants were very rich men. Participation in the slave trade, abolished in 1807, had provided an effective way of turning over very large sums of money, which became available for reinvestment in the town and its transport links. John Gladstone of Rodney Street (1764–1851), a member of the railway's Provisional Committee and father of the future Prime Minister, owned 2,508 slaves in 1833, on various plantations in the West Indies.[12] By no means all its supporters were in favour of slavery: Henry Booth, its dissenting Secretary and Treasurer, was opposed, and the Charleston and Hamburg's engineer, Horatio Allen (1802–89), recalled an uncomfortable dinner party in Liverpool at which conservatives and reformers lost their tempers with each other.[13] Manchester had grown wealthy on textiles, but was wary of investment in the new railway, subscribing only £12,000. The Exchequer Bill Loan Commissioners made £100,000 available to the Liverpool and Manchester in 1827.[14]

Civil engineering

This level of capital meant that the Liverpool and Manchester could be very substantially constructed. Although George Stephenson had done badly with the initial survey, he retained support within the Company, and the Rennies overplayed their hand in asking for complete freedom to choose their colleagues.[15] The result was that the railway was built largely under Stephenson's direct control through a system of direct labour and small contracts, with himself as one of the contractors. It ran through terrain that was mostly undemanding, on a near-straight alignment. The most challenging aspects of the work were the extensive area of peat bog known as Chat Moss, and the Wapping Tunnel under the streets of Liverpool, from the limit of locomotive working to the docks, on which waggons were hauled by stationary engines (chapter 7). The Baltimore and Ohio had to wait until October 1832 for direct rail access to the City Dock, threading its way through the streets, whereas the Charleston and Hamburg had to rely on horses and carts to connect it with the harbour.[16]

The 4.75-mile crossing of Chat Moss drew on established practice and long local experience, though one problematic section was only completed by sinking wooden and heather hurdles into the bog until they could provide a strong foundation. Cutting the tunnel was gruelling work, carried out by experienced contractors. Elsewhere, the line required 9 miles of embankment, 13 miles of cutting and a nine-span major viaduct over Sankey Brook, built of yellow sandstone and red brick.

Much of the supervision was in the hands of men with experience on the Stockton and Darlington, but the scale of work was vastly greater than any previous railway construction and much more akin to a major canal project. Labourers were drawn from the local population and from Ireland, from where gangs had been coming over to England for many years to bring in harvests and increasingly to build canals and docks.[17] Many were enticed across the Atlantic to work on canal projects and on the Baltimore and Ohio. They were described as:

> ... the finest workmen in Europe [who] dig out twenty-five cubic yards of heavy clay each day – but their desire to run to the public houses and get drunk is so great that many of them perform their day's work in a few hours. These dissolute men exert themselves so violently in their work that I have seen many powerful, muscular men with blood oozing out of their eyes and nostrils.[18]

Instructions were sent out that wages were to be paid in the offices, not in the public houses, and, though no doubt the drink trade did very well out of the men building the railway, there is little reference in any of Liverpool's newspapers to behaviour that offended respectable opinion or to court cases.

THE FIRST MAIN LINES 1824–1834

The Baltimore and Ohio had to follow the winding course of the Patapsco but still emulated the Liverpool and Manchester in several respects. Its brick and stone bridges were built from the outset for double track, designed for heavy traffic and with steam locomotives in mind. The federal government assigned the railroad three brigades of West Point engineers to survey the route, which was dug out by 2,000 men, 300 horses and 200 carts. Irishmen predominated among the labourers. However well they might have behaved in Lancashire and Cheshire, in Maryland they were constantly in trouble with the law. They fought among themselves, Ulstermen against Corkonians and the 'fardowners' (*fir donn*, 'brown men') of Connaught and Leinster against each other, as well as with African Americans and Germans.[19]

Other teams were put to building the bridges. The elegant single-arch Carrollton Bridge, designed by the architect James Lloyd and built by Caspar Wever, the Company's construction superintendent, is only the most notable of the 17 stone bridges the company built between 1829 and 1833, the first masonry railway bridges in the USA. The Thomas viaduct of 1834 on the Washington branch, still in use, and highly unusual in being built on a curve, resembles the Liverpool and Manchester's Sankey Brook viaduct in scale and ambition.[20] It was built to the design of Benjamin Henry Latrobe II (1806–78), one of several engineers responsible for the Baltimore and Ohio. This very thorough approach nevertheless made the construction of the Baltimore and Ohio a long-drawn-out matter. Its civil engineers quickly learnt that what might be advisable in Britain was too costly in the USA; after 1834 they tended to revert to the already well-established American tradition of wooden bridge building, and took this principle with them when they went on to work on other railroads in the USA.

The Charleston and Hamburg, by contrast, was underengineered. It was built for single track, using wooden trestles to carry it over the swamps and creeks of the Piedmont. Horatio Allen had already used this method on the Delaware and Hudson, and assured the citizens

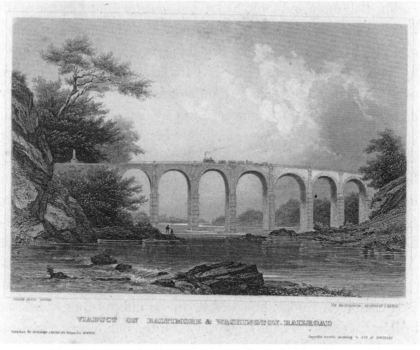

20. The Baltimore and Ohio's elegant Thomas viaduct was British-inspired but did not set an American precedent.

of Charleston that it would be safe from floods and would require no fencing since slave, freeman and beast could simply pass underneath the tracks. Lumber was cheap, and this method certainly allowed the 136-mile route to be built quickly. It was emulated on several later American railways. Some landowners offered to forego compensation for the right of way and offered their enslaved workers 'on the most reasonable terms' to grade the ground and clear the woods for the required opening of 200 feet.[21] The enslaved were cheaper than freemen as there were no wages involved, nor did they expect to be given whisky, but they were also considered more dexterous.[22] The 'rails' were southern pine, capped with flat iron bars, at 5-foot gauge. The English sociologist Harriet Martineau described it after it had been operating for a few years:

> I never saw an economical work of art harmonise so well with the vastness of a natural scene, as here. From the piazza of the house at Branchville, the forest fills the whole scene, with the rail-road stretching through it, in a perfectly straight line, to the vanishing point. The approaching train cannot be seen so far off as this. When it appears, a black dot, marked by its wreath of smoke, it is impossible to avoid watching it, growing and self-moving, till it stops before the door. I cannot draw; but I could not help trying to make a sketch of this, the largest and longest perspective I ever saw.[23]

This is perhaps the first description of a type of scene that was to become part of the American experience, the spectacle of a train making its way on a straight formation across a totally flat wilderness.

Novel it may have been, but travel on the Charleston and Hamburg in its early days was not a reassuring experience. Its locomotives were unreliable – Martineau and her party arrived at Charleston in the small hours instead of the late afternoon because of a boiler leak, having in the meantime been kept awake by a nocturnal frog chorus in the Carolina swamps while it was being repaired.[24] For Colonel 'Davy' Crockett (1786–1836), a trip on the Charleston and Hamburg in 1834 was his first sight of a train, 'a dozen big stages hung on to one machine, and to start up a hill. After a good deal of fuss we all got seated, and moved slowly off; the engine wheezing as if it had the tizzick. By-and-by she began to take short breaths, and away we went with a blue streak after us.' He relished the story of how, after it had grown dark, sparks from the locomotive frightened a waggoner's horses and made them bolt, smashing his waggon. 'He run to a house for help, and when they asked him what scared his horses, he said he did not jist know, but it must be hell in harness.'[25]

The locomotives also inflicted hammer-blows on the trestles with every revolution of their wheels, and it soon became clear that only tipping spoil between the rotting timbers would enable the railway to remain in operation.[26] By 1842–3, maintenance was demanding enough to require one gang consisting of a carpenter and nine enslaved men hired from their owners for each of 15 nine-mile sections of track, as well as an additional gang to which no one particular section was assigned. The railroad also owned 16 men outright. Wooden houses were built to accommodate them along the track, though they were also by this stage being accommodated in out-of-service passenger coaches.[27]

Motive power

All three railways made significant use of steam power. Trains on both the Liverpool and Manchester and the Charleston and Hamburg were hauled by locomotives, other than on short double-track sections down steep slopes to navigable water – the Mersey and the Savannah respectively – operated by fixed engines. The Liverpool and Manchester's inclined plane ran through the brick-lined gas-lit Wapping Tunnel below the streets of Liverpool, the first railway to run under a city.[28] It also used fixed plant to draw carriages up to Crown Street Station and considered, but ultimately rejected, proposals to use them on some of the stiffer gradients on the main line (chapter 7).[29] The Baltimore and Ohio initially installed four inclined planes on its Frederick branch, opened in 1831, two on each side of the Parrs Spring Ridge, operated by stationary steam engines, but found that locomotives and horses were able to haul trains up the gradients.[30]

The Rainhill trials on the Liverpool and Manchester not only confirmed the suitability of locomotive traction on a long railway and for what were then considered high speeds, but also defined the form that it would take for the next hundred years and more – though this was only because a radically different type was built and entered just in time, the Stephensons' *Rocket*. Adapted as the *Northumbrian* class, this design was the mainstay of the first few years of the new railway. In its further avatar as 'Planets' and 'Samsons', it was within a few years to be found on railways from Delaware to the Urals. Between 1829 and the end of 1834, the railway acquired, and in some cases already withdrew, 35 locomotives in this design series.[31]

After the tiny demonstration machine *Tom Thumb* had operated on the Baltimore and Ohio, the company sponsored its own trial, announced on 4 January 1831, offering a premium of $4,000 for the best design 'of *American Manufacture*' and $3,500 for the second best. This encouraged several speculators to submit locomotives, as at Rainhill, and, though little is known of them, they suggest an active spirit of experiment among American engineers. As chapter 8 describes, one specific type came to be developed as a consequence of the trial, the vertical boiler 'grasshoppers'.[32]

Although all three railways pioneered the use of locomotives over long distances, none of them committed to using steam at the time they were chartered. In 1828, the Charleston and Hamburg's engineer, Horatio Allen, confidently announced 'I have just returned to Liverpool, having been at Newcastle, visited the railroads in its vicinity and examined the operation of the locomotives with the closest attention. I have been completely convinced of their utility and superiority to horsepower.' After graduating from Columbia College, Allen had become agent to the Delaware and Hudson Canal Company, and had driven the *Stourbridge Lion* at Honesdale. His views were confirmed by the directors in 1830, who stated 'the locomotive alone shall be used. The perfection of this power in its application to railroads is fast maturing, and will certainly reach, within the period of constructing our road, a degree of excellence which will render the application of animal power a gross abuse of the gifts of genius and of science.'[33] The Charleston and Hamburg, in its own echo of Rainhill, stipulated that each locomotive should achieve 10 mph and pull three times its own weight. One of the directors, Ezra L. Miller (1784–1847), also strongly believed in locomotive traction, and had a 30-inch model made to help convince his colleagues; this operated on a circular track at the company's office.[34] The *Best Friend* was the first American-built locomotive to enter revenue-earning service, but its life was short – a mere six months until its boiler blew up, killing the enslaved African-American fireman who had suppressed the safety valve. It and the other West Pointers were joined in 1832–3 by the four unsuccessful eight-wheel articulated locomotives described in chapter 8, then in 1834 by the Baldwin 4-2-0, the *E. L. Miller*.[35]

Rolling stock

Much of the traffic carried by these railways – cotton, flour – could be carried in two-axle goods waggons, 'gondolas' in American parlance.[36] Passenger services, however, called for more imaginative solutions. Perhaps the least successful was the Baltimore and Ohio's short-wheelbase two-axle carriage vehicles, of which the first were built by Richard Imlay (1784–1867) of Baltimore and later of Philadelphia, a road-coach design body on a rail chassis. Imlay turned out what was familiar to him, even down to the leather thorough-brace suspension, the railway providing the underframe. As with road coaches, passengers could travel on the roof, though sparks and cinders from the locomotive were a novel hazard, made potentially worse by the light canvas awnings they carried.[37]

The Liverpool and Manchester also adapted road-coach practice but in a different way, by ordering longer conjoined compartment carriages, in which three such road-coach bodies were combined on a single frame. To complete the resemblance, they were painted black and yellow, a common stage livery, and were all given names, in part to enable passengers to recognise the carriages in which their seats were booked. Some were named after people in the public eye, others after emblems of speed or beauty. In an appropriate show of political impartiality, or a nod to the railway's own divided loyalties, one was named 'Conservative' and another 'Reformer'. There were also mail coaches, for which the tickets were more expensive, as well as some carriages with coupé ends, and some with curtains rather than windows. Second-class carriages, known as 'blue boxes' by their patrons, were simply that, some divided into three compartments with side doors and reversible seats enabling passengers to face the direction of travel, others with an entrance in the middle and seats parallel with the track. An experimental variation resembling an Irish jaunting car was not adopted. Illustrations also show a 'Chinese' carriage, reflecting a current interest in oriental decor.[38]

Although the Chinese carriage had a recessed upper body where the windows were located, it was otherwise an early attempt at a type of passenger vehicle which became very common, a roofed box construction. An illustration of the first passenger train on the Charleston and Hamburg shows two similar, though unadorned, passenger carriages, as well as a flat waggon carrying a recruiting party of Federal troops and a small cannon.[39] The Liverpool and Manchester ran two-axle flats on which the wealthy could travel in their own road carriages.[40]

All three railways adopted articulated bogie vehicles, though in different ways. For its opening day, the Liverpool and Manchester constructed a sumptuous 32-foot-long eight-wheel carriage with its axles grouped in pairs towards the end of the frame, and, though there is nothing in the archives that confirm details of its construction, this configuration strongly suggests that it ran on bogies.[41] It was a far grander affair than the Stockton and Darlington's 'Experiment', but may have been inspired by it.

American engineers and promoters were already aware of articulated vehicles and soon realised that they would offer a smoother ride on the rough track they would be laying.[42] They were also popularised by Ross Winans, a farmer in Sussex County, New Jersey, who had first turned up in Baltimore with a small working model of a car equipped with what he called 'friction wheels' – effectively a primitive form of roller bearing. It was the short-lived enthusiasm of the Baltimore and Ohio for this device that prompted its directors to send Winans to England to confer with the Stephensons and to witness the construction of the Liverpool and Manchester.[43] Here he again succeeded in interesting a railway management committee in friction wheels which he applied to a hand-operated trolley which was paraded at the Rainhill trials.[44] Whether or not he saw the Liverpool and Manchester's carriage under construction, or perhaps even had a hand in its design, on his return to the USA – by now an assistant engineer on the Baltimore and Ohio – he became an enthusiastic advocate for bogie vehicles. By late 1830 flour-cars derived from Liverpool and Manchester practice were being adapted as bogies in articulated eight-wheel firewood cars, soon followed by open-sided eight-wheelers carrying hay, barrels and cattle and one for a road carriage and horses all harnessed together, on which Louisa Catherine Adams, the President's wife, rode.[45]

In July of the following year, the inaugural steam-hauled passenger train included the 'Columbus' bogie carriage. Others followed – the 'Winchester' in 1832, made up of three stagecoach-style bodies, and the 'Dromedary' in 1834, in which four stage-coach bodies were carried on a drop frame with an overhead truss.[46] There was also a 'Sea-serpent' and a 'Washington'.[47] The cars built for the Washington branch in 1835 set the pattern for future American passenger transport, with end balconies and aisle seating.[48]

Engineering, repair and maintenance facilities

Just as steam locomotives were coming to be built by specialist firms rather than by the companies they were destined to serve (chapter 9), these first main lines had to establish effective means of ensuring that they remained in working order and basic repairs could be carried out. The Liverpool and Manchester's locomotives were maintained at 'Melling's shed' at Brickfield Station, Liverpool, where John Melling (1781–1856) held sway. By 1833 he was 'superintendent of the repairs of the locomotive engines' on a wage of three guineas a week, later raised to four pounds.⁴⁹ The Charleston and Hamburg seems also to have carried out repairs 'in house', under the supervision of Horatio Allen and with the encouragement of Ezra Miller. By contrast, the Baltimore and Ohio, once it finally decided on steam power in 1833, contracted out the construction and maintenance of locomotives and other equipment. Though the company constructed a forge, machine shop and other facilities for locomotives and carriages at Mont Clare at this time, this seems to have been under the superintendence of Phineas Davis, whom the railway contracted to build locomotives, until his death in an accident on the Washington branch on 27 September 1835.⁵⁰ Thereafter, as the Annual Report observed:

> ...the workshops at the Mount Clare depot are carried on by Messrs. Gillingham and Winans, independent of the Company. They are bound by contract to supply the Company with locomotive engines, and all other railroad machinery, at a stipulated price, and at all times to give precedence to the Company's demands for work. They have the use of the ground and buildings occupied by them, with the fixed machinery left by Mr. Davis, without rent, being bound to keep the same in repair and return them as they received them. In consideration of this, they manufacture the Company's engines so much below the market price for them elsewhere that the interest on the cost of buildings and fixed machinery, above mentioned, is fully paid; and, indeed, it would take but a little while, when the extension of the road westward required a larger number of engines, to reimburse to the Company the entire outlay for the shops at the Mount Clare depot.⁵¹

Organisation

Railway building and operation on this scale required close coordination, efficiency and timing on the part of a hierarchy of salaried and waged employees, from which in time a professional managerial class and a distinct working-class occupation type would emerge. The organisation of the Stockton and Darlington had been old-fashioned, even precapitalist, somewhere between a family bank and a Quaker meeting, whereas the Liverpool and Manchester, after a shaky start during the construction phase, became a formidably well-run and distinctly modern organisation.[52] This was no surprise. Liverpool's bankers and merchants had brought diligence and efficiency to commercial life, qualities they imparted to the town's public sector and to the railway.[53] There was also a touch of rational religious dissent in the person of Henry Booth, secretary, treasurer and effectively general manager from 1833, on a salary of £1,000 per annum. Booth was a Unitarian, heir to the intellectual traditions of his native Cheshire, and an exceptionally gifted organiser and publicist, the 'father of railway management'. He worked with an active board of directors, which met once a week, and with a management committee and specialist subcommittees within a clearly defined structure.[54] The railway employed no fewer than 706 men (and no women) by March 1832.[55] As it settled down, it had to develop rules and policies on matters such as the issuing of tickets, Sunday work, gratuities for staff, smoking on trains and the sale of food and drink.[56] It was the first to adopt full uniforms for the members of its workforce who were in the public eye.[57]

The Charleston and Hamburg went through the same process. As it candidly admitted in its report for 6 May 1833, 'In fact, the whole business of rail-roads and locomotives is new to all the world, and to this company in common with all others'.[58] Its published regulations specified that no guns or fowling pieces were permitted 'unless examined by the Conductor'. Feet could not be put on cushions. Servants were not permitted unless looking after children, and only with the consent of other passengers, presumably to minimise ethnic mixing, though 'Coloured Persons' could travel half fare, as could anyone below the age of 12.[59]

The Baltimore and Ohio did for the United States what the Liverpool and Manchester did for the United Kingdom: it established the way that these 'iron roads' would be managed. Not only did many who worked on its construction go on to find employment on other railroads, but it also published a detailed annual report from the beginning of the construction phase, causing it to be known as 'the railroad university of the United States'. Its approach to information management enabled its accounting procedures to move from simple bookkeeping to become a means of decision-making. Unit-cost comparison enabled an informed choice as to, for instance, the relative merits of timber and stone bridges, horses or locomotives or of one route compared to another.[60]

Architecture, design and philosophy

All three systems were built at a time and in countries where the practice of civic architecture was well advanced, yet neither of the two American railroads chose to make a statement through their public buildings. The Baltimore and Ohio clearly sought to impress with its well-engineered and attractively designed bridges but did not devote any significant resources to passenger accommodation or goods facilities. Depot buildings were knocked together out of boards salvaged from the workmen's shanties.[61] Early depictions of the Baltimore and Ohio show it as an almost unassuming presence in the busy but ordered environment of Ellicott's Mills or sharing the spectacular landform of Maryland Heights with the Potomac River and with the Chesapeake and Ohio Canal.[62] Its Frederick terminus consisted of a single loop and a siding into a freight depot; a photograph taken towards the end of its life shows this to have been a stone structure with a canopy to protect road carts.[63] Even when dedicated building for passengers came into being, they were not built to impress: the depot in the federal capital was tucked away where Pennsylvania Avenue met 2nd Street and was a three-storey single-fronted house with a tiled roof.[64] The Charleston and Hamburg was built on the cheap throughout, and also began life with minimal facilities. The Hamburg depot was little more than an enclosed yard with 'two spacious depositories ... with zinc roofs'.[65] The battlemented depot at Camden (Charleston) was not built until the 1840s. At the top of the inclined plane, the locomotive-worked section ended in a double loop where, alongside the track, the town of Aitken was laid out on a grid pattern.[66] The two American main lines resolved to establish operational railways and to worry about their public face only once revenue started to come in, confident that the vogue for 'internal improvements' would sustain them.

By contrast, the Liverpool and Manchester Railway carefully and deliberately set out to present itself not only as a lavish showcase of engineering capacity and investment potential but also as a philosophical statement. There are several possible and closely related reasons why this should have been so.

First, so much British capital had gone into canal building, and a mainline railway was such a novelty that this new transport system needed to justify itself. Conspicuous architectural display would bring home to canal trustees, and other vested interests such as uncooperative landowners, that railways could bring huge financial resources to bear, especially if they were funded from Liverpool. Furthermore, as the Liverpool and Manchester was being planned and built, the United Kingdom was undergoing one of its periodic spasms of cultural conflict. If there had ever been a consensus about 'improvement', it had certainly given way in the late 1820s to heated debate about the 'March of Intellect', so called, whether approvingly or derisively, when 'philosophic Whigs' were making the case not only for political reform but also for technical and scientific enquiry and for its dissemination at all social levels, while old-school Tories deprecated the mania for progress and for new-fangled ideas. These informed, or inflamed, the argument about some very specific political issues, which included giving the vote to Roman Catholics, conceded in 1829, and the extension of the franchise, granted in 1832. The cartoonist William Heath (1794–1840) depicted steam road vehicles as well as a vacuum tube for travel to India, air transportation and a bridge to South Africa, even a giant bat for transporting convicts to Australia, in his savage but ambivalent excoriations of everything he disliked in contemporary life or feared in prospect.[67] Ultimately, in the years that followed, the idea of 'progress' won the day, to become something of a Victorian truism.

To demonstrate that the railway was not a mechanical fancy but an undertaking of substance and worth, the board of the Liverpool and Manchester chose to present its corporate image in a number of different ways. They ordered medals to be struck, the first to commemorate a transport undertaking.[68] Railway-themed merchandise was encouraged. Rudolph Ackermann, one of the best-known lithographic publishers of the day, was commissioned to produce views of the railway, and he in turn commissioned the architect and illustrator Thomas Talbot Bury (1811–77) to produce the sketches and watercolours. Unlike the lithographs of Ellicott's Mills and Harpers Ferry, these place the railway centre-stage, its machines and its working men at the service of commerce and prosperity. Well-turned-out families admire locomotives and ride contentedly and safely on the trains. The track gauge is consistently exaggerated, in a way that makes the railway look like the conqueror of nature, its long straight formation running through deep cuttings and across the expanse of Chat Moss. They proved extremely popular; a reissue in 1832 was followed by Spanish and Italian editions, and the prints were produced separately in French and German. The artist Isaac Shaw was asked to make further sketches, which were engraved and published in *A New and Complete Work of the Liverpool and Manchester Railway*.[69]

Above all, it was in its architecture that the directors set out their vision for the railway. The two terminal passenger stations, at Crown Street in Liverpool and Liverpool Road in Manchester, were plain but dignified stone-built two-storey town-houses in a restrained classical style by John Foster, who specialised in civic Greek revival buildings.[70] They were innovative in railway terms in that they acknowledged the need for proper facilities to handle the arrival and departure of travellers, to shelter them and to keep their movement separate from the administration of goods traffic. Shelters were added at both stations in 1831.[71] They were the first of the many lavish railway stations built worldwide in the years to come. Coaching inns and canal hotels had been performing similar tasks for many years, and in terms of scale at least the Liverpool and Manchester's new stations can be compared with the depots built on the Budweis–Linz horse railway. They certainly represented a step change compared to the minimal facilities offered by the Stockton and Darlington. Liverpool Road Station at Manchester survives with very little change as part of the Museum of Science and Industry.

Goods facilities on the Liverpool and Manchester were no less impressive than passenger stations. Late in 1829 the company resolved to act as a carrier in its own right, and preparations were soon put in hand for a three-storey warehouse at the Manchester terminus, capable of accommodating 10,000 bales of cotton or other goods in proportion, completed shortly after July 1830, on a long, slightly curving plan. A further shipping shed and warehouse at the south-eastern corner of the site followed in 1831. These were built in the same design tradition as existing warehouses on canals and docks, of which several examples were to be found nearby on the Castlefield basin of the Bridgewater Canal.[72]

The most remarkable structures on the railway, however, were functional rather than commercial and were to be found at Chatsworth Street cutting at Edge Hill on the periphery of the built-up area of Liverpool, at the point where the passenger line from Crown Street Station joined the summit of the inclined plane from the docks emerging from the tunnel to form the main line to Manchester. It was here that locomotives were attached to the trains.

Foster gave his imagination free reign, designing one of the most extraordinary architectural concoctions on any railway of any era. A Moorish arch crossed the line, flanked by two battlemented houses for the stationary engines. These exhausted through two tall chimneys on either side of the tunnel mouths. This was the first time that a railway had emulated the triumphal arches and civic thoroughfares that had become a part of town planning, though the railway's 'public relations' man, John Scott Walker, made it clear that their significance lay even deeper:[73]

> Looking up from the area to the west, the visitor beholds, surmounting the two corners of the walls above, two beautiful Grecian columns, built of chequered brick, with pedestals and capital ornamented with stone. These columns rise in due proportion to the height of above 100 feet, and are conspicuous and decorative objects, marking the commencement of the open Rail-way, as the 'pillars of Hercules,' the entrance of the chief sea of the ancients. The stranger will appreciate the taste of the Engineer, when he learns that ornament is here combined with utility, that they are, in fine, the chimnies of the two engine furnaces below – the boilers of which he will observe placed in small tunnels cut laterally into the rock forming the area, and near to its eastern end. This area may be called the starting station of the Locomotive Engines, which will here be attached to the wagons and Rail-way coaches.[74]

Foster had evoked an ancient strain of symbolism that was still accessible to the well-read in the early nineteenth century. The open sea was what made Liverpool rich, but to classical antiquity the two pillars of Hercules, Gibraltar and Jebel Musa, had represented the limits of known or prudent navigation. Sailing beyond them in search of riches and advancement had become a symbol of heroic valour and imperial ambition in the sixteenth century, and of intellectual

21. The frontispiece to Bacon's 1620 *Novum Organum* inspired the Liverpool and Manchester's architecture.

discovery in the seventeenth. In 1620 the philosopher Francis Bacon adapted it for the title page of his call to intellectual regeneration and inductive methodology, the *Novum Organum*, showing the ship of knowledge passing through the columns of received understanding. Its suggestion that intellectual discovery might be the fruit of finding new lands and trading with their people (or, in Liverpool's case, in their people), was well-suited to nineteenth-century merchants dealing in overseas commodities and anxious to join the British social elite.[75] The Rainhill trials were further proof that knowledge was achieved by reasoning and practical experiment, and the railway as a whole symbolised this revived spirit of empiricism.

Ceremonial

Ceremonial was also part of the way these new main lines announced themselves, and was on a far more lavish scale than the speechifying and feasting that had marked the opening of previous railways. Construction of the Baltimore and Ohio began on 4 July 1828 with the laying of the first stone in a ceremony attended by Charles Carroll (1737–1832), one of the richest men in Maryland, and indeed the United States. In 1776 he had been the only Catholic to put his name to the Declaration of Independence, and by this time he was the last survivor of the signatories, a figure from another age in his knee breeches and buckled shoes. He sank a silver spade into the ground, whereupon the Masons went about their ritual of measuring the cornerstone, pronouncing it 'well formed, true and trusty' before anointing it with oil and wine and scattering it with corn. Carroll remarked to a friend 'I consider this among the most important acts of my life, second only to the signing the Declaration of Independence, if even it be second to that'.[76] A 'Rail Road March' (*maestoso con moto ben marcato*) was one of two songs composed for the occasion.[77]

The President of the United States, John Quincy Adams, broke ground the same day at a ceremony for the Chesapeake and Ohio Canal, a form of transport which he favoured over railways.[78]

The actual opening of the Baltimore and Ohio was quieter. On 22 May 1830, five months before the Liverpool and Manchester's traumatic event, Charles Carroll and the other directors sat down in Imlay's 'Pioneer' coach, which was pulled by a single horse to the initial terminus at Ellicott's Mills. The Charleston and Hamburg also opened with little fuss. On 3 October 1833 company president Elias Horry and state governor Robert Y. Hayne set off from Charleston at 5:45 a.m., reaching Aiken at 5 p.m., a total journey of 120 miles. This was not the end of the day, as the car with the Augusta mail and the passengers had to be let down the inclined plane. They arrived at Hamburg at about 8 p.m. The world's longest railway was now operational.

Even if the Liverpool and Manchester had not ended up killing one of its principal guests on its opening day, 15 September 1830 would have been remembered in British and world history. Previous transport undertakings in the United Kingdom had been inaugurated by local bigwigs, but, on that day of watery sun, Liverpool saw a gathering of continental European nobility such as had not been assembled since the Congress of Vienna rubbing shoulders with the British political elite. Not only was the guest of honour the Prime Minister and war hero, Arthur Wellesley, First Duke of Wellington (1769–1852), but four future British prime ministers were also present, and Sir Henry Brougham, the very embodiment of the 'philosophic Whig', was to be Lord Chancellor before the end of the year. Guests of rank, and in some cases of intellect and distinction also, included six earls, two marquises, six viscounts and over twenty other members of the peerage, though only one bishop. Some other guests were people in the public eye, like the writer and actor Fanny Kemble and the polymath Charles Babbage (1791–1871).[79]

From Russia came Count Pavel Nikolaievich Demidov (1798–1840), who had recently inherited his family's extensive holdings in Russia and their mines and ironworks, including the two engineer-serfs Yefim and Miron Cherepanov (chapters 5 and 8), whom he later sent over to study the railway. From Hungary came Count Lajos Batthyány de Némétújvár (1807–49), a future prime minster of his country, who was to be executed in the aftermath of the Hungarian Revolution, and Count István Szechenyi (1791–1860), an Anglophile and a moderniser.[80] The representative of the new world was Francis Barber Ogden (1783–1857), the United States consul in Liverpool, and friend and patron of John Ericsson.

The Duke was welcomed with Handel's chorus 'See, the conqu'ring hero comes', which came to be played at the opening of many a rural branch line for years to come, though increasingly the 'conqu'ring hero' was felt to be the locomotive pulling the inaugural train rather than the dignitary performing the ceremony. He was placed, along with the most distinguished of the other guests, in the eight-wheel carriage. The Duke was there to give a Tory endorsement to a Whiggish and philosophical enterprise, and a soldier's salute to a trade venture and a capital speculation.

A ceremony meant to be spectacular proved calamitous when a respected public figure was mortally wounded in full sight of the principal guests. It was William Huskisson's fate not only to be killed by a railway locomotive but also to be remembered for little else. Physically clumsy and already ill, the Member of Parliament for Liverpool had been advised not to attend the opening day but felt that it was his duty to his constituents to do so. At Parkside, while the locomotives were taking on water around midday, he was one of many who stretched their legs despite official warnings about the dangers. Seeing the Duke, and sensing an opportunity to repair a relationship which had become strained after he had left the government, Huskisson walked over to the carriage, only to be run over and injured by the *Rocket*, driven by Joseph Locke. He died nine hours later after enduring terrible pain. Liverpool's representative at Westminster, a former Under-Secretary at War, Secretary to the Treasury, Treasurer of the Navy and President of the Board of Trade, he was a prominent free trader and acknowledged as the leading representative of the commercial interest in Parliament. That he of all people should have been killed by a locomotive added a grotesque irony to the symbolic architecture of the railway and to the thwarted ceremonial of the day.

William Huskisson was not the first fatal victim of a railway accident, as is so often stated, and was by no means even the first person to be killed by a steam locomotive, but he remains to this day the highest profile railway fatality. As if that were not enough, the passengers on the opening day trains found that when they reached the far end of the line, they were greeted not by cheering crowds but by a resentful population lining the tracks despite the rain which had now begun to pour. Disagreements between gentlemanly Whigs and Tories were real enough, but the people of Manchester had ideas of their own. Revolutionary tricolours had been hung from some of the bridges, and there were rumours, untrue as it turned out, that men were marching from Oldham to attack the railway.[81]

Imitation and invention

It was once observed that the opening day of the Liverpool and Manchester was one of those epochal moments which 'divide, precisely, the past from the future, the old from the new, the historic from the pre-historic, and of which nothing that came after was ever quite the same as anything gone before'.[82] It certainly represented a shift in scale and ambition that surpassed both the earlier generation of iron railways and all but the longest canals and turnpikes. Not only was it entirely steam-operated, but its locomotives themselves were the design precursors of nearly all that followed. Another step change was the way that passenger facilities were set out and managed; its stations showed the way forward for railway companies in the years to come.[83] Above all, it broke with most predecessor railways in England in that it was built not to carry coal or some other mineral, but to serve the globalised economy of cotton. It connected two great industrial centres, one an ocean-serving port, the other a manufacturing town. Its architecture celebrated what the railway embodied, not only the empirical philosophy which identifies successful solutions to technical problems but also Britain's role as the 'mart of nations'.[84]

THE FIRST MAIN LINES 1824–1834

Highly influential though the Liverpool and Manchester indeed was, it was not the progenitor of all that immediately followed. The Charleston and Hamburg showed that a railway over 100 miles long could indeed be built on the cheap and operated at a profit, but it was also evident that its timber formation was unwise and the bizarre eight-wheel locomotives a failure. Considerable sums had to be spent correcting initial mistakes. The Baltimore and Ohio struggled with its granite rails and even revised its gauge twice but it was a better engineered system than its southern sister.[85] In a lecture to the Maryland Institute on 23 March 1868, John Hazelhurst Boneval Latrobe (1803–91), who had been the Baltimore and Ohio's lawyer for many years and who was the older brother of its engineer Benjamin Latrobe II (1806–78), gave an account of its design and construction, in which he remarked:

And yet, no one railroad was the prototype, exactly, of another. The Baltimore and Ohio Railroad set men to thinking, and gave them the benefit of its experience. But originality was everywhere aimed at; and improvement was the consequence everywhere. The Chevalier Von Gerstner, a distinguished German engineer, who had constructed a short railroad in Russia, from St. Petersburg to Tsarskoseloe, came to America while the railroad fever was at its height, for information, as he said. At this date England was well under way with the system, and the speaker expressing surprise to the Chevalier that he had not remained there, instead of coming to the United States, the latter answered at once: 'That is the very thing I want to escape from – this system of England, where George Stevenson's [sic] thumb, pressed upon a plan, is an imprimatur, which gives it currency and makes it authority throughout Great Britain; while here, in America, no one man's imprimatur is better than another's. Each is trying to surpass his neighbor. There is a rivalry here out of which grows improvement. In England it is imitation — in America it is invention.'[86]

Perhaps as Latrobe recalled the conversation, von Gerstner was endeavouring to be polite to his transatlantic hosts, yet he had met George Stephenson, no doubt listened to his advice and nevertheless had followed his own judgement when he came to build what had been, briefly, the longest railway in the world, between Budweis and Linz (chapter 10), before it yielded the palm to the Charleston and Hamburg. The Stephenson system offered much, as the Liverpool and Manchester showed, but it would be subject to careful scrutiny and to adaptation as the iron railway came of age between 1830 and 1834.

In June 1833 President Andrew Jackson travelled on the Baltimore and Ohio between Relay and Baltimore, only the second head of state to make a journey on the iron road. A citizens' committee met him when he was on tour, and the trip seems to have been a spontaneous gesture. Whereas the Austrian Emperor had presided at the opening of the Budweis–Linz railway with due Habsburg ceremony, the first citizen of the American republic, so far as is known, seems simply to have bought his ticket just like everyone else and to have sat down in a carriage with the other passengers.[87]

12

Coming of age
The public railway 1830–1834

The high-capacity mineral carriers, the internal communications and the first main lines described earlier evolved in parallel with a new generation of public railways.

By 1834 well over 600 miles of chartered railroad were operating in the USA, compared to about 50 in 1829. Fewer schemes were initiated in the United Kingdom in the immediate aftermath of the opening of the Liverpool and Manchester because of the reform crisis. The febrile 'days of May' in 1832 brought the country within sight of revolution, but by the summer changes to the political fabric made it possible to complete a number of new systems as well as to begin some very ambitious projects.

Turbulent France was not in a position to invest in major railway systems until the dust had settled after the revolution of 1830. In the German Confederation, it was not enough that the 'educated world' considered Stephenson railways necessary and desirable, or that the commercial interest should advocate them (chapter 5); nothing much was likely to happen until civil servants came to the same conclusion, and they tended to favour canals and roads. In Russia, the age-old and never-ending fight between bureaucrats and ministers who favoured Western practices and Slavophiles who rejected them meant that no swift decision could be made about railway building. Many thought that the diversion of capital away from agricultural improvement would benefit neither nobleman nor serf.[1]

One country which gave serious thought as to how public railways might be planned, financed and built was Belgium. Shortly after it secured its independence from the Netherlands in 1830, its government sought to safeguard the revolution by initiating a comprehensive survey of potential routes to serve the country's well-established industrial and commercial economy, and also, in liberal circles at least, to forge a national identity.[2] Many of its engineers and statisticians were closely linked to the French utopian socialist Saint-Simonian movement (chapter 9), which called for a meritocratic 'industrial class' of workers, scientists, industrialists, engineers, managers and bankers to construct a new society.[3] It created the world's first planned public state system. Initially a railway was proposed from the Prussian border to the Scheldt at Antwerp, but the lower reaches of the river lay in the Netherlands, and successive revisions of the plan therefore also ensured access to the Belgian coast at Ostend.[4]

Scope and purpose

Many of the railways actually opened between 1830 and 1834 were short in length and answered only local needs. A British example might be the Bodmin and Wadebridge, which carried sea sand from the Camel Estuary for use as an agricultural fertiliser on the acid Cornish soil, an unusual example of an early iron railway where the bulk of the traffic went inland.[5] Others were still being built to carry coal to navigable water (chapter 9), as Tyneside waggonways had been since the seventeenth century. Such were the Room Run, the Little Schuylkill, the Mine Hill and Schuylkill Haven and the Mount Carbon in Pennsylvania; a railway connecting the Épinac pits in the Saint-Étienne coalfield with the Burgundy Canal (the only significant French system built in this period); and the Clarence, the Leicester and Swannington and the Stanhope and Tyne in England.[6]

As we have seen (chapter 8), in south Durham, the Clarence Railway ran from Coxhoe Colliery to staithes on the River Tees but also offered an alternative outlet for coal from Bishop Auckland by means of a junction at Simpasture with the Stockton and Darlington, itself extended beyond its original termini to pits at Hagger Leases and to the Tees at Middlesbrough in 1830. Another regional network came into being in 1831, when the Garnkirk and Glasgow was opened, ensuring a supply of domestic fuel to Glasgow itself and of coal and iron ore to Charles Tennant's chemical works at St Rollox, just outside the town, one of the largest in Europe, by connecting with an earlier railway, the Monkland and Kirkintilloch. This had been in use for a number of years, but it ran from Palacecraig and Cairnhill collieries to the Forth and Clyde Canal, and its operators seemed more interested in the Edinburgh market. Further tributary railways in the Lanarkshire coalfield were the Ballochney, completed in 1828, and the Wishaw and Coltness, of which the first section opened in 1834.[7] The four lithographs of the Garnkirk and Glasgow's opening day on 27 September 1831, by David Octavius Hill (1802–70), confer on it the dignity of a main line and are meant to make it clear that this development was greeted enthusiastically by the people of Scotland – all ages and all ranks of society have turned out for the great day, and mothers are pointing out to their children, girls as well as boys, all that is going on. The view of two locomotive-hauled trains passing each other in the cutting at Provanmill with the Clyde Estuary in the background shows an artery of trade built to connect with the wider world.[8]

The Leeds and Selby Railway, which opened on 22 September 1834, with only one of its two tracks complete, was the outcome of several otherwise abortive projects of 1825 to connect the Lancashire coast of England with its Yorkshire, North Sea, coast. Since 1816 Leeds had had a canal outlet to Liverpool, but ships of up to 200 tons could reach Selby on the River Ouse. Hull merchants, anxious to retain the woollen trade between Leeds and Hamburg, persuaded the Leeds interest to sponsor a railway only as far as Selby and a jetty to the new steam packet services, thereby bypassing the Air and Calder navigation and the new riverport at Goole. Though it was designed to be a goods-carrying 'land bridge', it also catered for passengers, who were entrusted to guards in suits of green livery, with brass plates on their hats bearing the name of the company.[9]

In Lancashire, the directors of the Liverpool and Manchester were content with their revenue-earning route, but the Kenyon and Leigh Junction connected it with the Bolton and Leigh, and independent branches were constructed to Wigan and Newton. Existing short-haul colliery systems were also adapted to connect with the main line.[10] The St Helens and Runcorn Gap Railway, by contrast, was in no hurry to establish a physical junction with the Liverpool and Manchester, which it crossed on an elegant stone bridge. A well-known lithograph shows a Braithwaite and Ericsson locomotive hauling coal waggons over the bridge towards the docks on the Mersey at Widnes while the *Northumbrian* or one of its sister locomotives hauls a passenger train towards Manchester.[11]

In America it was still broadly accepted that inland waterways would continue to be the main arteries of internal communication and that other modes of transport would serve them. The Ithaca and Owego opened a route connecting the Erie Canal and the Great Lakes with the Pennsylvania coalfields by linking the canal-served Lake Cayuga with the Susquehanna River. The Tuscumbria, Courtland and Decatur and the Mohawk and Hudson were both built to cut out long stretches of difficult river. The Allegheny Portage Railroad formed a link between two separate sections of the Pennsylvania Canal, thereby creating a through route from Philadelphia and the Delaware River to the Ohio.[12] The Pontchartrain Railroad in Louisiana connected an estuary of the Gulf of Mexico with the lower stretches of the Mississippi at New Orleans, so shipping could avoid having to negotiate the shifting sandbanks of the river mouth. It made for a short but delightful journey for well-heeled white visitors:

> It was odd to be passing through a gay garden on a rail-road. Green cypress grew out of the clear water everywhere; and there were acres of blue and white iris; and a thousand rich, unknown blossoms waving over the pools. A negro here and there emerged from a flowery thicket, pushing himself on a raft, or in a canoe, through the reeds. The sluggish bayou was on one side; and here and there, a group of old French houses on the other. It was like skimming, as one does in dreams, over the meadows of Sicily, or the plains of Ceylon.[13]

Elsewhere, networks were already discernible in the way that railways were beginning to connect with each other. The two first railways in New York State joined end-to-end at Schenectady, forming a through route 37 miles along. These were also among the earliest to be built primarily for passengers. One of these was the Mohawk and Hudson, though this did also carry on a brisk trade in flour and other commodities.[14] What was effectively its extension, the Saratoga and Schenectady, ran to a fast-growing resort town, and also derived most of its revenue from passengers, though traffic was only substantial during the season, and all trains ran 'mixed', with both passengers and goods.[15] Philadelphia, on the Delaware River, became a railroad hub very early on, with lines going west into Pennsylvania and to the Susquehanna (1833), north to Germantown (1832) and to Trenton along the Delaware valley, linking with the Camden and Amboy (1834), itself built to connect the vicinity of Philadelphia with the Raritan River and so with New York.

Another type of passenger railway was being constructed in the United Kingdom. The Dublin and Kingstown was not only the first steam public railway in Ireland when it opened on 17 December 1834, and the first to serve a capital city, but it was also the first commuter railway, carrying only passengers and parcels.[16] It ran from Dublin to Dunleary (Dún Laoghaire), a fishing village which had been chosen by Thomas Telford to be the site of his Irish harbour, and which had been renamed 'Kingstown' following George IV's visit there in 1821. For many years this was the point where bleary-eyed and queasy voyagers disembarked from the Holyhead packet to make the onward journey by rail, but it was also near enough to Dublin to allow for daily return travel (it was slightly less than 8.5 miles long), encouraging the prosperous middle classes to forsake the city for villa dwellings facing the bay. The elegant suburbs which sprang up along its route – Booterstown, Blackrock and Salthill – were a far cry from the squalid tenements which garlanded plateways in the valleys of South Wales (chapter 3).[17]

Dublin Bay and Saratoga were both testament to the effects of railways in urban promotion. A related development was the building of the first intra-urban system, a 'streetcar' or city tramway. The New York and Harlem was promoted by the owners of property in the northern part of the promontory between the Hudson and the East River, around the small village of Harlem, which was then still set amid farms, to increase its value. As Gerstner pointed out, 'the Harlem Railroad actually functions as an *omnibus service within the city of New York*'.[18] It proved popular with nannies looking after small children who would take them out to enjoy the country air in good weather.[19]

Legislation

Some of these systems could be built by local agreement. Quite a number in the north-east of England coalfield were built under wayleaves just as wooden waggonways had been, even including the ambitious Stanhope and Tyne.[20] However, most railways of any length and ambition sought authorisation by the state. French railway proposals of this period were still governed by the procedure of 1810 set out in chapter 9; it was not until 1836 that royal decree was replaced with concession by law.[21] In Britain, incorporation by an Act of Parliament was a costly business in itself, and the stipulation that four-fifths of a company's share capital must be subscribed before the bill could proceed to committee meant that promoters were under pressure to underestimate construction costs. As a result many systems began life with a burden of debt that could only be allayed by further borrowing.[22] In the USA, state governments could charter corporations by legislative act or through a registration process. The first Belgian railway law, passed on 1 May 1834, set up the national system by decree: 'A railway system shall be established in the Kingdom, having as its central point Malines, and proceeding to the east to the border with Prussia via Louvain, Liège and Verviers; to the north to Antwerp; to the west to Ostend via Termonde, Ghent and Bruges; and to the south to Brussels and the frontiers with France via Hainaut'.[23]

Finance

Railways were costly, but there was considerable variation in the amount of capital needed for each one. The Dublin and Kingstown cost £60,000 per mile, much more even than the Liverpool and Manchester; the Leicester and Swannington a mere £7,740.[24] The New York and Harlem was the costliest per mile in the USA at $141,333, a consequence of having to build a very solid road using stone sleepers through the middle of a built-up area.[25] Otherwise the most expensive American railway for its route length had been the Pontchartrain in Louisiana, at $72,000 a mile; it was only 4.5 miles long but was double-track throughout and ran through a swamp.

Almost as expensive were the Boston and Lowell at $70,000 and the Mohawk and Hudson at $63,568. The soundly built Baltimore and Ohio cost $38,232. The Tuscambia, Cortland and Decatur, making its way over more than 45 miles of Alabama, along a single line of strap rails, was built for no more than $8,840 per mile.[26]

Both private finance and public funds went into railway building in the United Kingdom and the United States. Some promoters paid for short mineral lines out of their own pockets, such as the Pentewan in Cornwall and the Épinac in France.[27] However, most needed borrowed capital. Following the success of the Liverpool and Manchester, British railway companies had little problem selling their shares and scrip directly to the public. In some cases, investors were local people, often in quite humble circumstances. The Bodmin and Wadebridge owed its existence to an improving landlord, Sir William Molesworth, who cajoled his tenants and dependants into subscribing, no doubt to their subsequent regret.[28] The Leicester and Swannington attracted subscribers from within the county, as well as many from Liverpool, who had seen what level of return railways offered.[29] Provincial joint-stock banks invested in newly formed railway companies, and their bills were discounted by London banking houses. Quaker finance played an important part: the Dublin and Kingstown was a Quaker initiative, as the Stockton and Darlington had been. In the north-east of England, where coal ownership and political power had always been virtually synonymous, Joseph Pease's election to the reformed House of Commons in 1832 meant that the influence of the Society of Friends now extended to parliament.

The situation in the United States was different. Capital was very limited. Some of the relatively inexpensive railroads in New York depended on local stock subscriptions. The New York and Harlem, costly though it was to build, could pay for construction by the sale of stock in the metropolis.[30] However, most railroads raised capital by using the services of an intermediary to sell bonds to the money markets of London. Quakers were again particularly active in this respect, through the banking houses of Philadelphia, which goes a long way to explaining how the city of brotherly love became a railroad hub so early on. The Camden and Amboy, for instance, floated sterling bond issues secured by mortgages on its property, convertible to stock at the holder's option. Even the shorter mineral-carrying lines, which could be practically certain of regular traffic, issued bonds in order to minimise investment risks for promoters who were primarily interested in the benefits they would confer on the industries they served rather than the return these railroads themselves offered.[31]

Governments also took a direct role in railway finance. In the UK, this meant offering money from the Exchequer Bill Loan Commission.[32] The Clarence Railway, which lay outside 'the charmed circle of Quaker investment', was lent £110,000. £160,000 went to the Newcastle and Carlisle, of which the first short section opened in 1834, and £8,000 to the Bodmin and Wadebridge.[33] In the USA, loans came from the states, though the situation was not uniform by any means. Several states accepted that costly projects of wide interest 'belong to the nation'.[34] The state of New York planned and built canals but not railroads, though it anticipated acquiring the Utica and Schenectady. Massachusetts made loans available to transport undertakings, and some legislatures, Indiana and Ohio among them, understood from an early stage that states with a very small population would neither generate the private finance to support transport infrastructure projects nor readily submit to increased taxes. In 1825 the Erie Canal had shown how public works could be funded by sovereign debt repaid with toll revenue, rather than by potentially unpopular tax increases, prompting some states as well as corporations to issue debt for transport infrastructure, all of which British investors eagerly purchased. American railroad building was already laying up problems for itself in this respect and, though it was not until 1835 that the boom began, followed by the inevitable bust two years later, the signs were already apparent. 'The whole country looked to England for capital to carry out its system of internal improvements.'[35]

France, by contrast, did not even move beyond the planning stage in this period, despite its long-standing traditions of centralisation and direct state promotion of infrastructure, as well as its monarchy committed to commercial values and to progress, the quality of its engineering corps and, neither last nor least, the wealth of James Mayer de Rothschild (1792–1868). Though the Banque de France's suspicion of joint-stock finance houses meant that infrastructure projects were slow to develop, the 20 or so wealthy family banks including Rothschild's who formed the so-called *haute banque* participated in the railway planning which only began several years after the regime change of 1830.[36] Vested and regional interests, as well as distrust between government, the legislature and financiers, slowed the process down, and though the end result was a unified system, it was late in coming.[37]

The Rothschilds typically underwrote government bonds rather than invested directly in transport or industry. The only railway in the United Kingdom that Nathan Meyer Rothschild ever directly sponsored, though it was not built, would have connected slate quarries in Merioneth in North Wales with the sea; he abandoned the scheme, and it was left to Dublin bankers to promote what became the Ffestiniog Railway, in 1832.[38] However, his brothers in Paris and Vienna paid careful attention to the development of the British railway system, as well as to steamships, and the three of them were evidently prepared to finance continental European schemes once the growing possibilities had been tried and tested in Britain. The public railway networks constructed in Belgium, France and in some of the South German countries were financed by the sale of state bonds underwritten by the Rothschilds. In these circumstances, where governments licensed and subsidised private companies, the brothers became drawn into railway finance. It took time for a railway to be built, and longer for it to become profitable; bankers who marketed railway shares could not ignore the management of the concerns whose shares they sold, any more than they could ignore the stability of the states whose bonds they marketed.

Another, indirect, way in which Nathan Meyer Rothschild promoted railway building was by lending £15,000,000 to the British government in 1835, in conjunction with the Anglo-Sephardi banker Moses Montefiore, to finance compensation to those who had formerly owned enslaved workers, under the terms of the Slavery Abolition Act.[39] The total bill amounted to 40 per cent of the Treasury's annual income, or approximately 5 per cent of British gross domestic product. Not only did the loan swell the Rothschild coffers for future reinvestment, but many of the people who were compensated put the money into railways, in some cases leading to the payment of commission to the Rothschilds and to Barings Bank.

Engineering

In the years from 1830 to 1834, a young engineer wishing to know more about how railways were being built would have been better advised to visit the United States than Britain or anywhere else in Europe. Not only was the mileage under construction remarkable but so was the geographical extent. As well as the mineral-carrying systems in Massachusetts and Pennsylvania, and the first main lines in Maryland and South Carolina described in earlier chapters, railways were now to be found in New York, New Jersey, Ohio, Michigan, Virginia and Alabama. They were being built west of the Alleghenies, where the Pontchartrain Railroad in Louisiana was the first. Thanks to the detailed research which von Gerstner carried out a few years later, detailed statistical information is available for all of these systems. Most were to the Stephenson standard gauge, now 4' 8½" (1435 mm) with the addition of the extra half-inch, but the states of Ohio and New Jersey favoured 4' 10" (1473 mm), and the Charleston and Hamburg had already set the southern standard at 5' (1524 mm), which lasted until after the Civil War. The great majority used wrought-iron straps on timber rails, as had the Charleston and Hamburg. Others used 'flat-bottom' rail, generally spiked directly to wooden sleepers, devised by Robert Livingston Stevens (chapter 6).[40] The use of chaired track on stone or timber sleepers, though recorded

22. Trautwine's 1834 Columbia bridge carries a Philadelphia and Reading train over the Schuylkill.

on the Baltimore and Ohio for assessment purposes, seems to have been limited. As noted in chapter 6, the Clarence type of chaired track was emulated on the Allegheny Portage Railroad.[41]

Whereas bridges in the United Kingdom were more often built of brick or masonry, those in the United States were generally timber, either on piles or on stone abutments, following lessons learnt on the Baltimore and Ohio.[42] The Paterson and Hudson River Railroad built a timber drawbridge over the Hackensack.[43] Some made use of the lattice designs developed by Ithiel Town (1784–1844), which could be built quickly by relatively unskilled workers from readily available material.[44] The Columbia Bridge over the Schuylkill at Philadelphia was one of the most impressive, designed by one of Strickland's pupils, John Cresson Trautwine (1810–83), a seven-span covered pine structure on stone abutments and piers. It survived long enough to be photographed in the 1850s.[45]

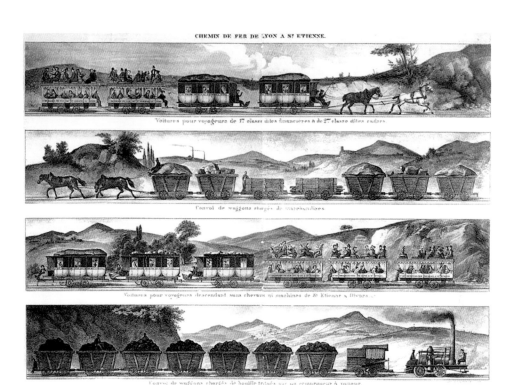

17. Horse-drawn, gravity and locomotive-hauled passenger, goods and coal trains on the Saint-Étienne to Lyon railway.

18. The people of Scotland, young and old, have turned out to see the Garnkirk and Glasgow's two 'Planets' and their trains pass each other on the opening day in 1831.

19. The opening of the Liverpool and Manchester Railway. Isaac Shaw's sketch of Edge Hill shows the Moorish arch and the Duke of Wellington's ceremonial carriage.

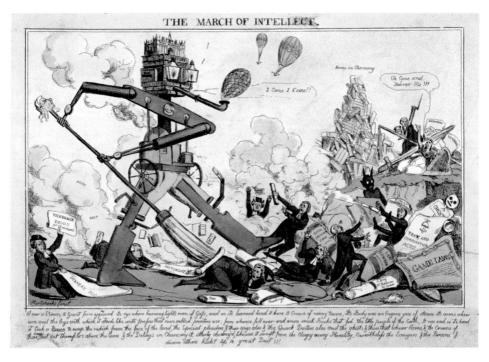

20. Steam power was welcomed by many and feared by others. Cartoons satirised both the 'march of intellect' which made it possible and the 'old corruption' which it threatened to sweep away.

21. 'Samson' locomotive *Jupiter* hauls first-class carriages and the mail on the Liverpool and Manchester Railway, whilst *North Star* hauls open carriages.

22. Liverpool and Manchester trains are about to pass each other whilst a Braithwaite and Ericsson locomotive hauls coal waggons to the River Mersey on the St Helen's and Runcorn Gap Railway.

23. The *Maria Anna* has steamed upriver from Vienna to Linz, and a passenger train on the horse railway sets off to cross the Danube on its way to Bohemia.

24. A 'grasshopper' draws a Baltimore and Ohio train into the rudimentary original terminus by Ellicott's flour mill on the Patapsco River.

25. The Baltimore and Ohio reached Harpers Ferry at the confluence of the Potomac and Shenandoah in Virginia in 1834. Maryland Heights dominate the river, the railway and the Chesapeake and Ohio canal.

26. Edward Lamson Henry's meticulous recreation of the opening of the Mohawk and Hudson in 1831 shows girls and boys from very different backgrounds watching David Matthews rake *DeWitt Clinton*'s fire before the train sets off from Albany to Schenectady.

THE MOMENTOUS QUESTION.

"Tell me, oh tell me, dearest Albert, have you any Railway Shares?"

27. *(left)* Unwise investment: Leech's 1845 *Punch* cartoon shows Prince Albert hanging his head – underneath one of his much-disliked shako helmets – in embarrassment, as Queen Victoria asks if he has been foolish enough to buy railway shares.

28. *(below)* A revolution in communications: in 1845 the mail coach from Louth in Lincolnshire is loaded onto an Eastern Counties train at Peterborough East Station to complete its journey to London.

29. Influence from the USA: Philadelphia-designed but Russian-built locomotives and carriages cross the wooden 'American bridge' over the Obvodny Canal on the Nikolaeskaya at St Petersburg.

30. Pride of place goes to an 'American type' locomotive in this view of José Eusebio Alfonso's well-equipped sugar mill on the Matanzas to La Union railway in Cuba.

31. Turner's *Rain, Steam, and Speed* shows a 'Firefly' locomotive hauling a train over the Thames at Maidenhead.

32. Thomas Cole's *River in the Catskills* is the earliest American oil painting to show a train, an unobtrusive presence crossing a tributary of the Hudson.

The one railway tunnel in the United States built before 1834, on the Allegheny Portage Railroad, was unusual in its ornamental entrances in Roman Revival style, with a low relief lintel supported by Doric pilasters on each side.

Horse traction

Chapter 7 describes the development of the steam locomotive in the early 1830s, but horses became far more numerous in absolute terms in railway service in the 1830s. They were commonly used as a stopgap on some systems, and as an ancillary form of motive power on new railways that had also invested in locomotives. In addition, existing railways continued to make use of them, and completely new systems were still being designed for horse operation.

One which began life making extensive use of horses as an initial measure was the Allegheny Portage Railroad, where operations were constrained by two factors – the 10 inclined planes and the intervening 'levels' along its 36-mile route, and its charter of incorporation which permitted the state to provide power only for the inclined planes, in the form of fixed engines. Though the railroad and the canal between them cut the journey time from Philadelphia to the Ohio River from weeks to just four days, its operations initially resembled a rural British plateway from 20 years before, even down to the post in the middle of each 'level' which conferred the right to proceed on the haulier who reached it first. Trains were short, both in order to negotiate the inclined planes and because they were all horse-drawn, so they were also frequent. Because most hauliers worked for local farmers who treated the railroad as a glorified turnpike, the railroad could neither dismiss them nor make them follow a timetable. It was only after an appeal to Thaddeus Stevens (1792–1868), later famous as a Radical Republican during the Civil War but in 1834 no more than a member of the Pennsylvania House of Representatives, that the law was changed to allow the Commonwealth to operate trains and to provide locomotives. The *Boston* arrived that year to work the 'long level', 13 miles altogether, and others soon followed.[46]

Some railways used horses on lighter trains. The Mohawk and Hudson concluded that any train that required three horses on its main line was less cost-effective than using a steam locomotive. It nevertheless employed nine horse drivers and four stable boys.[47] Gerstner calculated that using steam locomotives was 30 per cent cheaper than horses on the New York and Harlem, with the advantage of double the speed, assuming a life span of eight years for a locomotive.[48] Horses were used where traffic did not justify locomotives or where mechanical traction was forbidden, such as in built-up areas, either absolutely or during the hours of darkness or through covered bridges.[49] Short-haul movement and shunting was often carried out by horses.[50]

Many well-established railways had no need to convert to locomotive operation if traffic did not increase. The independent carriers who operated the trains on many systems often had neither the means nor the need to use them. These include many which are described in earlier chapters; the Surrey Iron Railway used only horses until its closure in 1846, and, on systems in Wales and the border country, their use commonly extended into the 1860s, sometimes much later.[51]

Railways specifically designed for horse operation in the early 1830s and for some time afterwards included some quite ambitious systems, such as the slate-carrying Ffestiniog in North Wales, not completed until 1836. This Stephenson-inspired system adopted dandy waggons for the downward journeys like the Stockton and Darlington.[52] Another was the Bratislava–Trnava railway in the kingdom of Hungary, built between 1839 and 1846, though with locomotive clearances with eventual conversion in mind. In 1838 proposals were put to Tsar Nicholas I for railways from St Petersburg to Moscow with separate tracks for swift locomotive operation and for slower goods trains hauled by horses. One was to have been a double line of rails via Volochek and Tver, another no less than a six-track route via Novgorod, which it was argued could ultimately be extended to the Prussian border, the Black Sea and to Mozdok in the Caucasus (chapter 13).[53] No such thing was ever built, but horses were typically used underground in mines, on contractors' lines and on other industrial systems, often into the late twentieth century, as well as on intra-urban passenger systems, where their employment has not yet come to an end. There were many circumstances in which the locomotive was unnecessary or an irrelevance, and the horse the only rational choice of motive power, just as wooden rails survived long and late on forestry systems.[54]

Rolling stock

Passenger carriages on these new railways evolved the basic designs which had already been used on the first main lines in the United Kingdom and the USA (chapter 11): two-axle types with single short road coach bodies or longer conjoined compartment carriage, and four-axle bogie articulateds.

Richard Imlay, who had turned out the tiny road coach body types for the Baltimore and Ohio, now also built them for the Little Schuylkill, the Newcastle and Frenchtown and the Philadelphia, Germantown and Norristown, complete with thorough-brace suspension. James Goold (1790–1879) of Albany, New York, built similar cars for the Mohawk and Hudson but with longer wheelbases and larger wheels to improve tracking. Conjoined-compartment carriages, made up of three such bodies, first introduced on the Liverpool and Manchester for the opening in 1830, were now to be found on the Camden and Amboy (in 1831), on the New York and Harlem (1832) and the Dublin and Kingstown (1834). The American examples all had facilities for passengers who were prepared to ride on the roof. The Irish ones had a guard's perch, but no more; these ran on substantial long-wheelbase frames and were equipped with spring buffers.[55]

Compartment carriages built as a plain wheeled box with side doors, which had also first been seen on the Liverpool and Manchester in 1830, were introduced by John B. Jervis for the Mohawk and Hudson in 1832. It was clear that two-axle carriages would run well enough on well-laid British or Irish permanent ways but were uncomfortable and dangerous on rougher American track. There was a limit to how long the wheelbase could be made on a sharply curved route. Furthermore, a broken axle would cause a derailment and might pitch anyone riding on the roof a considerable distance. Before long some other American two-axle types may have been built with end balconies and aisle seating but by this stage the better-riding articulated bogie or 'swivel truck' design was becoming more common. There were British precedents, including the 'Experiment' on the Stockton and Darlington (chapter 9) and the Duke's carriage on the Liverpool and Manchester, but they were much discussed in the USA. Richard Price Morgan (1790–1882) of Stockbridge, Massachusetts, inspired by seeing work under way on the Baltimore and Ohio and by learning about the Liverpool and Manchester, proposed a massive two-floor articulated 'land-barge', with 'all the convenience and comfort which belong to the best steam boats', to run on the proposed Boston and Hudson Rail Road. It would have had berths on the lower floor, a promenade covered by an awning and a cupola lookout for the 'captain'.[56] Morgan, more a visionary than a practical engineer, anticipated both the American caboose for train crews and the 'dome' cars introduced on some railroads after 1945, but articulated carriages were widely adopted once the Baltimore and Ohio had shown the way forward (chapter 11).[57] Richard Imlay appreciated their potential early on and constructed his first in 1832.[58]

For a while, some systems continued the habit of naming carriages and of painting them in different colours. The New York and Harlem ran a 'President'. Though the Leeds and Selby named its locomotives after admirals (Chapter 8), its first-class carriages bore the names of

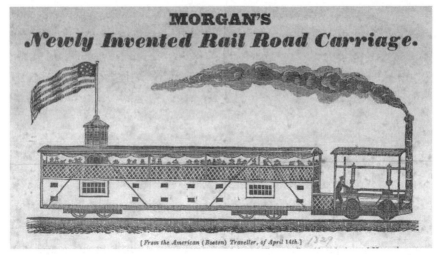

23. Morgan's articulated passenger coach promised the comforts increasingly found on American steamboats.

goddesses – 'Juno', 'Diana' and 'Vesta'.[59] The Dublin and Kingstown colour-coded carriages by class: purple lake for the first class, yellow for closed seconds, green for open seconds, Prussian blue for thirds.[60] In the United States, class distinctions in carriages were unknown, so the lawyer and the merchant supposedly sat down with the labourer and the artisan, though efforts were made to ensure that female passengers were not offended by men's behaviour. Ethnic affiliation was another matter entirely: African Americans and Ibero-Americans experienced discrimination from the start, in both northern and southern states.[61] Trains mixed together people from very different backgrounds. One which came to grief on the Camden and Amboy in November 1833, and which nearly claimed the life of former President John Quincy Adams, involved what Adams described as an 'accommodation car, a sort of moving stage, in a square, with open railing, a platform, and a row of benches holding forty or fifty persons', which was no doubt carrying the Irish-speaking immigrants whom he had encountered earlier on. Passengers in the compartment coaches included service officers, clergymen and academics.[62] If they were regular passengers and were undeterred by the accident, they might on another occasion have beheld Davy Crockett, always the frontiersman, calculating the train's speed, which he estimated to be 25 mph, by spitting out of the carriage window and letting the spittle hit him in the face.[63]

Stations and depots

The Liverpool and Manchester had set the standard for well-designed and well-built passenger and goods stations in 1830, but in the immediate aftermath of its opening only two railways in the United Kingdom showed any interest in following its lead. The Leeds and Selby constructed compact 'depots' at both ends of the line, with adjacent goods and passenger facilities, locomotive sheds at Leeds and a street-facing 'building containing offices'.[64] The other was the Dublin and Kingstown, which could draw on a long-established Irish tradition of imposing civic architecture. Its city terminus was on Westland Row, immediately adjacent to both Trinity College and Merrion Square, both of them strongholds of the Protestant Anglo-Irish community – though a sign of the times was that its neighbouring building was a new Catholic chapel, with a splendid Doric portico. Here the engineer, Charles Blacker Vignoles, set out arrival and departure platforms, covered by a double-pile slated roof, and a street-facing classical facade.[65] In order to oblige the wealthy landowners through whose lands it passed, the railway's directors consented to some remarkable architectural effusions on the way to Kingstown, most notably Lord Cloncurry's bridge, near Blackrock, in fact twin granite towers with a connecting footbridge to enable his lordship to retain access to the sea for bathing.

On other railways, facilities, if they existed at all, were less lavish. The Garnkirk and Glasgow erected a 'commodious Inn' for excursionists at its junction with the Monkland and Kirkintilloch in 1832, but otherwise did not put up station buildings until a few years later.

A 'Railway Coach Office' in Glassford Street, Glasgow, issued tickets from 1834, and horse-drawn road vehicles ran from the city hotels to the terminus.[66]

In Britain places where passengers joined or left a train or where goods were transferred on and off the rails were generally known as a 'station', but in the USA the word 'depot' became common very early on. American depots were neither architecturally ambitious nor capacious. The well-known painting *The First Railroad Train on the Mohawk and Hudson Road* by Edward Lamson Henry (1841–1919) is an imaginative and much later but meticulously researched recreation, and perhaps captures the essence of early rail passenger travel in the USA, particularly the lack of any dedicated buildings other than the conveniently located tavern. It depicts the 'grand opening excursion' on 24 September 1831, when various dignitaries set off from Albany behind the *DeWitt Clinton*.[67] Schenectady's historian quotes a 'time schedule' which remarks 'Passengers may be secured at the office of Thorp's and Sprague's in Albany and Schenectady'. They were road coach proprietors and provided the services to the trains from the town centres, much as on the Garnkirk and Glasgow.[68] The New York and Harlem had 'waiting rooms' at each end of the line.[69] Facilities which were probably fairly rudimentary did not serve the 'corporate image' well enough to be commemorated, and the time had not yet come when they could be depicted for their homespun American values. Contemporary illustrations of American depots are few. An engraving entitled *Railroad Depot at Philadelphia*, undated but evidently early, as it shows a *Rocket* lookalike pulling some four-wheel passenger cars, includes as background a two-storey balconied dwelling lettered 'Philadelphia Germanstown & Norristown Railway Depot'.[70] The architect Joseph C. Hoxie (1814–70) was responsible for the railway's buildings.[71]

The Pontchartrain Railroad's terminus on the New Orleans riverfront consisted of 'an extensive shed on cast-iron pillars', built in 1832.[72] A sketch made a few years later by the artist George

24. This Philadelphia depot was well designed but the artist fancifully included an outdated *Rocket*-type locomotive.

Washington Sully shows a timber overall roof, nestling decorously between the houses of the French Creoles, German immigrants and free African Americans who settled this area from the 1830s.[73] From it, a locomotive built in Bolton, Lancashire, hauled its train along Elysian Fields Avenue towards Milneburg.[74] Later architect-designed passenger facilities are to be found at Lowell, Massachusetts, and Richard Upjohn's designs for the Boston and Lowell terminus at Boston and at Norton, Massachusetts.[75] Goods such as flour or raw cotton required warehousing, but dates and details are elusive. Gerstner refers to a depot on the Tuscambia, Cortland and Decatur.

One surviving structure from this period is the 'Lemon House', a substantial double-fronted two-storey stone dwelling in the Pennsylvania Georgian style, built by 1834, facing the tracks of the

Allegheny Portage Railroad, but also adjacent to the course of the Northern Turnpike. It was built and run by Samuel Lemon (1793–1867) and Jean Moore Lemon (1797–1880), who operated this and several other taverns in the area.[76]

Workshops and maintenance

To anyone who recalls the last generation of steam-worked secondary and industrial railways, it seems evident that locomotives required storage under cover, at or near a terminus of the line, with fuel and watering facilities and some means of carrying out basic repairs, even if only a smithing hearth. However, evidence is sorely lacking for the period 1830–34.[77] The Leicester and Swannington's 'gate houses, engine house and warehouse at West Bridge' (the Leicester terminus) cost £4,601; they included 'an engine shed for two locomotives, workshop and smithy'.[78] The Dublin and Kingstown begin life with a store and workshops for locomotives and carriages near Serpentine Avenue, several miles out of Dublin.[79] The *American Railroad Journal* noted in 1834 'A new locomotive Engine has been built at Schenectady, at the workshop of the Mohawk Railroad Company, which bids fair to excel even those of Mr Stevenson', leaving it unclear whether it was the machine or the facility which was considered to outdo Newcastle.[80] The Camden and Amboy had 'car and locomotive sheds' by 1842.[81] Little else can be said, but the impression is that such buildings were few on American railroads at this time.

The 'Padorama' of the Manchester and Liverpool Rail Road

One other railway from the 'coming of age' deserves mention, though it was a model, not a full-size system. This was an entertainment which became popular in London's Baker Street in 1834, a 'Padorama [*sic*]; of the Manchester and Liverpool Rail Road containing 10,000 square feet of canvas'. The published guide breathlessly assured 'the inhabitants of the metropolis, whose avocations may prevent them from seeing the *real* rail-road itself', that it had been accomplished by 'men of science and enterprize', that 'the locomotive steam-engine ... has been brought to a degree of perfection not previously thought of' and that land given over to horse fodder could henceforth be devoted to feeding the people.[82] Furthermore, the model locomotives and carriages were faithful delineations of the originals and the depictions of the scenery 'had been executed by artists of acknowledged talent', though the 'dull portions' were omitted. A brief article in the *Spectator* suggests that the background scenery was either projected on the outside of a large rotating cylinder or scrolled between two spindles but makes it clear that the 'models' were also somehow set in motion.[83] It may have worked like an orrery, a model of the solar system which typically was operated by clockwork and had the sun at the centre, with the locomotives and the trains rather than the planets at the ends of a series of radiating arms. Whether the locomotives and rolling stock were 'flats' like toy tin soldiers or three-dimensional is unclear but this seems to have been the world's first scenic model railway, and was a step removed from the model of the *Rocket* and its carriages which Goethe bought for his grandchildren.[84]

The *Athenaeum* noted: 'every one of our juvenile friends ought in particular to see it, as it is very instructive for youth'.[85] In the aftermath of the Great Reform Bill of 1832, useful knowledge was emerging victorious over unthinking veneration for a time-honoured past in Britain's culture wars, and was to be passed on to the next generation. As its symbol, the iron road had truly come of age.

13

'The new avenues of iron road' 1834–1850

... each of the new avenues of iron road ramifies like the bough of a tree ...[1]

By 1834 it was clear that the roads and canals in which so much money had been invested, and which had carried the hopes of improvers, would now become the tributaries of the iron road. Experiment, calculation and experience had also confirmed that its Stephensonian form was the most promising, and the dissemination of scientific knowledge ensured that its possibilities were becoming known throughout Europe and over much of America. Finance would be made available, and the legislative preconditions put in place to ensure that it would grow rapidly. By mid-century, recognisable national networks were becoming evident in some countries, connected with seagoing ships carrying textiles and foodstuffs across oceans.[2]

How this happened has been told many times, in many different ways and often in very great detail, but all too frequently as the story of a transformative and sudden 'invention', one that was made bigger, better and faster by a new generation of heroic engineers. As will by now be clear, the process of identifying, proving and using new railway technologies had begun much earlier. Though the years from 1834 to 1850 were exceptionally innovative and inventive, and railways were being built at an extraordinary rate in both North America and Europe, they reflect the much longer evolution which this volume describes, as well as the broader changes in economy and society which made them both possible and necessary.[3]

Expansion

The pace of construction in these two continents was far from uniform. As the civil engineer Ludwig Klein remarked in 1842 in his preface to von Gerstner's *Die Innern Communicationen* (chapter 10):

> Railroads were first introduced and perfected and introduced on a large scale in *England*, the cradle of industry and of so many useful inventions. It was in the *United States* of America, however, that the railroad found its best reception and most extensive use. On the *European* continent, railroads were also introduced early – much earlier, in fact, than in the United States. Yet they have made but slow progress in Europe.[4]

In the USA they did seem to 'ramify like the bough of a tree'. One reason was that successive presidents managed until 1861 to postpone conflict over enslavement, preserving a fragile peace which required transport links but which tolerated the 'peculiar institution'. Construction was slower during the panic of 1836 to 1844, largely caused by the decline in the Bank of England's reserve through speculation in American transport systems, which also made cotton more easily available and lowered its market value. However, the 600 miles and more of chartered railway which were operating in the USA in 1834 had become 9,030 by 1850, of which 2,401 were in the southern states.[5] So far only one railway operated west of the Mississippi, the Alexandria and Cheneyville in Louisiana, and not even the Baltimore

Map 5. Railways in the USA in 1850.

and Ohio had as yet connected the Eastern Seaboard with the river, though far more ambitious schemes had been under discussion for some time. Dr Hartwell Carver memorialised Congress in 1847 on the subject of a railway from Lake Michigan to the Pacific.[6] An alternative scheme was proposed by the New York trader and China merchant Asa Whitney for an 'iron path' to the Pacific, with the support of Senator Thomas Hart Benton of Missouri.[7]

Canada had no more than about 100 route miles by mid-century but the Railway Guarantee Act of 1849 soon led to a rapid expansion, as well as to the near bankruptcy of the province, since it assured interest up to 6 per cent on loans to be raised by any railway at least 75 miles long.[8]

As the threat of major political disturbance receded in Britain, the ambitious schemes which had hung fire for a number of years (chapter 12) were granted parliamentary approval and work could begin. The Grand Junction, which connected Liverpool with the Midlands, ran its first trains in 1837, while the London and Birmingham, which had been under discussion since the speculative year of 1825, opened the following year. These two amalgamated in 1846 to become the London and North Western Railway, which ultimately became one of the world's largest corporations. The London and Bristol Railway, the future Great Western, was founded in 1833, received its act in 1835, and opened its first length in 1838. London's first steam-hauled public railway was the London and Greenwich, operational from 1836.

These particular railways were made possible by a tight-knit, hard-working and exceptionally talented group associated in one way or another with Robert Stephenson – above all Joseph Locke (1805–60), Isambard Kingdom Brunel (1806–59) and George Parker Bidder (1806–78). Of these, Stephenson and Locke came from the inventive milieu of the north-east of England coal industry (chapter 5). Locke had worked with George Stephenson on the Stockton and Darlington and on the Liverpool and Manchester and had surveyed and partly built the Grand Junction, before going on to construct much of the British mainline system.⁹ George Stephenson himself continued to carry out survey work, prudently choosing wherever possible routes along river valleys with easy gradients, and found work in Belgium and Spain as well as in the United Kingdom.¹⁰ Brunel was the son of a remarkably able engineer from a Norman farming background who had fled the French Revolution, married an Englishwoman and become a Protestant, whereas Bidder had had a bizarre childhood as a fairground attraction, 'the calculating boy', until his mathematical abilities came to the notice of patrons who sent him to Edinburgh University, where he met Robert Stephenson.¹¹ These men were respected members of early Victorian society and enjoyed the rewards it had to offer – membership of the Royal Society, of professional associations and of London clubs, a country estate, even, in Stephenson's case, a lavish steam yacht. A second 'railway mania' of 1844–6 (a stock-market bubble associated with the financier George Hudson) created a public network 6,085 miles long by 1850, the densest in the world, but at the price of duplication and inefficiency.¹² Westminster and its civil servants took a more direct role in railway promotion in Ireland, creating a sparser but more rational system.¹³

In France, railways had long appealed to progressively minded engineers and bankers (chapter 9) and the idea of a national system had been under discussion since 1814 (chapter 10). After the revolution of 1830, the Saint-Simonians (chapters 5 and 12) kept this ambition in the public eye. They were closely allied both with a forward-looking group of engineers and the financier, journalist and politician Émile Pereire. The country's strong traditions of an able bureaucracy and centralisation favoured railway building, but lack of capital, political rivalries and its largely agricultural economy told against it. Outside Paris, only Lyons and Rouen had bankers with sufficient resources, and other provincial towns were disinclined to support schemes which did not benefit them directly.

With the backing of James de Rothschild and of the idealistic financier (and school friend of Brunel) Adolphe d'Eichthal, Pereire and his brother Isaac undertook the construction of the short but lavishly built line from Paris to the Seine at Le Pecq, which was opened on 24 August 1837 and was extended across the river to St Germain 10 years later. A joint British–French venture obtained the concession for the country's first main line, from Paris to Rouen along the valley of the Seine, in 1840. The engineer was Joseph Locke, the contractors William Mackenzie and Thomas Brassey and the work was undertaken by British labour. Locke persuaded two engineers based at Crewe, William Allcard and William Barber Buddicom, to superintend the building of locomotives and rolling stock at Rouen.

Work began on this system while France was still in the throes of a recession, and the building of new lines only resumed on a serious scale after 1842, when King Louis Philippe promulgated the 'law relating to the establishment of mainline railways', an event commemorated by a weighty medal showing France enthroned, dispatching both Mars and Mercury to do her bidding while trains make their way across the countryside.[14] The law stipulated that railways would be government property, but might be leased to operating companies, and that the land would be supplied or bought by the national government, although the departments and communes through which each line passed were to pay two-thirds of the cost. The government was to prepare the formation and build structures and bridges, while the lessees were to supply all equipment and maintain the railway in good condition. By mid-century Paris lay at the centre of a network which extended north beyond Rouen to the channel at Le Havre, west to the Loire and eastwards to the Marne. There was as yet no link to the Mediterranean, though short lines ran inland from the coast at Marseille and Sète. Isolated systems also operated along the valley of the Moselle between Metz and Nancy, and from Strasbourg to Basel.

The Belgian government's survey of potential cross-country routes (chapter 12) bore fruit with the opening of the Brussels–Malines railway on 5 May 1835, in the presence of the king and the prime minister. By 1843 the planned 'railway cross' was in place, one route extending from the North Sea at Ostend through Ghent, Bruges, Malines, Louvain and Liège to the German Confederation, another from Antwerp through Malines to Brussels and so on to France through the industrial region around Mons, with branches to Tournai, Namur and St Trond. This was the first centrally planned and funded national railway system brought into being to serve the public interest rather than to augment private capital.[15]

The first Stephenson railway in the German Confederation opened on 7 December 1835. Promoted not by a state bureaucracy but by a group of liberal-minded merchants, the Bavarian Ludwig Railway (*Bayerische Ludwigseisenbahn*) was envisioned, like the Budweis–Linz horse railway, as a means of connecting the Danube basin with northern Europe. King Ludwig preferred his own scheme for a canal to the Rhine and did not attend the ceremony. It ran the four-fifths of a *meile* (6.04 km) between Nuremburg and Furth on a practically straight route alongside the *chaussée*, single track and to British standard gauge. Not only did its one locomotive come from Newcastle but so did the driver, William Wilson, though the cost of carting coal from Saxony to this isolated system limited the employment of both machine and man for some time.[16] Each German state was responsible for the lines within its own borders, but by the 1840s trunk lines linked the major cities. Chancellor Metternich's dislike of railways was quietly forgotten once their commercial possibilities became clear, to say nothing of their potential role in keeping the diverse and far-flung Austrian Empire in being (chapter 10). Work began on the *Kaiser-Ferdinands-Nordbahn* to connect Vienna with Prague in 1837. The next major project of the empire was the *Südlichen Staatsbahn*, the Southern State Railway, from Vienna to Trieste, the first to connect the central European heartlands with the economy of the

Mediterranean. This found work for the unemployed of the capital during the troubled times of 1848, and its Semmering section, completed in 1854, was the first railway to be driven through the Alps. Its engineer, Karl von Ghega (1802–60), a mathematical prodigy who had gained practical experience as a road builder, accepted steep gradients, sharp curves and locomotive operation throughout, making effective use of tunnels and viaducts to take best advantage of topography. This was a step change from the Cromford and High Peak and the Allegheny, with their mixture of inclined planes and level sections (chapters 7, 10 and 12). The result was one of the greatest feats of civil engineering from this period of railway history.

By mid-century it was possible to travel by train between Paris, Berlin, Brussels, Vienna, Prague, Warsaw and Budapest, and the major northern European ports were all rail-connected.[17] Isolated public systems, mostly short, were also operating in Sweden, Denmark, Holland, Spain, the Italian peninsula, Switzerland and Russia. There were as yet no railways in the Ottoman Empire's European territories.

The building of railways in Russia reflected the ambitions and priorities of Tsar Nicholas I (r. 1825–55, chapter 5); deeply conservative and suspicious of most Western ideas, he nevertheless presided over a period of territorial expansion, economic growth and industrial development. Initially he favoured road building but had retained an interest in steam traction since his trip to Leeds and to the Middleton Railway in 1816 with Dr Hamel (chapter 5) and was prepared to grant an audience to Franz Anton von Gerstner, builder of the Budweis–Linz horse railway (chapter 10), to listen to his plans for a Russian network. The Tsar's ministers considered Gerstner's proposal far too costly and ambitious but agreed that he should build a 6-foot gauge demonstration line 25 *versts* (27 km) long, from St Petersburg to Tsarskoe Selo, 'the Tsar's village', and to the pleasure grounds named after London's Vauxhall, thereby bequeathing the word *vokzal* to the Russian language as a generic name for all stations

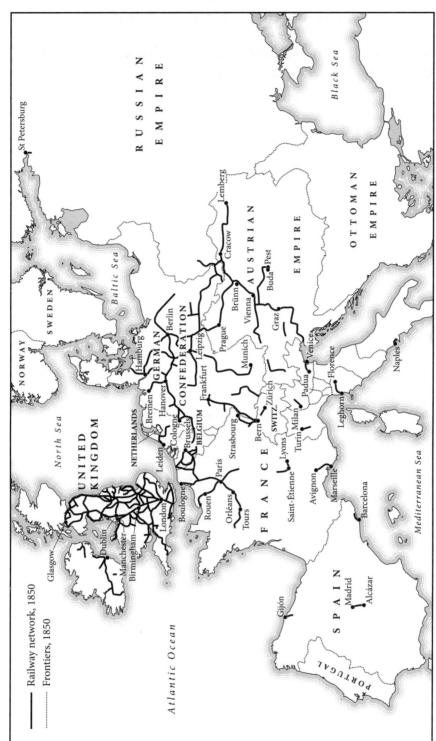

Map 6. The European railway network in 1850.

of any size.¹⁸ The first locomotive came from Timothy Hackworth at Shildon in the north-east of England, accompanied by his son, a youth of 16. He was present when it was blessed according to the rites of the Orthodox church, and listened to the Tsar marvel at how much locomotives had improved over the 20 years since his trip to Leeds as a grand duke.¹⁹

On 30 October 1837 Nicholas I travelled on the first train in his private carriage hoisted on a flatbed waggon. The railway had a very limited capacity and next to no commercial value, since its passengers were mainly sightseeing nobles, though it came to be used for training and to carry out locomotive tests, and it accustomed people to seeing trains.²⁰ Over the next few years, ambitious proposals passed back and forth for multitrack systems connecting the Baltic with the Black Sea and with the Caucasus (chapter 12). American, French, Belgian and British precedents were discussed, as were the relative merits of railways, roads and canals.²¹ The engineer-Tsar initially gave his support to a railway from Warsaw to Vienna to strengthen Polish economic ties with Austria, then on 1 February 1842 he passed to the senate the ukase for the Nikolaevskaya, a railway from St Petersburg to Moscow, which would not only be the longest in the world when it was completed in 1851 but also the greatest work undertaken by the Russian state since the building of St Petersburg itself.²² There the parallels did not end, for, just as Peter the Great sent to Holland for experts, Nicholas also looked abroad, but this time to the USA.

The result was a near complete transfer of technology from the North American continent to Russia. The Tsar's emissaries, Colonels Nikolai Osipovich Kraft and Pavel Petrovich Melnikov, had been sent on a tour of US railways in 1839 and recommended as project leader the experienced Major George Washington Whistler (1800–49), who had a background in both survey work and mechanical engineering. As locomotive builders their choice fell on Eastwick & Harrison of Philadelphia, and Ross Winans, who declined but sent his son Thomas instead. The firm of Harrison, Winans & Eastwick secured a five-year contract to supply the Nikolaevskaya with rolling stock and wood-burning locomotives for $3,000,000, later increased to $5,000,000, and the former state arsenal at Alexandrovsky was adapted as a workshop to build them. Harrison also performed an important service to the Nikolaevskaya in securing financial backing for locomotive building, from William Crawshay II (1788–1867) of Cyfarthfa Ironworks near Merthyr Tydfil.[23]

The public railway was no longer purely a European or North American phenomenon. The expansion of sugar cultivation in Cuba, a colony still faithful to Spain but open to international trade, led to the island's first railway in 1837, approved by Madrid, financed from London, designed by Yankees and built by Irishmen.[24] In 1850 work began on a railway connecting the oceans across the Isthmus of Panama, a proposal first considered by Simón Bolívar in 1825 (chapter 10).[25] Though British empire builders had already drawn up ambitious railway schemes for both Egypt and Madras, only a few short industrial lines had as yet been built.[26] Africa and Asia were otherwise innocent of railways, and Australia had only some tiny industrial systems.[27]

Whereas European and North American countries and their colonies could learn about each other's new technologies with comparative ease, the empires of the East were a different matter. Knowledge of the 'western fire-wheel carriage' had reached Canton (Guangzhou) in China by 1841, just as the opium wars were forcing state officials to take European technology seriously while at the same time making it impossible to invest for the future.[28]

In a similar way, reformist Japanese engineers and officials within the shogunate were discreetly making themselves familiar with the operation of the *reirote* ('railroad') and with what they called 'Dutch knowledge' generally.[29] When Commodore Perry of the United States Navy arrived in 1854, his hosts were delighted with the miniature railway he laid out at Yokohama, but they were slightly more knowledgeable on the subject than the Americans had anticipated.[30]

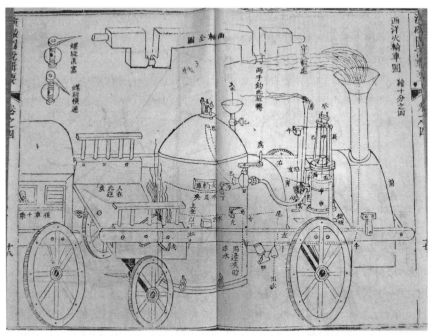

25. Ding Gongchen's model 'western fire-wheel carriage' was built at Guangzhou (Canton) in 1841.

Mechanical engineering

The widespread adoption of the railway between 1834 and mid-century meant that some established technologies could now actually be discarded or relegated to specific circumstances. Wooden rails did not disappear, but they ultimately came to be confined to logging systems in America and Australia. Plateways remained only where there was sufficient traffic to justify operation but not enough to justify replacement. It took many years to convert the busy and extensive system serving the collieries and ironworks in the Monmouthshire valleys.[31]

Rope haulage remained a feature of many industrial lines until the late twentieth century, but for some time it was also a necessity on public railways wherever locomotives were too weak for gradients, were too noisy or posed a fire risk. In 1837 Robert Stephenson installed an enlarged version of his Liverpool tunnel system on the first section of the London and Birmingham, between Euston Station and Camden, and recommended a similar one for the Düsseldorf and Elberfeld, before adapting the design for the level, 3.75-miles long London and Blackwall, opened in 1840.[32] As late as 1850 he recommended rope haulage to the Swiss government, in conjunction with improved roads, steamboats and locomotive-worked railways, for passenger transport across the Jura Mountains, recognising that tunnelling techniques were still insufficiently advanced to contemplate an integrated rail route.[33] Brunel included a steam-powered inclined plane on his Taff Vale Railway in South Wales in 1841.[34] They were, however, never anything other than an operating headache, and had become a rarity on public passenger systems by the late nineteenth century. Von Ghega felt no need to resort to rope haulage on his mountainous and steeply graded Semmeringbahn.[35]

Conversely, some other components were only added in the Victorian period – for instance, mechanical signalling was introduced as systems grew more complex, and as accidents became both more lethal and more subject to public scrutiny. The most common method was a fixed semaphore, derived from the 'optical telegraphs' widely used by governments and armed forces from the late eighteenth century, which used pivoted indicator arms to convey information.

Signalling lessened the likelihood of collisions but could not prevent disasters caused by material failure. Boiler explosions occurred and bridges collapsed. The world's worst railway accident before mid-century, and for many years to come, took place near Versailles on 8 May 1842 when a derailment caused by a broken locomotive axle caused the fire to spread over the track and ignite the wooden carriages. It was never established how many died, whether only the 52 recognised casualties or as many as 200.[36] Railway travel could undoubtedly be dangerous, but must be seen in context. Other forms of transport such as road carriages and steam riverboats could also be perilous, and the worst railway accidents in terms of fatalities took place in the twentieth century.

Otherwise, for the most part, existing technologies could be developed. In the late 1830s rails assumed the form they retain to this day. On the Grand Junction, Joseph Locke adapted the T-section rails into a double-head type which theoretically could be turned over and reused; this was superseded by a more practical bullhead section, also held in chairs.[37] The Vignoles flat-bottom rail also became widespread and could be spiked directly to wooden cross-tie sleepers, which were increasingly used in preference to stone blocks once it was appreciated that some degree of flexibility was preferable to an entirely solid road bed. Most rails came from England or Wales. By 1850 Dowlais Ironworks near Merthyr Tydfil alone had supplied Cuba, Prussia, Saxony, Bavaria, Silesia, Hungary, Russia, the Netherlands and the United States.[38] In the USA, lack of capital and the lengthy banking crisis meant that strap rail, aptly described as 'the greatest wooden nickel in American railroad practice', predominated into the 1840s, despite its high maintenance cost and its potential for accidents.[39] Only two major European systems felt it necessary to adopt this technology, the Leipzig–Dresden (1837–39) and the connecting Magdeburg–Leipzig line (1838–40), where it was soon replaced.[40]

In passenger rolling stock design, rigid-wheelbase vehicles began to give way to the articulated bogies which had been pioneered in the 1820s. Locomotives became more powerful, and could be built to make more economical use of steam by the introduction of features such as cut-off, yet the basic configuration did not change greatly for many years – a two-cylinder engine operating cranks at right angles, near-horizontal cylinders, a smoke-tube boiler and exhaust draughting. In the USA, John B. Jervis's leading bogie design led the Norris Locomotive Works in Philadelphia to develop a 4-2-0 type with a Bury boiler, which proved popular at home but which also came to be exported to Europe from 1837, influencing design at the newly established Borsig works in Berlin, the first major continental

European manufacturer.[41] The Jervis bogie also led that year to the 'American' type 4-4-0, which became standard throughout the continent for many years, as the need for increased power required larger boilers and fireboxes. This was the design which truly won the west, and the very last was built as late as 1945.[42] Harrison, Winans & Eastwick's locomotives for the Nikolaevskaya reflected current American practice. The demands of the Semmeringbahn led to an Alpine Rainhill in which four different designs were trialled and their most promising elements combined.[43]

Standardisation became a priority: Daniel Gooch of the Great Western derived his 'Firefly' class from the 'Patentee' type, and 62 were built by 7 different manufacturers between 1840 and 1842.[44] Robert Stephenson & Co.'s long-boiler patent of 1841 combined improved heat transfer with a smaller firebox placed behind the rear axle. This proved popular in France because it burnt less imported coke and many were built under licence. Other successful European designs of the 1840s were the 'Buddicombe' or 'Crewe' type, widely adopted in Britain, France and Spain, with outside cylinders and a double frame to obviate the failure of crank axles, and the 'Crampton', which combined speed with a low centre of gravity by placing the single driving axle behind the firebox, so that the wheels could be very large.

New engineering concerns competed with established firms to build and maintain locomotives and rolling stock on behalf of railway companies. The Borsig works was set up in the 'Feuerland' district on the edge of Berlin, where there was a long-standing tradition of metalworking, just as Hunslet, near Leeds, not far from where Murray's foundry was situated (chapter 4), became home to Kitson's, E.B. Wilson and Fenton, Craven & Co.

Steam locomotives required complex infrastructure, not only sheds and workshops but also facilities for fuelling and turning them, so they could work chimney-first.[45] For a number of years their range was still limited, so longer railways provided accommodation for them at the termini, as well as change-over locations at a mid-point, where workshops might also be located, such as Crewe on the Grand Junction Railway, strategically situated where lines from Liverpool, Manchester and Birmingham all met, Swindon on the London and Bristol and Wolverton on the London and Birmingham.[46] The Paris–Rouen Railway established its first workshops near where coal and equipment could be landed from England, initially on the site of a Carthusian monastery, then at a purpose-built establishment at Sotteville-lès-Rouen. The first 'round-house', in which locomotives were stalled under cover around a central turntable, was built at Birmingham in 1839 and was soon emulated on other railways in Britain, the USA and continental Europe.[47]

Renewed attempts were made to devise alternatives to steam. Compressed air locomotives, first suggested by Joseph von Baader in 1822 (chapter 8), proved impractical. The atmospheric railway of the 1840s used a fixed engine to generate differential air pressure in a continuous pipe to propel a piston attached to a vehicle through a resealable slot, and was claimed to be able to cope with gradients beyond the capacity of steam locomotives. It was trialled around

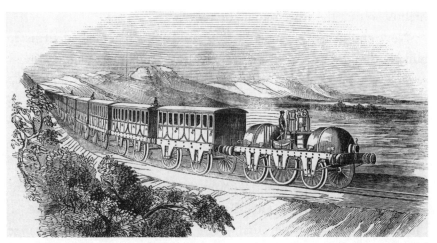

26. Compressed air locomotives were one of several unsuccessful attempts to replace steam.

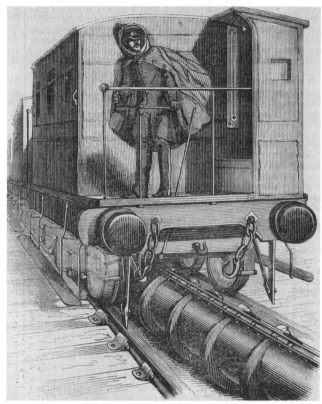

27. Atmospheric propulsion enjoyed a brief vogue but soon proved impractical.

London, at Dalkey in Ireland, between Vésinet and St Germain near Paris and on the South Devon Railway, but came to nothing because of technical problems and did not find favour with investors.[48] Von Ghega decided against it for the Semmeringbahn, just as he rejected inclined planes, even though he knew when he began work that locomotive technology was not yet capable of the task.[49]

Fr Nicholas Callan of the Maynooth seminary in Ireland devised an electric battery which he estimated could power a locomotive on the Dublin and Kingstown and be far cheaper than steam, but could not take it beyond a laboratory experiment. More immediately successful were Thomas Davenport, a Vermont blacksmith who built a demonstration railway at Springfield, Massachusetts, the electro-magnetic version of Stevens's Hoboken steam track (chapter 4), and Robert Davidson of Aberdeen, whose *Galvani* zinc-acid battery locomotive was trialled on the Edinburgh and Glasgow. It appeared promising enough for a group of steam railwaymen to damage it beyond repair at Perth. The American physician and experimenter Charles Grafton Page and his assistant Daniel Davies built an experimental electric locomotive on which they set off from the federal capital along the Baltimore and Ohio on 29 April 1851, intending to make an 80-mile demonstration round trip, only to turn back after five miles, returning to Washington on the down-grade reeking of nitric acid.[50] Electric traction did not move beyond its experimental phase until the 1890s, and was initially restricted to intra-urban systems and to a few suburban services. Steam was the only practical form of mainline traction for the 1830s and 1840s, and for many years to come.

Civil engineering

Construction of these systems was achieved by good organisation as well as by a large workforce. In any sort of hilly country, routes were planned so that 'cut' from excavations became 'fill' for embankments, drawn in end-tipping waggons along temporary tracks by horses or, if the distance were more than about two miles, by locomotives. 'High-end tipping', as opposed to building up in successive shallow layers, had been the method used to create embankments on both the Stockton and Darlington and the Liverpool and Manchester, and was the most common on the main lines built thereafter.[51] Little use was made of machines to aid construction, though pile drivers and steam excavators were put to work on the Nikolaevskaya.[52]

Bridging techniques mostly followed established practice, though the scale might be larger; the London and Bristol's Maidenhead viaduct was at the time the largest brick arch span in the world. The Semmeringbahn included four two-storied masonry viaducts.

European viaducts were mostly built in the Roman style of multiple-stone arches and piers, whereas American railroads made greater use of wood, and later iron, bridges. Wooden truss bridges on the American plan were used on the Nikolaevskaya, making use of Russia's abundant timber reserves. Cast-iron's potential for catastrophic collapse was brought home with the failure of Robert Stephenson's bridge over the Dee at Chester in 1847. The wrought-iron tubular box-girder bridges designed by him and by William Fairbairn to cross the Conwy River and the Menai Straits on the Chester and Holyhead were derived from contemporary shipbuilding practice. These two bridges were innovative, but very few others were built once steel became available as a constructional material in the 1860s.[53]

Tunnelling remained largely a matter of careful survey, low explosives and pick-and-shovel, as it had been at Wapping on the Liverpool and Manchester and on the earlier canal systems. Several of the tunnels on the London and Birmingham presented Robert Stephenson with particular geotechnical problems, in particular Kilsby, resulting in the first major application of pumps for groundwater lowering. John C. Bourne's engravings of the railway under construction show the Cornish engines and flatrods installed to operate them, as well as the horse whims which raised the spoil from the headings. Stephenson shared his experience of tunnel construction with Brunel, who did not experience the same challenges on the Great Western, though the Box Tunnel was a huge undertaking, involving several thousand men, of whom a hundred are said to have died on the contract.[54] Locke avoided tunnels wherever possible; there are none on the routes built according to his survey between Birmingham and Glasgow. The 15 Semmeringbahn tunnels together measured 2254.09 *klaftern* (4.275 km), over one-tenth of the total route.[55]

Fundamental departures from the Stephensonian system, and from its American and continental variants, proved unnecessary or met with failure. The most radical was Isambard Kingdom Brunel's decision to build the London and Bristol (the Great Western) to

7′ 0¼″ (2140 mm) gauge on a 'baulk road' (rails laid on longitudinal timbers). Brunel was always temperamentally inclined to return to basic principles to see if a better scientific understanding could offer improved specifications, better performance and more economic operation, not always with successful results.[56] Within a fortnight of the railway obtaining its Act, he persuaded the directors that the broad gauge would improve stability and that larger locomotive wheel diameters would provide less resistance, thereby offering both lower operating costs and higher speeds. His broad gauge lasted until 1892, but proved costly, led to complexity at interchange points and conferred no advantages. Whistler persuaded the Tsar and his ministers that von Gerstner's 6-foot gauge for the St Petersburg–Tsarskoe Selo Railway was unnecessary and most subsequent Russian systems were built to 5-foot gauge.[57] The Dutch government soon regretted its decision to adopt 2-metre gauge on the advice of Bernard Goudriaan, a government engineer who had been impressed by the Great Western, and the Erie Railroad in the USA abandoned its 6-foot gauge in the 1870s.[58] Brunel-gauge railways only ever justified themselves in building and maintaining breakwaters, where their ability to move and tip large rocks proved useful. One such, at Ponta Delgada in the Azores, remained in use into the 1960s.

Function

Railways had transported ores and coal for many years before 1750, and these have remained a staple ever since. New railways such as the Durham and Sunderland in the north-east of England, Brunel's Taff Vale in South Wales and many others were still being built to connect mining regions with the sea.[59] Cuba's sugar railway similarly served an extractive industry which relied on an export market.

Some purposes to which railways could be put had barely been anticipated by 1834. In its first year, the Liverpool and Manchester had carried troops on their way to Dublin to suppress rioting, and it became clear that the speed and capacity of railways would make them decisive not only in subduing civil unrest but also in warfare between sovereign states.[60] In 1846 the Prussian VI Corps arrived by train in Kraków to put down an insurrection.[61] The Tsar considered reducing the size of his army, partly to finance the railway but also to make it a more flexible and responsive force, capable of being applied where needed by train.[62]

Some significant changes in function reflect developments in other sectors. The Baltimore and Ohio was the first major system built for the food industry, moving live animals as well as flour from Ellicott's Mills, but by the 1840s some American railways were attempting to transport fresh fruit.[63] The Nikolaevskaya enabled the agricultural surplus areas around the Upper Volga to supply the more densely populated and industrialised St Petersburg region with grain and cattle.[64]

'Labourers on the rail road, strangers to the parish'[65]

Railway building on this scale was an exceptionally labour-intensive activity. There were 6,250 men employed on building the Great Western in July 1837, but 40,000 were at work on the Nikolaevskaya nine years later.[66]

Workers changed the landscape of Europe and eastern North America as never before, with only hand tools and explosives. Popular legend depicts these 'navvies' (from 'navigators', canal builders) as muscular outlaws with nothing to sell but their strength, making their homes in unsanitary huts near the railhead and frightening women and officers of the law in nearby villages and towns. Without a doubt, many were working a long way from where they had been brought up, were constantly on the move, sometimes differing in language, religion and outlook from their host communities, and living a way of life which valued endurance and physical strength, but this was not the whole story. In Britain, wherever census enumerators were brave enough or sufficiently conscientious to venture into navvy camps and to ask questions, the answers suggest that many were in fact from the region in which they worked, that they were the strongest and brightest of the agricultural labourers and were very apt to desert the contract to bring in the harvest. Irishmen were not particularly evident – for instance, they only contributed 2 per cent of the workforce completing the Great Western in 1843. Complaints about them in newspapers may also reflect local nervousness about the lack of an effective, or any, county police force. Some navvies were peripatetic specialists who moved from one job to another, and might find themselves following the contractor for whom they worked to another country – English, Welsh, Scottish and Irish navvies on the Paris–Rouen Railway amazed the local population with their capacity for hard work.[67] They were treated to a 'Homeric feast' including an entire roast ox when the Tourville tunnel was completed.[68]

Irishmen were heavily involved in railway building in the USA. Two-thirds of a million crossed the Atlantic in the 1830s, most of them labourers, and more still arrived after the famines of the following decade.[69] Some also made their way to Cuba as railway navvies, where the sugar producers initially suggested that entire parishes from Ireland might wish to emigrate directly to join those who were recruited from the USA, in the hope that they would settle to create a white, Catholic underclass. They soon preferred the services of Canary Islanders and – a sign of times to come – Chinese contract labourers.[70]

Many countries had a surplus population by the 1830s, labourers and poorer farmers driven from their land, or artisans who could no longer practise their trades, but all of them anxious to find any way of making a living. The unskilled men on the Nikolaevskaya were drawn from Belorussia, Lithuania and the adjoining provinces. They experienced, and expected, extremes of wretchedness, and were frequently in debt to the contractors who managed them.[71] In 1847 it was reported that 'upwards of a thousand women and girls are labouring as "navvies"' on the Royal Württemberg State Railways between Stuttgart and Ulm, and a 'large number' on the Prague–Dresden Railway.[72] Women

'THE NEW AVENUES OF IRON ROAD' 1834–1850

28. Enslaved men are set to work on track maintenance in Tennessee.

also worked on the building of the Semmeringbahn, in a labour force drawn from the many different nations ruled by Vienna.[73]

In the USA, free and enslaved African Americans worked as railway builders. Coerced labour was generally hired rather than purchased for construction work, and shareholders could pay their subscriptions in labour if they were awarded a contract. Enslaved women and children were put to work as well as men; the Montgomery and West Point in Alabama owned 53 men, 7 women and 11 children in 1850. Slaves had no choice but to be sober, industrious and punctual, as their owners were well aware; the Central Railroad of Georgia fired white workers in favour of slaves for this reason. Their railway-building work in the South sustained the cotton economy, and prolonged their servitude.[74]

Free black workers were the main competitors of the Irish in the labour market, and fights between them were bitter.[75] Ethnic conflicts were also exacerbated by religious differences. Catholic Irishmen fought on, and for, construction contracts with Welsh Methodists, rarely a pugnacious affiliation, and with 'natives' (Protestant Americans).[76] They also fought with each other; Corkonians and 'fardowners' were still slugging it out in Virginia in 1850, just as they had on the Baltimore and Ohio 20 years before (chapter 11).[77]

Railwaymen

Railways were becoming major employers in their own right. They offered job security, since, as a service rather than a productive sector, it was difficult to dispense with staff at a time of economic depression. The pay was mostly good. Working on the railways implied literacy and the ability to tell the time and to work as part of a team. Footplate crews were quick to develop an affinity for their locomotive, especially if it was the one to which they were regularly assigned. Crewe, Swindon and Wolverton were designed from the start as railway communities as well as workshops, with housing, places of worship and educational institutions, just as places as otherwise different as Derby, Sotteville-lès-Rouen, Utrecht and Baltimore were becoming home to significant numbers of railwaymen and their families. The introduction of uniforms on European railways, particularly for public-facing staff, encouraged them to identify with their employing company. In the United Kingdom, the cut and colour variously resembled a servant's livery, or the everyday wear of a naval officer or policeman. Official clothing on French and German railways reflected the stronger continental tradition of a civil uniform.[78] American conductors (guards in British parlance) were not put into uniform until the 1850s, since there was a feeling against dressing them like the 'liveried lackeys of European noblemen', but they were identified by a badge and by their distinctive cry of 'All o' board!'.[79]

Coerced labour also played a major role. Serfs worked in the Alexandrovsky works on the Nikolaevskaya, superintended by British

'THE NEW AVENUES OF IRON ROAD' 1834–1850

or American foremen.[80] Slaves and free African Americans in the southern USA worked as firemen or tended the wood and water stops, though it was noted in the 1830s that 'an intelligent slave occupied the position of assistant superintendent' on the Charleston and Hamburg. Proposals to allow African Americans to drive locomotives on goods trains led nowhere. Again, enslaved men were generally hired rather than owned outright to avoid capital outlay. Contracts with their owners contained clauses about how they were to be clothed and fed, how much time they were allowed for home visits and who would be liable if they absconded.[81]

Freemen, serf or enslaved, these were jobs for men. Once the construction phase was over, women were rarely employed by mainline railways.

Corporate organisation and public face

A sustainable system was not only a matter of identifying a successful technology. The corporate structures put in place by Britain and the USA enabled railways to become big businesses in their own right, in some cases the largest private-sector employers in their respective countries. State participation in finance, whether in the form of loans as in the United Kingdom and the USA or direct sponsorship on the Belgian model, provided alternative models of operation in the public interest. In countries where railways were run by private enterprise, organisations came into being to co-ordinate their activities, such as the Railway Clearing House in Britain (1842) and the German Confederation's Verein Deutscher Eisenbahn-Verwaltungen (1848). In Britain, Gladstone's Railway Regulation Act of 1844 provided a minimum standard of comfort and affordability for passenger travel, set out the compulsory services that companies must offer and reserved the powers of nationalisation should private enterprise fail.[82]

Railway companies also proved adept at how they presented themselves, whether to investors, to the travelling public or to political elites. Station architecture ultimately took its cue from the ambition of the Liverpool and Manchester and of the Dublin and Kingstown, rather than from the wooden shanties and zinc sheds of the Baltimore and Ohio and the Charleston and Hamburg. For years to come, American depots were less impressive than those of Europe, and those of the southern states less impressive than those of the north.[83] European railways, on the other hand, spared no expense. The London and Birmingham copied the columnar device used at Liverpool for the chimneys on the inclined plane winding house at the summit of Camden Bank (chapter 11), and received passengers at Euston itself through a propylaeum of the Doric order.[84] The classically inspired station building at the Place de l'Europe on the Paris–St Germain Railway with its large windows recalls an orangery in a nobleman's park. The architect William Tite, whose masterpiece was London's Royal Exchange, designed stations in England and Scotland as well as between Paris and Le Havre, where classical, neo-Tudor and neo-Byzantine styles were all to be found. Konstantin Andreyevich Thon (1794–1881), a favourite of the Tsar, was responsible for the Renaissance terminals on the Nikolaevskaya.

The Liverpool and Manchester had also shown how helpful a lavishly illustrated promotional publication could be (chapter 11), an example not lost on the London and Birmingham and the Great Western, both of which commissioned John C. Bourne, nor on the Le Havre–Rouen, where Jean-Baptiste-René Viau set to work.[85]

'A perfect passion for railroads'

In chancelleries and palaces across Europe, outright opposition to railways was giving way to worries about their cost, the probable return on investment and their likely social impact. Even conservative statesmen like Metternich and cautious ones like Adolphe Thiers accepted that these new systems were inevitable. Some monarchs like Louis Philippe, *'le roi bourgeois'*, and William I of the Netherlands, 'the merchant king' (r. 1815–40), identified with the commercial interests of their subjects and saw them as the handmaids of trade, whereas Tsar Nicholas believed they would strengthen autocratic rule. Pope Gregory XVI called them not *'chemins de fer'* (iron roads) but *'chemins d'enfer'* (roads of hell), whereas his successor, Pius IX, elected in 1846, enjoyed train travel, which greatly encouraged British ministers and free traders to believe that he would permit a railway from Bologna to Brindisi across the papal states.[86] Priests would now bless the opening of a new railway and a choir sing a *te deum*.[87] Only Muhammad Ali, the *Wāli* (Ottoman viceroy) of Egypt (r. 1811–48) held out, shrewdly suspecting that the railway he was being urged to build from the Mediterranean to the Red Sea would benefit the British by providing a faster route to India but would do no good to his own country.[88]

Elite fear that allowing ordinary folk to travel would weaken social cohesion (chapters 5, 10 and 11) had to be set against the advantages these systems conferred on investors, and the benefits they offered working people. Sir Robert Peel equated the free movement of labour not with potential disorder but with shared prosperity for all classes: 'whatever improvement in communication will enable the poor man ... to carry his labour, perhaps the only valuable property he possesses, to the best market, and where it is most wanted, must be a decided advantage, not only to him but to the community at large'.[89] The Tsar and his successive ministers of the interior continued to worry that railways would lead both serfs and the middle classes to entertain dangerously progressive notions but also appreciated that the Nikolaevskaya would increase Russian food production by enabling seasonal rural labourers to move around the country, and that it meant that factories could be dispersed along the route in order to avoid creating large concentrations of industrial workers in any one area, particularly Moscow.[90]

Opposition was now more likely to come from vested interests and from those excluded from prosperity. In 1842 'attempts were made to injure the railroads' during the 'Plug Riots' in the north of England, and carters, lock keepers and bargees attacked stations and bridges on the Paris–Brussels and Paris–Rouen railways during the February Revolution of 1848.[91] Even in the USA, the Saint-Simonian socialist Michel Chevalier heard of objections from canal and turnpike operators, stagecoach companies and innkeepers, though on the whole he believed that forward-looking US citizens had 'a perfect passion for railroads', and loved them 'as a lover loves his mistress'.[92] Men and women could now travel by train to go to work or to market, to attend a religious or political meeting, emigrate or even to enjoy themselves.[93]

In this sense, railways not only embodied modernising trends like industrialisation, the growth of secular society and challenges to social norms, but also enabled people to understand what these might mean for them in practice. They might find a train journey an exhilarating experience, but feel overwhelmed by a station that exuded the new confidence of private capital or the established power of an authoritarian state. In Britain and the USA there was little agreement as to whether trains should operate on a Sunday.[94] Women experienced some new freedoms on the railway but were also exposed to new dangers.[95] Saint-Simonians hoped that Enlightenment rationalism and universal brotherhood would be spread by the railway, and in 1839 the rackety old plateways down the Monmouthshire valleys in Wales provided the marching routes for Chartists raising the banner of rebellion.[96] In the USA, segregation was an issue for the railways. The enslaved might be sold in the depot and transported as goods, but they might also use a train as a means of escape, and it is telling that the clandestine escape routes and safe houses to the free states and to Canada were known as 'the underground railroad'. A nursemaid might travel with her owner and with the children in some comfort, and sometimes enslaved men might be expected to accompany a plantation owner as part of his retinue, generally at a reduced fare. In southern states, enslaved African Americans could in some

'THE NEW AVENUES OF IRON ROAD' 1834–1850

places even pay the full price to sit in a carriage on their own account. A Connecticut man travelling on the Richmond, Fredericksburg and Potomac was struck by the ease and friendliness with which 'a white girl ... and a bright and very pretty mulatto girl' chatted and shared sweets out of the same bag on a train journey.[97] It was generally agreed that travellers here were more relaxed about ethnic mixing than in the supposedly more progressive north, where free African Americans were more likely to be relegated to the baggage car. A white man who wished to smoke might be made to join them.[98]

Other modernising factors were time-keeping and standardisation, and military forms of work discipline. At arrival and departure times, station masters, signalmen, footplate crews and train guards would pluck watches from the pocket of their uniform waistcoats to make sure the timetable was being kept. In rural parishes, the locomotive's whistle now divided up the day instead of the church bell. The train it pulled might carry not only goods and passengers but also the mail and newspapers, and the tracks on which it ran might share a right of way with the telegraph.[99] The railway was becoming a revolution in communications as well as in transport.[100]

It had now more than come of age; it had become the dominant transport system of the industrialising world, perhaps even its defining technology. It brought swift, safe and affordable travel to many people, and required natural resources to be exploited on an unprecedented scale. Very different human economies and societies could now pursue their own versions of modernity.[101] Within a tumultuous period of human history, and from the Moskva to the Mississippi, it had innovated and strengthened both government and trade, and changed the appearance of towns, cities and countryside. Was it also already even changing the way people looked at the world in which they lived?

14

'You can't hinder the railroad'

Speed – despatch – distance – are still relative terms.[1]

The frontispiece to Bourne's 1846 *Great Western Railway* shows *Acheron*, one of the 'Firefly' class, steaming out through an ornamented tunnel portal in the rock face, as if thrusting itself into the modern world. Its driver and fireman are absorbed in their tasks. Men who may be navvies or permanent way labourers ignore the locomotive, but a top-hatted figure welcomes its arrival from darkness into daylight.

Only slightly more than 60 years separate Bourne's lithograph from Fittler's engraving of the wooden Parkmore way and the Newcastle Tyne (chapter 1), a mere 36 from Cartwright's aquatint of the Risca 'long bridge' (chapter 3), yet not only has the technology undergone a dramatic transformation in the intervening period, but so has its imaginative and intellectual context. The railway is no longer the accoutrement of trade depicted by Fittler, nor the sign of 'improvement' as it was for Cartwright.[2] Even Bury's illustrations of the Liverpool and Manchester now seem strangely dated. Bourne made the railway a sublime moment where nature, technology and humanity cohere, with the locomotive as its symbol.[3] Watching the train go by became an opportunity to rhapsodise or reflect.[4]

29. *Acheron* steams into the light of day on the Great Western Railway.

The coming of the railway was not the least of the many changes that characterised the long and tumultuous period of modernisation we call the 'Industrial Revolution', which in turn paralleled convulsive alterations in political order across the world in the late eighteenth and early nineteenth centuries. The close, often complicated, relationship between mechanical capacity and governmental, military, economic and social developments has formed a theme of this study but what is also evident is that the railway also had a profound imaginative impact. It was Bourne's genius as artist and as publicist to suggest that the railway would not blight but fulfil the English landscape, and that it would harmonise past and future.[5]

He was called upon to do so because many considered railways to be disruptive. An indirect tribute to their effect, in Britain at least, was a spate of novels set in the recent past celebrating the brief golden age of the stagecoach, from 1815 to around 1834. In the imagination of Walter Scott, William Hazlitt, Charlotte Brontë, George Eliot and Charles Dickens, road transport had united the kingdom without eroding distinctions of identity or locality.[6] William Wordsworth, for whom the local was sacred, fired off a sonnet to the *Morning Post* (which published it a few days later) and to Gladstone, President of the Board of Trade (who sent an emollient reply), when he learnt of plans to extend the railway from Kendal to Windermere, lamenting the danger to 'Schemes of retirement sown/ In youth'. Yet even Wordsworth felt compelled to state that he was not opposed to railways as such, reminding his readers of a sonnet 'published in 1837, and composed some years earlier' entitled 'Steamboats and Railways' in which he conceded that 'Nature doth embrace/ Her lawful offspring in man's Art', one of his more leaden offerings.[7]

Bourne's *Great Western Railway* was published in the same year as the first instalment of Dickens's *Dombey and Son*, which famously recalls the building of the London and Birmingham through Camden. Here 'the first shock of a great earthquake had, just at that period, rent the whole neighbourhood to its centre', bringing 'dire disorder' in its short term but opening a 'mighty course of civilisation and improvement'. Narrative events reflect Dickens's ambiguity; the defeated Carker is killed by a train whereas Mr Toodle finds a steady job which he loves as a locomotive stoker, and then driver.

Another who saw tension between environment and modernity, and between the past and the future, was J. M. W. Turner, whose *Rain, Steam, and Speed – The Great Western Railway*, exhibited in 1844, also depicts a 'Firefly', at the head of a west-bound train crossing the Thames

'YOU CAN'T HINDER THE RAILROAD'

30. Construction work: the London and Birmingham Railway tears its way through Camden.

at Maidenhead viaduct through a summer rainstorm. The river is reduced to indeterminacy, whereas the railway is urgent and direct. Bourne showed a locomotive responding to its driver and fireman, but Turner's 'Firefly' is demonic and self-sufficient, red with heat.

Rain, Steam, and Speed is contemporary with Thomas Cole's *River in the Catskills*, the earliest known American oil painting to depict a train, also crossing a river on a bridge, but an unobtrusive, almost insignificant, presence in a serene landscape. Another sign of the times is the axeman in the foreground and the felled trees. Perhaps the wild and the utilitarian could coexist, yet Cole wrote that the railway made the human body 'merely a sort of Tender to a Locomotive Car; its appetites & functions wait on a Machine which is merciless & tyrannical'.[8]

Whether or not humanity was becoming an adjunct of a machine, it was evident that speed, space, landscape and time would never be the same again. Henry Booth, the secretary of the Liverpool and Manchester Railway, reflected on what 'the March of Intellect' had accomplished:

> Notions which we have received from our ancestors, and verified by our own experience, are overthrown in a day, and a new standard erected, by which to form our ideas for the future. Speed – despatch – distance – are still relative terms, but their meaning has been totally changed within a few months: what was quick is now slow; what was distant is now near; and this change in our ideas will not be limited to the environs of Liverpool and Manchester – it will pervade society at large.[9]

For Booth, the regular packet ships crossing the Atlantic had brought an 'energy and despatch' to ocean sea trade which was now emulated by the railway; for him speed above all represented commercial advantage, enabling more effective investment in assets such as carriages and locomotives. 'The saving of capital, therefore, in this department of business is considerable, from expedition alone.'[10]

Speed, despatch and distance fed the imagination as well as the bank balance.[11] The French writer Victor Hugo (1802–85) had only seen an 'ignoble' industrial railway before he made his first train journey, from Antwerp to Brussels. He was enchanted by the iron horse, '*une bête veritable*', a machine which for him became a living thing, unlike Thomas Cole's locomotive which mechanised humanity. He was delighted by the way its speed turned flowers and cornfields into swathes of colour and made nature dance before his eyes.[12] For the American essayist and philosopher Ralph Waldo Emerson (1803–82), this change in perception and environment was not pleasurable but disturbing, as 'oaks, fields, hills, hitherto esteemed symbols of stability, do absolutely dance by you' – the railroad had evidently eroded and reordered nature. Though the cerebral Emerson had little in common with Davy Crockett, he repeated the colonel's tale of the Charleston and Hamburg locomotive frightening the horse (the countryman called it 'Hell in harness', chapter 11).[13] Yet he was drawn to this new technology; as he was on his way home to the USA in 1833, he had filled an idle hour in Liverpool by visiting the railway, where he 'saw Rocket and Goliath and Pluto and Firefly and the rest of that vulcanian generation'. He even listened patiently to Jacob Perkins (chapter 8) expounding on his locomotive proposals.[14] When he rode behind a 'teakettle' on the Boston and Worcester the following year, like Booth he sensed 'hitherto uncomputed mechanical advantages'.[15] If he deprecated the way the railroad had coarsened the fabric of American life and contributed to its materialism, he nevertheless came to hold bonds or stock in at least six American concerns, affording him the financial security to develop and expound his philosophy of a universe composed of nature and of soul.[16]

When Emerson's friend and neighbour Henry David Thoreau (1817–62) undertook his two-year exercise in simple living in the woods of Concord, Massachusetts, he did not remove himself from the modern world entirely, as the railway ran nearby. Thoreau took to wandering by the track so much that train crews thought he was a platelayer. He observed freight cars carrying the goods 'of foreign parts, of coral reefs, and Indian oceans, and tropical climes, and the extent of the globe' on their way from Long Wharf in Boston to Lake Champlain and on to Montreal.[17] They were drawn by locomotives whose whistle 'penetrates my woods summer and winter, sounding like the scream of a hawk sailing over some farmer's yard, informing me that many restless city merchants are arriving within the circle of the town, or adventurous country traders from the other side'.[18] These unnatural, mechanical sounds depleted body and mind as well as degrading the environment. Yet he also wrote that 'When I hear the iron horse make the hills echo with his snort-like thunder, shaking the earth with his feet, and breathing fire and smoke from his nostrils ... it seems as if the earth had got a race now worthy to inhabit it'.[19]

In similar vein, the English critic John Ruskin (1819–1900) loathed intrusive modernity and the laissez-faire economics that lay behind it, but felt compelled to admit that 'I cannot express the amazed awe, the crushed humility, with which I sometimes watch a locomotive take its breath at a railway station, and think what work there is in its bars and wheels, and what manner of men they must be who dig brown ironstone out of the ground, and forge it into THAT!'[20]

Ruskin spoke for all who have ever wandered to the end of a platform on a steam-worked railway and marvelled at what they saw as a train was about to leave – the well-tended furnace in the firebox, the rising needle in the pressure gauge, the sense of latent power in the machine and the concentration on the faces of the driver and fireman. Ruskin's fascination is all the more eloquent because of his ambiguity about the locomotive. This 'gracious instrument', which moved the goods of the world along its iron road, could only have been assembled by the votaries of a tenth muse, of mechanical art, yet it was also an 'unreasonable thing' like an ox that could only pull or push, and it would all too soon banish 'pastoral song' from England.[21] It seemed as if a new age had dawned, whether for good or ill or for both, defined by the railway.

'Now, my lads, you can't hinder the railroad'

Ruskin never came to a conclusion about 'the real dignity of mechanical art' and its relation to soul and spirit, any more than Emerson and Thoreau. Others asked simpler questions and came up with more pragmatic answers. In the early 1830s surveyors for the London and Birmingham Railway were at work in rural Warwickshire, where their responsibilities took them on to lands belonging to the Newdigate family, managed by their capable Welsh agent, Robert Evans (1773–1849). Their Arbury estate formed part of the ancient forest of Arden and lay in an area that many have liked to call 'the heart of England'. Roman roads had avoided it, and the Reformation of the church had almost passed it by, but Arden had not been entirely impervious to outside influences. The Newdigates had developed collieries on their estate, and Sir Roger Newdigate (1719–1806) built elaborate private canals together with some short tributary railways to export their output.[22] Robert Evans would have been familiar with these iron railways, and neither he nor his employer objected to the arrival of surveyors drawing up their plans for a vastly longer system.[23]

Whether anyone else offered them violence or whether the surveyors encountered any other trouble is unrecorded, but Evans's daughter Mary Ann (George Eliot) turned their visit into a clash of old and new in her novel *Middlemarch* (1871–2).[24] Recalling or imagining events of 40 years previously, she relates how the surveyors are set upon by Hiram Ford the waggoner and his 'smock-frock' friends and are rescued by Caleb Garth, a literary reimagining of her father, and by his protégé Fred Vincy. Caleb in 'his everyday mild air' tells them, 'Somebody told you the railroad was a bad thing. That was a lie. It may do a bit of harm here and there, to this and that; and so does the sun in heaven. But the railway's a good thing.'

There it might have been easy to leave the matter, but George Eliot puts into the mouth of an old man who has observed the fracas the rejoinder – '"... it's been all aloike to the poor mon. What's the canells been t'him? They'n brought him neyther me-at nor be-acon, nor wage to lay by ... An' so it'll be wi' the railroads ... This is the big folks's world, this is. But yo're for the big folks, Master Garth, yo are".'[25]

For George Eliot, the Radical-turned-Tory, the railway came to Middlemarch at the same time as parliamentary reform and cholera, and she understood that the unknown was rarely welcome. Princes, ecclesiastics and philosophers variously welcomed or feared the coming of the railway, but she also sensed a profound if barely articulate concern that it meant no good to the waggoner or the labourer. All that Caleb Garth can do is persuade Hiram Ford and the smock-frocks that they shall do no more 'meddling', because 'you can't hinder the railroad'. On that, at least, all came to agree.

A NOTE ON SOURCES AND TERMINOLOGY

Remarkably, there has been no attempt to write a global history of the early iron railway. This book attempts to fill that gap. In 1938 Oxford University Press published *A History of British Railways Down to the Year 1830* by Chapman Frederick Dendy Marshall, and Bertram Baxter's *Stone Blocks and Iron Rails* appeared in 1966. Both of these books attempted to deal with the entire sweep of early British railway history.[1] They have both been superseded. Very many authoritative and informed studies have been undertaken of individual railways and locomotives, as well as of other aspects of the early iron railway, and this book offers a synthesis of them.

As much as anything, this reflects the work of the International Early Railways Conferences, which first met in 1998, and the Early Main Line Railways Conference, which met in 2014 and 2018, and which have been bringing together a varied group of people with diverse interests. In each case the proceedings were published, which has led to information about these systems becoming much more easily available, to say nothing of offering a more informed conceptual approach to this field of study. Sir Neil Cossons, who was then Director of the Science Museum in London, in his keynote address to the 1998 conference, expressed the hope that serious and conceptually informed scholarship in this area might be revived, one that the conferences have entirely realised.[2] In 2001 Professor Colin Divall of the University of York offered a 'constructive provocation', asking historians of the early railway, whether academics or lay scholars or enthusiasts, to engage with specialists in other, related disciplines as well as with more public forms of history and heritage, and with interpretation more generally.[3] This study attempts to respond to both these challenges and to draw on the work that these conferences made available, as well as on other recent research. It also makes use of the many contemporary publications and regional historical studies which have recently been digitised.

Both Cossons and Divall questioned the assumption, widespread if rarely explicit, that railway histories should mainly, or only, be about technology. My experience of railways, which led me to the lineside as a rapt child, and has since taken me to the footplate, the library and the board room, makes me believe that if we are truly to understand this type of transport system, we certainly need to comprehend the machine – and a railway is, after all, such a thing, or perhaps a machine ensemble – but we also need to grasp the human communities which called it into being, their economic imperatives, their readiness or reluctance to contemplate the new and the men and women who came to the fore as a consequence. The iron railway is a cultural artefact: it was made not only by engineers but also by bankers and entrepreneurs, by those who earned their living on it or in the industries it served and by the great mass of men and women who chose to travel on it. For this reason, readers will have found much here on political circumstance, finance, corporate promotion, intellectual debate, even religion – it is hard to imagine the iron road coming into being without the support of the Quaker community, whether in Darlington or Philadelphia. They will, on the other hand, have found comparatively little here about locomotive drawbar power or grate area, though I have provided references to sources where these matters are discussed in detail. I have dwelt less on the physical and mechanical properties of iron railways in the period 1830 to 1850, since these have already been researched in detail.

As an attempt at synthesis, this study draws on the work of many scholars who have published on aspects of early railway history. Some are book-length, which include the wide-ranging thematic works of Dr Michael Lewis, Dr Michael Bailey and John H. White; others are histories or archaeological studies of individual systems or of mechanical items such as locomotives. Many shorter studies have appeared in the published proceedings of the International Early Railways Conference and the Early Main-Line Railways Conference. I have also drawn on much earlier historical studies, such as William Weaver Tomlinson's 1915 work on the North Eastern Railway and Ernest Leopold Ahrons's 1927 *The British Railway Steam Locomotive*. Wherever possible, I have used secondary sources which are fully referenced.

What has revolutionised study of the modern past over the last few years has been the digitisation of books, journals and newspapers, which has meant that a search which would once have required a trip to the British Library or to the Bibliothèque Nationale can now be carried out in five minutes. It is easier for a reader who might wish to follow up these sources to access the original, so wherever it is helpful I have referenced primary sources rather than a secondary document but I also include in the bibliography all the publications which have led me to the originals and which have helped me understand them. For the period 1834–50 (chapter 13), there is an embarrassment of riches in terms of source material, so I have referred only to documents which are immediately relevant or which offer an informed overview.

My native proficiency extends to English and Welsh, and I have a good reading knowledge of French. I have done what I can to struggle through German, Dutch and Spanish texts, and I am very grateful to my former pupil Dr Alexander Zaslavsky for assistance with Russian.

A basic chronology of the early iron road has certainly been established by the distinguished authors whom I mention above, but it is important to note that there will still be many unread documents preserved in archives all over the world which could alter our understanding. To give one example from near my home: in 1915, Tomlinson stated that the first use of what proved ultimately to be a breakthrough technology, wrought-iron rails, was at Walbottle Colliery 'about the year 1808', and that three and a half miles of the Tindale Fell Railway, also in the north-east of England, were laid with them between 1808 and 1812.[4] There is no reason to doubt the thoroughness of Tomlinson's research. However, documents in the archive of a legal firm preserved at Bangor University in North Wales strongly suggest that a nearby slate quarry, Cloddfa'r Coed in the Nantlle Valley, started using them around 1811, a detail which challenges neither the primacy of Tindale Fell nor the accuracy of Tomlinson's account, but does prompt the thought that a chance discovery elsewhere might do both.[5] Cloddfa'r Coed's engineer was John Hughes, a local man who died in 1845 and who never learnt to speak English. Though his name remained one to conjure with locally until very recently, in railway history he is an unknown, a home-grown artisan who neither sought nor achieved any recognition.[6] Further search in archives may well alter the established chronology of the iron railway and perhaps tell us more about the contributions made to it by those who have no other memorial.

A NOTE ON SOURCES AND TERMINOLOGY

Railway terminology has the potential to confuse, and I have tried not to complicate matters.

A railway is defined here as a prepared way for wheeled vehicles which offers them both support and guidance. Waggonways (wooden-rail systems), plateways (L-section rails for flangeless wheels) and railroads (cast-iron rails for flanged wheels) are all forms of railway. However, in the United Kingdom the word 'railway' also came specifically to mean the type of system making use of wrought-iron rails for flanged wheels which emerged in the 1820s, which led to the networks we know today. Further to confuse matters, in the United States the equivalent word is 'railroad'. I hope it is clear from the context exactly what type of system is being described.

To this end, I have avoided the word 'tramway' entirely, except where it formed part of an official title. The word was often used at the time to signify a plateway but was also sometimes used for edge railways, and its meaning became even less precise in the course of the nineteenth century, encompassing an intra-urban surface railway (a 'streetcar' in the USA) and any lightly laid industrial or mineral line.

It is partly because the word 'railway' also means so many different things, as well as to emphasise the move from wood to iron and from animal to mechanical traction, that I have frequently used the phrase 'iron road' in this study – it was a phrase used at the time, and also translates the French *chemin de fer*, Russian *zheleznodorozhny* and the German *Eisenbahn*, as well as underlying the Irish *iarnród*. Modern Welsh usage has settled on *rheilffordd* (railway) but *cledrffordd* (again, railway) and *ffordd haearn* (iron road) were used until quite recently.

I have only capitalised the names of systems if they had an official name – if they were chartered or otherwise established by law – so the Surrey Iron Railway and the Baltimore and Ohio Railroad but the Lambton waggonway and the Hetton Colliery railroad.

Although a self-propelling machine on rails which pulls a train is often called an 'engine', I consistently refer to such a thing as a 'locomotive' to distinguish it from a 'fixed engine' which moves rail vehicles by ropes or other means either along the level or on a gradient, which is the departure from the absolute level on any formation. An 'inclined plane' is a stretch of railway on a significant gradient which is worked by ropes.

The period from 1750 to 1850 was one of extraordinary political upheaval. Countries and jurisdictions changed their names, even their calendars. For this reason, I refer to them by the names they bore at the time, such as 'the Ottoman Empire' and 'the German Confederation'. I distinguish (for instance) between the Kingdom of Great Britain (1707–1800), which encompassed England, Scotland, Berwick-on-Tweed and Wales, the Kingdom of Ireland (1542–1800), and the United Kingdom created in 1800–1, which also embraced a less-than-willing Ireland. It is assumed throughout that 'Britain' refers to the island within which England, Scotland and Wales are located, but that it does not include the island of Ireland. It is also assumed that both Britain and Ireland are part of Europe, and that continental Europe extends as far east as the Ural Mountains. Monmouthshire, which had the densest network of iron railways in the world by 1834, is assumed to be in Wales; although its legal status was ambiguous for years, Bedwellty and Blaenavon cannot be considered English.

I have used the place names which were in official use at the time, partly in order that readers may carry out internet searches if they wish, but have put its present-day official equivalent in parentheses if there is any danger that not doing so will confuse. An example is the harbour, town and railway terminus of Dún Laoghaire in Ireland, often spelt Dunleary, but which was officially known as Kingstown from 1821 to 1920, and still referred to as such by some people out of habit or because of Unionist leanings for many years afterwards. The much-contested territory of Silesia is another case in point: Germanic and Slav languages and dialects have all been spoken there and it seems easiest simply to refer for instance to 'Königshütte' but to add 'Królewska Huta' in parentheses and in this case to add 'Chorzów', the overall urban conurbation, for good measure.

A NOTE ON SOURCES AND TERMINOLOGY

I have put the word 'improvement' in inverted commas throughout as a reminder to myself and to the reader that this was a specific if vaguely defined concept and that it 'improved' in the interests of some individuals and not necessarily of society as a whole – ever since Jane Austen penned *Mansfield Park*, it has been impossible to take the idea entirely seriously.

Finally, many of the railways and locomotives described here were built in countries which used the traditional English units of measurement, the 'Winchester measure', further codified as the Imperial System by the Weights and Measures Act of 1824. These are still used in the United Kingdom and are still understood in the Irish Republic. The USA customary system derives from the English model and remains widely used. For this reason I have used these units throughout when discussing the historic period 1767 to 1834, though I have expressed statistics and lengths in their metric forms where occasionally I have introduced contemporary matters or details from remote antiquity for comparative purposes. Where historic units of measurements which are no longer in use are referred to, such as the French *lieue* (league), the Prussian and Austrian *meile* and the Russian *verst*, I have mostly expressed them in their original form with the metric equivalent.

NOTES

Introduction

1. A portage railway (one that connects two water bodies) across the isthmus of Corinth, running on carved stone ruts, was built in the sixth century BCE and lasted until 67 CE; it certainly carried warships, and probably marble and timber as well (Lewis 2001).
2. The definitive study of these systems was published in 1970, by Dr Michael Lewis of the University of Hull (Lewis 1970; see also Bennett, Clavering and Rounding 1990). A 'staithe' is a fixed structure where vessels – keels or ships – are loaded.
3. An early suggestion for an electrostatic telegraph appeared in *The Scots Magazine* in 1753 (volume 15, 73–74). Subsequent experiments led to working systems by 1816, and commercially viable telegraphy by the 1830s.
4. The publications of Sir George Cayley (1773–1857) set out the basis of the modern aeroplane but the first confirmed powered flight was not to be made until 1903.
5. Coleman 1992, 3–4.
6. Trinder (2003, 14–15) describes surviving examples of the first known iron rails as 'crudely cast in open moulds and their width and depth varies with the point at which measurements are made. They raise questions about the interpretation of the history of innovations. To substitute castings for wooden rails as the top layer of a double system seems on reflection to have been a small step for mankind ... Perhaps the most important thing we learn from the rails is that we gain more understanding from the study of historical contexts rather than by compiling lists of "firsts".'
7. Sir Herbert Butterfield (1900–79), Regius Professor of History and Vice-Chancellor of the University of Cambridge, was critical of the ways in which history was often written as a retrospective creation of a line of progress towards a glorious present (Butterfield 1931).
8. Guy 2001, 117–18.
9. Freeman 1999, 9–18.
10. Egypt, Argentina and India would all later learn the hard way that British investment in their railways came at a price (see El Sayed and Gwyn 2016; Damus 2016; Nehru 1936).
11. White 1976. Some even used unflanged wheels with separate guide wheels running against the side of each rail, in the manner of a sixteenth-century mine railway. It was said that there was no lumber company so rich that it could afford a pole road.

12. See Edgerton 2006 for discussion of technology-as-use and technology-as-invention.
13. At the time of writing (2022), a colliery at Sandaoling in the Xinjiang Uygur Autonomous Region of north-west China uses a number of steam locomotives on a daily basis.
14. This is not a figure of speech. Attempts to calculate the economic benefit of railways by analysing historical data with statistical and econometric methods ('cliometrics') have used different assumptions, statistical models and data, and have led to very different and sometimes very unpersuasive conclusions. Counterfactual analysis based on the difference between the actual cost of movement by rail and the hypothetical cost of the same services by other means presuppose an inelastic demand for transport. Transport is not an isolated entity in any economy – for instance, railways conferred huge benefits on the iron and coal industries, and railway capital and management structures may have shown other industries the way to greater productivity. Discussion of the British context has also been vitiated by a failure to understand the role (or even appreciate the existence) of railways before 1825.
15. Bradley 2016; see also Freeman 1999.
16. Union Inter Des Chemins de Fer: Railway statistics – synopsis, 2014 (https://uic.org/IMG/pdf/synopsis_2014.pdf).

Chapter 1 Trade, transport and coal 1767–1815

1. Hair 1988, 5.
2. Banks, *Journal of an Excursion to Wales &c.*, Cambridge University mss, Add. 6294; Tucker 1981; Trinder 2005, 40. The first of these rails had been cast in 1767 (Lewis 1970, 260–62).
3. Less energy is required where neither the wheel nor the road is prone to deformation through applied force. In scientific terms, this is expressed as the concept of rolling resistance (otherwise called rolling friction or rolling drag), and is crucial to understanding the relative advantages of different types of railway and road. This is the force which resists motion when a wheel rolls on a surface. Any coasting wheeled vehicle will gradually slow down due to rolling resistance, but a waggon with iron wheels will roll further on iron rails than it will on wooden rails, and considerably further than a road vehicle running on tarmac. An iron railway therefore requires comparatively less tractive effort to move a wheeled load, whether it is exerted by an animal, a locomotive or a fixed engine, than a wooden railway, less still than on a road. On canals heavier loads still were possible for the same tractive effort, since the only resistance is a fluid on a solid.
4. Cantle 1978–9.
5. Clark and Jacks 2007. It has been suggested that Britain accounted for about 85 per cent of European coal consumption in 1800 (Parthasarathi 2011, 161).
6. Borsay 2001, 167.
7. Trinder 2013, 221.
8. By the mid-eighteenth century, coalmining already had a long history in the Principality of Liège, in France, in Silesia and in China. Ambitious monarchs like Frederick the Great of Prussia and Tipu Sultan of Mysore had identified its potential for their dominions.
9. Thirteen per cent of production was the estimate of a coal viewer in 1830 (Trinder 2013, 219).
10. Nef 1966, 201; McCutcheon 1980, 333–4; Flinn 1984; Magennis 2014.
11. Hair 1988, 5.
12. *Hund* is German for dog, and the word was often explained as a reference to the barking noise it made when being pushed, though, as Lewis (1970, 7–8) points out, it is far more likely to be derived from a Slav verb meaning to chase or drive fast and to be cognate with

the Magyar *hintó*, 'carriage'. Lewis (2001) suggests that the stone railways of antiquity such as in the Roman adit at the Três Minas gold mine in Portugal, which preserves V-section channels cut into the rock at a gauge of about 1.2 metres, might have lingered on in Byzantine territories, to re-emerge in the late medieval period in mines in the eastern Alps or the Tirol. See Allison, Murphy and Smith (2010) for the Caldbeck system.

13. No railway systems, for instance, were identified in the excavated portions of the technically advanced Coleorton deep collieries in Leicestershire, active from 1460 to 1600, the only ones from this period to have had the benefit of detailed archaeological study (Hartley 1994).
14. Trinder 2013, 229.
15. Smith 1957, 1960; Lewis 1970.
16. The flanged wheel may have been another import from continental Europe, or it may have been an English invention. The Augsburg financiers at Caldbeck had a stake in copper mines in Neusohl in Hungary, where ore and deads (rock containing no ore) were being moved in horse-drawn boxes guided by small vertical rollers against the inside of the oak rails, but which ran on larger wooden horizontal rollers. As these wore down, they would eventually produce a flange and perform the guiding themselves (Lewis 2006, 17–22).
17. Lewis 1970, 267.
18. Angerstein 2001, 338.
19. Alfrey and Clark 1993, 71.
20. Lewis 1970, 107, 195–8; Jones 1987; Trinder 1996, 103.
21. Lewis 1970, 264.
22. Lewis 1970, 256.
23. Bennett, Clavering and Rounding (1990 vol. 2 plan 26) illustrates Derwenthaugh staithes in 1752, the best known example of their operation, with their off-gates for empty waggons and their 'turnrails' (turntables).
24. Bennett, Clavering and Rounding 1990, 58, 61–2.
25. Bennett, Clavering and Rounding 1990.
26. For the role of the viewer in the later development of the railway, see Stokes 2006.
27. See Carlton, Turnbull and Williams 2019 for an archaeological study of the Willington waggonway.
28. Lewis 1970, 184–90, 268; Darsley 2006.
29. Sometimes the 'main way' was called the 'waggonway' and the 'bye-way' was the 'sideway' (Lewis 1970, 144).
30. The scale of their civil engineering is conveyed by the regional terms for embankments – 'batteries', 'mounts' or bulwarks' (Lewis 1970, 148–50).
31. Jars 1774, 204–5.
32. Jars 1774, 205. 'Planche V' is an engraving of a chaldron waggon, a horse, a driver, a wooden way and a turntable.
33. *Newcastle Courant*, 8 July 1727. It was certainly the longest single-span bridge in Great Britain for the next 30 years.
34. Lewis 1970, 168–9, 194–8.
35. Lewis 2003.
36. The (Edinburgh) Speculative Society was founded in 1764, the (Dublin) Royal Irish Academy in 1795. The Literary and Philosophical Society of Newcastle upon Tyne (the 'Lit and Phil') was founded in 1793 as a 'conversation club', though discussion of religion and politics was prohibited. Women were first admitted to its library in 1804. Bruton (2015) discusses the 'Shropshire Enlightenment' and Uglow (2002) the Birmingham Lunar Society.
37. For Sinclair, see *Dictionary of National Biography; Oxford Dictionary of National Biography*. Jane Austen's 1814 novel *Mansfield Park* is the classic critique of 'improvement' as meddlesome novelty, from a Tory perspective.

38. Lewis 1970, 201.
39. Skempton and Andrews 1976–7.
40. Bye 2010, 142. The Middleton estate at Leeds and the railway were owned by Charles John Brandling (1769–1826), who owned land and collieries in Northumberland and Durham.
41. Lewis 1970, 295.
42. Bennett, Clavering and Rounding 1990, Vol. 2, 2, 4–5.
43. Turnbull 2019, 86–99.
44. Skempton 1973–4.
45. Bennett, Clavering and Rounding 1990, Vol. 2, 3.
46. *Morning Advertiser* 1 May 1810; Sykes 1833, 55; Mackenzie and Ross 1834, 43.
47. Hair 1988, 64, 145.
48. Patent 653, dated 9 February 1750 to Michael Menzies esq., anticipates inclined planes operated by gravity counterbalance or by water balance, whereby tanks filled with water running on rails hauled up loads of coal. These would variously connect the coalfaces with the foot of the shaft, enable coal and waste to be tipped at the head of the shaft, and transport coal to the river staithe. Those underground could run on wooden rails, or operate as rutways cut into stones or the coal of the tunnel floor. These and the tipping systems were intended to be easily movable as work progressed. Menzies also advocated water 'tubs' underground and surface canoes to move the coal. See Woodcroft 1854, 122 and HMSO 1856, entry for AD 1750, patent 653 *English Patents of Inventions, Specifications* volumes 626–759.
49. Faujas-de-St-Fond 1797, 163–5; Wood 1825, 93–108, 104n; Dunn 1844, 52; Powell 2000, 28. Wood states that Barnes's incline made use of a 'plummet' sunk in a well at the summit to counterbalance the empty waggons being raised up the incline.
50. Sykes 1833, 43–4. Matthias Dunn dates the first steam-powered inclined plane in the region to 'about the year 1805', from Bewicke Main pit at Birtley across the high grounds of Black Fell, followed in 1808 by a series of inclines from Urpeth Colliery over Ayton Banks, some of which were operated by fixed engines (Dunn 1844, 52). Mountford (2013, 25) states that the first Black Fell engine was installed 'about 1805', that from its summit the railway made its way to the River Wear, and that the events of 1809 reflect its rebuilding to give access to the Tyne. Sykes is specific that it took place on 17 May 1809.
51. Accession no: Tyne and Wear Archives and Museum 2011.3209. The Wearmouth bridge was the second large cast-iron bridge to be built in the world after Ironbridge in Shropshire but was over twice as long.
52. It was described by the *Globe* newspaper on 11 December 1815 as 'an easy Lead on a Iron Rail-Way to extensive Staiths at the Port of Sunderland, where the Coal is shipped by Spouts – an Advantage enjoyed by no other Colliery on the River Wear'.
53. *Cumberland Pacquet, and Ware's Whitehaven Advertiser*, 28 March 1815; *Tyne Mercury; Northumberland and Durham and Cumberland Gazette*, 8 August 1815; the two men who stood trial were acquitted.
54. Sykes 1833, 92–3; Guy 2001, 126–7. Tomlinson (1915, 27) gives the number of fatalities as 16. Hopkin (2014, 236–7) gives the date as 13 July. For examples of press reports, see *Durham County Advertiser*, 5 August 1815; *Gloucester Journal*, 14 August 1815. See Richardson (1843, 150–1) for the pit explosion.
55. In 1818 George Hill of Gateshead prepared a report proposing the installation of rope haulage on further sections – Northumberland Records Office 3410/East/1/141.
56. These are described in Winstanley 2014, 121, 123, 130–1.
57. Sykes 1833, 74; Dendy Marshall 1953, 43–7.
58. Guy 2001, 122–3.
59. Turnbull 2019, 102–8.
60. Guy 2001, 119–22.
61. Duffy 1982–3 discusses the technomorphology of the Stephenson traction system.

Chapter 2 'Rails best adapted to the road': cast-iron rails and their alternatives in Britain 1767–1832

1. Stevenson 1824, 6.
2. Often also referred to as a tramway or a 'tram-road'. Plate rails bequeathed the word 'platelayer' to the English language, still used for the railway employee who puts the track down and maintains it.
3. Lewis 2003, 103, 114; Patel 2020.
4. Lewis 1970, 293; Rynne 2006, 359–61.
5. Wood 1825, 43–75, 157–68 is an overview of the issues as they appeared to a particularly well-informed engineer of the period.
6. Northover 2014, 144–9. Other secondary sources that provide a context to the choice of track materials include Paxton 2001; Wilmott 2001; Brotchie and Jack 2007, 8–10; Boyes and Lamb 2012, 93–6; Hartley 2014; Waterhouse 2014, 2017; Grudgings 2019.
7. Van Laun and Bick 2000; Van Laun 2001a; Van Laun 2001b, 249 records plateway gauges of 12", 1' 6", 1' 10½", 2', 2' 3¾", 2' 4", 2' 6", 2' 8", 2'9", 2' 9½", 2' 10½", 3', 3' 3", 3' 3½", 3' 6", 3' 8", 4' 2" and 4' 4".
8. Trinder 2003, 14–15.
9. Lewis 1970, 253, 265, 320–1; Hughes 1990, 172; Boyd 1991, 18.
10. From Thomas Wilson's *The Pitman's Pay*; quoted in Lewis 1970, 316.
11. Skempton 2002, 163–4.
12. Farey 1817, 161, 295.
13. Lewis 1970, 293.
14. For Buddle, see Hiskey 1978.
15. Curr 1797.
16. Curr 1797, 5.
17. Curr 1797, 23–9, plate 2.
18. Skempton 2002.
19. Van Laun 2001b, 19–22.
20. Outram 1801; Riden 1972, 1973, 37; Schofield 2000; Patel 2018.
21. Lewis 1970, 293.
22. Lewis 2020, 141–6
23. Lewis 2020, 64.
24. Clarke 2001, 219–24; Lewis 1970, 346–7; Henaux 1843, 279; Caldwell, Campbell and Brougham 2014, 65; Anon 1831; Morris 1841, 312–13. The Swedish examples may have been cast locally and the American ones certainly were. So far, no evidence has emerged for plateways in the ironworks built by Rhys Davies of Tredegar in South Wales for Marshall August de Marmont and King Louis Philippe in France, nor at the Tredegar Ironworks in Richmond, Virginia, with which Davies was subsequently involved (Madison 2015; *Monmouthshire Merlin*, 30 November 1867; Reid 1983, 15–17). Thought was given to building the Ithaca and Owego in New York state as a plateway (Lee 2008, 3). Remarkably, plate rails of some description, perhaps of wrought iron, were initially proposed as a money-saving exercise for the Warsaw to Vienna railway as late as 1839 (chapter 13) (Haywood 1969, 198).
25. Lewis 2020, 81, 134.
26. Boyes 2001, 195.
27. Lewis 2003 is a comprehensive summary of the development of this type of rail.
28. Van Laun 2001b, 17; Van Laun 2003.
29. Wyatt 1803; Patel 2020. Anon (1804) is a French account of the Penrhyn rails.
30. Davies 1943–5, 144–5 and plates 14 and 15; Lewis 2003, 105–6.
31. William Losh and George Stephenson, patent No. 4067, 30 September 1816.
32. Rees 2001, 159–60, 162.

33. Oeynhausen and Dechen 1829, 77, 85; Mountford 2006, 86.
34. Oeynhausen and Dechen 1829, 54–5.
35. Liffen 2006, 53.
36. Cotte 2003b, 8; preserved example in the *mairie* of Andrezieux-Bouthéon; Caldwell, Campbell and Brougham 2014, 62–5.
37. Riden 1973, 39, 41, 49.
38. Van Laun 2001b, 22–3, 81–2, 202, 211–12.
39. Excavations in 1996 at Lambton D pit in Sunderland uncovered a considerable length of wooden railway (Ayris, Nolan and Durkin 1998).
40. Lewis 1970, 296.
41. Rees 2001, 160.
42. These were: two construction railways at Boston built in 1795 and 1805; a short experimental line at Philadelphia in 1809; a stone-carrying line at Crum Creek in Pennsylvania and at an explosives factory in Chesterfield County, Virginia, both in 1810; at Kiskiminetas Creek in Pennsylvania in 1816; from Bear Creek furnace to a river navigation in Pennsylvania in 1818; at the Niagara portage, of uncertain date; and initially also the Granite Railway of 1826 at Boston, discussed further in chapter 3 (Gamst 2001, 256–9).
43. Caldwell, Campbell and Brougham 2014, 55.
44. Lewis 1970, 301.
45. Curr 1797, 23–9, plate 2.
46. Lewis 1970, 164, 219, 260
47. Riden 1973, 38; Hopkin 2014, 226.
48. Outram 1801, 474–5.
49. Bangor University mss: Penrhyn Further Additional 12/14. For Jessop, see Skempton and Hadfield 1979.
50. Edgeworth 1820. He invented a machine for measuring land, devised an early form of caterpillar track, a carriage steering system and an optical telegraph, and became convinced of the possibilities of steam locomotion very early on. The novelist Maria Edgeworth was one of his 22 children.
51. Edgeworth 1802; Lewis 1970, 297–8.
52. Parliamentary Papers 1808, 42–44 (Jessop's letters and evidence), 48 (Edgeworth's theories on iron roads), 95 ('Observations by a Country Gentleman'), 110–11 (stone and iron railways running on either side of a main road), 116–17 ('Plan of a species of Single Railway; by Mr. A. Walker, of Hayes') and 118–20 ('Letter from Mr. Henry Matthews, on Roads and Railways').
53. Robertson 1983, 34–5.
54. Smith 2014.
55. Cort 1834, 27.
56. Geraghty 2010. A variation, which cannot be called a railway as it included no elements of guidance, was the street tramway system used in some Italian cities. One had been installed in Milan by 1836 (Mazzoni 1836, 3; Le Neve Foster 1875).
57. Lewis 2001 provides the context for stoneways in antiquity.
58. Ewans 1977, 22. Pointwork in the early iron railway is discussed in Lewis 2019a.
59. The patent is 3959/1815, 15 November 1815 (Boyes 2001). For Joseph von Baader, see Baader (1857, 5–23). For the mono-rail, see Baader 1822, plate 3 figure 3, plate 5 figure 2. '*Wagen mit Seitenrädern und neben den Schienen laufenden Zugpferden*' ('Waggons with side wheels and draught horses running alongside the rails').
60. Lewis 1970, 304–6; Boyes 2001; 'Suspension railways', *American Railroad Journal and Advocate of Internal Improvements* 2, 1833, 421.
61. Allen 1953, 100.
62. Northover 2014, 154.

Chapter 3 Canal feeders, quarry railways and construction sites

1. Smiles 1857, 62.
2. Phillips 1803, 598 claimed that 90 of the 165 canals in acts passed after 1758 had collieries nearby and a further 47 were close to iron, lead and copper mines.
3. Hughes forthcoming, a detailed study of the Swansea Canal and its tributary railways, conveys the variety and extent of the systems serving just one canal.
4. Skempton 1953–5, 25.
5. McKeown 2006, 102.
6. In China from very early times, boats were hauled up and down slipways by capstan. The sixth-century BCE Corinth *diolkos* involved rope haulage along a stone portage railway, and a well-known example was constructed in 1437 at Lizzafusina (Zafosina) on the River Brenta at Fusina near Venice. See Tew (1951–53).
7. The Sirhowy's act of 1802 sanctioned a road, not completed until *c.* 1825 (Hughes 1990, 166, 315; Cossons 1972, vol. 1, 392). On the Grosmont, the same toll gates and houses served both (Cook and Clinker 1984, 21, 24–6, 61, 76–7). The Brecon Forest formed part of a network intended both to 'improve' the uplands and to facilitate travel between Swansea and the English West Midlands, its proprietors running 'a light private omnibus' along the road to connect with the Hereford coach and so on to Birmingham and Liverpool (Hughes 1990, 60; *Cambrian*, 23 August 1834). In addition, a railroad serving Dinorwic slate quarry in North Wales partly ran along an earlier industrial cart-road, whereas at Dowlais and Oystermouth, turnpikes made use of existing rail corridors (Gwyn 2015, 231–2; Rattenbury and Lewis 2004, 13–15; Lee 1954, 9).
8. Rattenbury and Lewis, 2004; Lewis 2020.
9. Ellison 1936–38, 1937; Cook and Clinker 1984; Rattenbury and Cook 1996.
10. Cumming 1824, 17.
11. Hughes 2010.
12. Boyd 1981, 5–115; Gwyn 2001a, 2014, 2015.
13. Gwyn 2015, 251–2.
14. Business entities such as these, whereby all shareholders own company stock in proportion, as evidenced by their shares, had existed in England since the mid-sixteenth century. Van Laun (1976) discusses the passing of such an act.
15. Erskine May 1844, 383–460. Private bills originated in medieval petitions to Parliament and were well established by the late eighteenth century as a means of promoting not only canals, turnpikes and railways but also dock and harbour works and water-supply systems.
16. Lewis 1970, 281.
17. Rattenbury and Lewis 2004, 30–44.
18. In the primary market, the investor can purchase shares directly from the company. In the secondary market, investors buy and sell the stocks and bonds among themselves. For the Lake Lock, see Goodchild (2006) and Dawson (2020a).
19. NRM: Image No. 10562739. The Surrey Iron Railway's act was granted on 21 May 1801, and operations began in September 1802 (Gerhold 2010, 193).
20. *Morning Chronicle*, 'we do not despair to see either iron railways, or at least paved causeways, generally adopted ...' (6 January 1801); 'one horse will do the work of ten ... will in all probability be extended to Portsmouth' (25 May 1801); 'We rejoice to see that the modern improvement of Iron Railways is very generally adopted throughout the whole Country ...' (10 November 1801).
21. It nevertheless limped on until 1846 (Gerhold 2010). A Russian visitor in 1802 published a description, together with an account of plateways on the London docks, dedicated to Count Nikolai Petrovich Rumyantsev, the Minister of Commerce (Vaxell 1805).

22. The Carmarthenshire's act was granted on 3 June 1802, and the first section was opened in May 1803. The *Ipswich Journal* for 26 November 1803 records the opening as far as Brondini and Cynheidre on 7 November. The *Cambrian* newspaper for 13 October 1804 records the completion of the embankment at Mynydd Mawr.
23. Robertson 1983, 22–3.
24. Bick 1987, 14.
25. See for example Dendy Marshall 1938, 72; Gwyn 2003.
26. Van Laun 1996–8.
27. Dendy Marshall 1938, 64.
28. 57 Geo III, 'to authorise the issue of Exchequer Bills and the Advance of Money out of the Consolidated Fund, to a limited Amount, for the carrying on of Public Works and Fisheries in the United Kingdom and Employment of the Poor in Great Britain'; Boyes 1978; Skempton 2002, 690.
29. Dendy Marshall 1938, 61. Women hauliers were common in the Forest of Dean.
30. Boyes and Lamb 2012, 21
31. Skempton 2002, xxix–xiv.
32. Skempton 2002, 554–69; Lewis 2010; Paxton 2014; Goodbody 2010; McCarthy 2014.
33. Skempton 2002, 120.
34. Hughes 2000, 144–6; Skempton 2002, 337–9; Northover 2014, 151–3.
35. Skempton 2002, 496–7; Lewis 2005–7.
36. For the Ashby-de-la-Zouche area, see Clinker and Hadfield 1958. For Leicestershire generally and South Derbyshire, see Palmer and Neaverson 2002.
37. For William Reynolds (1758–1803), ironmaster and scientist, of Madeley Wood and Ketley Ironworks, promoter of the Ketley inclined plane and of the Coalbrookdale locomotive project of 1802–3, see Trinder 2008.
38. Tub boats, also known as compartment boats or container boats, were typically run in trains.
39. Boyes and Lamb 2012, 32.
40. Beatty 1955, 209–14.
41. Fulton 1796, plate 2 'The Double Inclined Plane' (counterweight double plane); plate 3 'The Single Inclined Plane' (counterweight single plane); plate 4 'The Medium Plane for a small scent' (waterwheel plane); plate 5 'Parts of the Machinery'; plate 6 'The First mode of passing wide and deep Valleys'; plate 7 'The mode of passing Rivers and gaining height at the same time'.
42. Brunton 1939, 18–20; Hughes 1990, 116–27.
43. Gwynedd archives; XM/5171/1; Paxton 2001, 88.
44. Boyes and Lamb 2012, 97; Boyes 2014.
45. Northover 2014, 151.
46. Rattenbury 1964–5; Jones and Hankinson 2004.
47. Lewis 2010, 180.
48. Davies 1992, 85–7.
49. Pevsner and Williamson 1979, 174: Vanags 2000, 14–15.
50. Boyes and Lamb 2012, 100–1.
51. *Edinburgh Encyclopaedia*, Vol. 15, 246–7.
52. McCutcheon 1980, 58–65. Ducart was also known as Daviso de Arcort, Davis Dukart, Duchart, and Duckart.
53. Trinder 2005, 85. The artist was William Armfield Hobday. Lord (1998, 53–4) discusses the painting but ascribes it to Richard Wilson.
54. Trinder 2005, 76–80, 105–6; Van Laun 2001b, 186–7. The drawing is dated 1797. The incline may have been the work of Thomas Dadford senior (1730–1809), engineer to the Glamorgan Canal, whom Crawshay also admired (Skempton 2002, 164–6).
55. Notable later installations using traveller carriages include the Khojak inclined planes in what is now Pakistan and the Siclau forestry plane in Romania (Mountford 2013).
56. Trinder 1981, 76–82.

57. Egerton 1812. The eighth earl, a son of the Bishop of Durham, a clergyman, scholar, naturalist and antiquarian, lived in Paris from 1796, despite the state of war between France and Britain, where he attracted the attention of the police by dressing his dogs and cats as ladies and gentlemen and taking them for rides in his carriage from the Hotel Egerton at 335 rue Saint-Honoré.
58. Trinder 2008, 22.
59. Dickinson 1913, 40.
60. Clarke 2001, 219–24.
61. Bibliothèque École des Ponts Paris Tech: ms 1558 (Dessin de la machine pour faire monter et descendre les Bateaux d'un canal inférieur à un supérieur ... Levé et dessiné par M. Betancourt Ingénieur Espagnol). Betancourt's skills encompassed steam engines, telegraphy, road building, aeronautics and urban planning, and he was successively employed by his own government and by those of France and Russia. He had already come to England in 1788 to learn about Boulton and Watt engines.
62. Betancourt y Molina 1807.
63. Maillard 1784.
64. Other members included the mineralogist Karl Heidinger (1756–97) and Captain Johann Swoboda of the Austrian Military Academy. I am grateful to Mike Clarke for these references.
65. Dutens 1819, 42–7, planche 3 et 4.
66. Fulton 1796, 10.
67. Chapman 1797, 22.
68. He did publish a French translation of his 1796 volume carried out by François de Recicourt, a French military and civil engineer (Fulton 1799, though dated as the revolutionary year 7) which added an appeal to Nicolas-Louis François de Neufchâteau (1750–1812), who initiated the French system of inland navigation.
69. Millin 1807, 373–5.
70. Skempton 2002, 267–71.
71. Waterhouse 2017, 493–7.
72. Skempton 2002, 804; Wyatt 1803.
73. Hughes 1990, 116–21; *The Engineer*, 11 February 1870.
74. The steam locomotive era in South Wales lasted over 170 years, until May 1985, when *Menelaus*, a 1935-built saddle tank, was retired from service at Marine Colliery in Ebbw Vale (Owen 2001, 34). This was also the last use of a steam locomotive in a British colliery.
75. Lewis 2020, 61–73, 120.
76. Lewis 2020, 73–91, 105, 121–3; Ince 2001, 63–8.
77. Ewans 1977, 25.
78. Cook and Clinker 1984, 17, 38.
79. Boyes and Lamb 2012, 62–3, 86–7.
80. Hughes 1990, 184–93.
81. Guy (2014) discusses the various different functions these railways fulfilled.
82. Lewis 2010, 182.
83. Van Laun 2001b.
84. Boyes and Lamb 2012, 60–9.
85. *Shrewsbury Chronicle* 19 February 1808, 3 August 1810; *Edinburgh Encyclopaedia* 15, 246–7; Gwyn 2008.
86. Gerhold 2010.
87. Dendy Marshall 1938, 58–64; Hughes 1990, 290–2.
88. Hughes 1990, 39–92; Lewis 1911.
89. Hughes 1990, 291–2; Northover 2014, 141.
90. Lewis (2014 201–8) speculates that there had been other, unadvertised, services in South Wales.

91. Hughes 1990, 292; Lewis 2014.
92. Ripley 1993, 11, 23–5.
93. Farey 1817, 295–6.
94. Lewis 2010, 185–6.
95. Oeynhausen and Dechen 1829, 131.
96. Lewis 2010, 179–80.
97. Buchanan 1811, 16. Phillips (1972, 217) shows a sketch of the Oystermouth carriage in August 1809 from University of Chicago MS 280 f 65, reproduced in Lewis (2014, 203).
98. Chapman 1813, 24.
99. Lewis 2014.
100. Lewis 2014.
101. Chapman 1813, 24.
102. Gwyn 2004.
103. A reference in an American lawsuit of 1853 to 'freight cars' on the 'Killeyney-kill and Dalkey' can only be to this system (Whiting 1853, 124, 125; Winans 1854).
104. Mike Clarke, personal communication. A road had been built through the pass in 1735 and the Semmering Railway was completed in 1854 (chapter 13; Tusch 2019).
105. Davies and Moorhouse 2003, 221; Muhle 2015, 179.
106. Russeger 1837, 222; Hoffmann and Zrnić 2012, 388. Their descriptions are confirmed by the surviving ruins.
107. The engineer Héron de Villefosse, who managed the Ruhr collieries during the French occupation, was familiar with British practice and with the use of railways (Villefosse 1810–19, 531–59).
108. Lewis 1970, 344–5.
109. Henaux 1843, 279. So far, no evidence has emerged for plateways in the ironworks built by Rhys Davies of Tredegar in South Wales for Marshall August de Marmont and King Louis Philippe in France, nor at the Tredegar Ironworks in Richmond, Virginia, with which Davies was subsequently involved (Madison 2015; *Monmouthshire Merlin*, 30 November 1867; Reid 1983, 15–17).
110. Lewis (1970, 330–55) summarises English influence in the German-speaking lands.
111. Lewis 1970, 347.
112. Anon 1831.
113. Lewis 1970, 304–6; Boyes 2001.
114. See Gamst (2001, 256–61) for early wooden railroads in America.
115. Durnford 1863, 106–7; Gamst 2001, 258; Skempton 2002. Durnford mentions both a plane 350' long at an angle of 45° operated by a four-horse gin, and one 'about 500 feet long' operated by a steam engine.
116. This obelisk commemorated the battle of the Revolutionary war and took 18 years to complete.
117. Stuart 1871, 120–7.
118. Gamst 1992. Bryant was adamant that 'All the cars, tracks, and machinery are my original inventions' (Stuart 1871, 125).
119. Morris 1841, 312–13 states that at the Lonaconing works in Alleghany Co., Maryland, they were used in the coal and iron mines, each weighing 45 or 36 lb per yard, and 3' 9" in length, other than plates for the 10-foot radius curves, which were cast separately; for the Ithaca and Owego, see Lee (2008, 3).
120. Lewis 1970, 304–6; Boyes 2001; Suspension railways, *American Railroad Journal and Advocate of Internal Improvements* 2, 1833. 421.
121. Lewis 2020, 161–2.
122. For the longevity of some of these railways, see Ripley 1993, 44–6, Gwyn 2010; Boyes and Lamb 2012, 178–9. The Little Eaton gang road operated from 1795 until 1908, and the Peak Forest Railway only carried its last loads in 1922, when the London and

NOTES to pp. 79–84

North Eastern Railway realised it was about to become the owner of an eighteenth-century plateway. The British canal network remained busy until the 1920s, and short industrial systems continued to be constructed to serve it. When the pioneering industrial archaeologist Tom Rolt went to work in Shropshire during the Second World War, he was amazed to find plateways still in use around Horsehay works, near where Banks had seen the iron railway in 1768, where they were delivering materials to construct steel invasion barges (Rolt 2005, 44). As noted, a plateway at Bicslade in the Forest of Dean operated into the 1950s, and some very short systems even later.

Chapter 4 'Art has supplied the place of horses': traction 1767–1815

1. Faujas-de-St-Fond 1799, vol. 1, 142–5.
2. Hair 1987, 1.
3. The Mines and Collieries Act of 1842 forbade women and girls, and any child below the age of 10, from working underground.
4. Sykes 1833, 43–4.
5. Bye 2010, 143; Lewis 2010, 185. See also Gamst 2001, 258 for the use of winches on a portage system at Niagara in Ontario.
6. They outlasted the steam locomotive in commercial railway service in the United Kingdom, which ended in 1990, at Castle Donnington power station in Leicestershire. The last 'pit ponies' were only retired in 1999, at a colliery in South Wales (Thompson 2008, 66). Horse-drawn pleasure railways operate in a number of locations across the world.
7. Wood 1825, 90–1.
8. The equivalent figures for a pack horse are one-eighth of a ton; for a horse pulling a stage waggon on a 'soft' road, five-eighths of a ton; on a macadam road, 2 tons; on a river, 30 tons; and on a canal, 50 tons (Skempton 1953–4, 25). Abraham Rees describes how 'an indifferent horse' could draw three tons on a wooden railway (Cossons 1972, 1, 323). See also for instance, *Aberdeen Journal*, 21 January 1807: '... a trial was made to ascertain what a horse could draw on the iron rail way, from the harbour of Ayr to Newton coal-pits. Six waggons were loaded with three tons each, the six waggons exceed two tons, making in all fully 20 tons. A cart horse was yoked, but, in starting, the chain which bound the fifth waggon to the fourth gave way, and the horse proceeded with four waggons with ease; thus pulling a load of nearly 14 tons weight.'
9. Bruton 1955–6, 77; van Laun 2001a, 28–9; Wood 1825, 90–2, 230–9; Lewis 1970, 209, 311; New 2019; see also Local Studies Collection, Newcastle City Library: Early History: Letters and Cuttings 1, 5, letter 3. The price of a horse went up from about £14 for a good three- to six-year-old horse, mare or gelding in the late eighteenth century to about £40 in 1821, though less powerful animals could be bought for half the price and a pony would be cheaper still.
10. Lewis 1970, 201.
11. *Morning Advertiser* 1 May 1810; Sykes 1833, 55; Mackenzie and Ross 1834, 43.
12. This practice was noted in Wales by the French engineer Louis-Georges-Gabriel de Gallois-Lachapelle (De Gallois 1818, 132).
13. Hyde Hall 1952, 105.
14. Boyes and Lamb 2012, 97; Ewans 1977, 24–5.
15. See Lee 1954, plate 2; Rattenbury and Cook 1996, 79; Boyes and Lamb 2012, 97.
16. Lewis 1970, 307–29; Hughes 2000, 84.
17. De Gallois 1818, 135.
18. Wood 1954, 59.
19. *Liverpool Mercury*, 23 August 1816.
20. Beazley 1967, 147.
21. House of Lords, *Eighth Report of the Commissioners for Making and Maintaining the Caledonian Canal*, 22 May 1811, 23.

22. The anonymous 'gentleman' whose description of the Whitehaven Colliery was published in the *Liverpool Mercury* on 23 August 1816 clearly experienced a pang when he noted that water from the lowest level was removed in casks on horse-drawn rail waggons 'attended by females: one of whom, an healthy lass of about 18, passed us several times during our visit to the mines, sat with great composure on the car, and though not as graceful as a Venus drawn by Peacocks, yet is not uninteresting, while she occasionally warbled out some wild but cheerful notes, which appeared intended to excite the tender passion in some of her fellow labourers around'.
23. Durnford 1863, 106–7; a four-horse gin operated an incline plane at an angle of 45°.
24. Price 1982, 158; Hughes 2000, 90; Lewis 1970, 276; Williams 1908, 180–1. Zulu oxen were used on a wooden-railed construction system in South Africa from 1856 (Darsley 2010, 226–7).
25. Williams 1908, 189.
26. Specifically, to Derbyshire by 1793 (Outram's Butterley plateway), to South Wales by 1794 (at Cyfarthfa Ironworks, to lower iron and limestone from the casting house level to the canal head), to the north-east of England in 1797 (at Benwell staithes) and to North Wales by 1801 (on the Penrhyn Quarry railroad) (Boyes and Lamb 2012, 85; Hughes 1990, 108; van Laun 2001b, 30, 186; Wood 1825, 104; Powell 2000, 28; Gwyn 2015, 231–2).
27. Egerton 1802. Egerton (1812, 15) refers to *'lames de fer'*, which suggests plate rails.
28. Mullineux 1971.
29. Fulton (1796, plate 2) shows just such an arrangement, though the text does not enlarge on its mechanical advantages. See Chapman 1813, 25–7.
30. Boyes and Lamb 2012, 87.
31. Bye 2003, 134–5, 2010, 146.
32. Karlsson 2001.
33. This was one of the earliest railways in the United States. An account and description by Sir William Strickland (1753–1834), a friend of Thomas Jefferson, is preserved in the Henry Luce III Center for the Study of American Culture in the New York Historical Society collection, object no: 1958.70 (*Technical Study of the Incline Plane Mechanism of the South Hadley Canal, South Hadley, Massachusetts*), dated 1794. The use of a cradle raises the possibility that it was influenced by Ketley. Sir William Strickland should not be confused with William Strickland (1788–1854), the Philadelphia architect and civil engineer (see chapter 5).
34. Gainschnigg studied mathematics at the Benedictine University of Salzburg under Professor Dr Ulrich Schiegg, mathematician, astronomer and land surveyor. It is possible that either or both had access to the notes and drawings prepared by Maillard and his colleagues Haidinger and Swoboda after their 1795 visit to Britain or to a copy of Fulton 1796. I am grateful to Mike Clarke for this observation. Anon (1805) is a French account of inclined plane in technology in the United Kingdom.
35. Waterhouse 2017, 439–64, figs 12.19, 12.20; Gwyn 2015, 140–3, though the author retracts a suggestion that the earliest date from *c*. 1810, as they are more likely to have been introduced in the 1820s.
36. Trinder 1981, 76.
37. Trinder 2005, 75–7. These were presumably the engines recorded in 1796 by the geologist Charles Hatchett (Trinder 1981, 103). Since these were the first steam engines to work railed vehicles, details of Heslop's patent are appropriate. It involved two open-topped cylinders, equidistant from the beam's pivot, one 31 inches (79 cm) diameter (the 'hot' cylinder) and the other 24 inches (61 cm) diameter (the 'cold' cylinder), in a cold water jacket. On nearing the top of its piston's stroke, water was injected from the cistern. Steam at 7 PSI was admitted under the piston of the 'hot' cylinder until it reached the top of the stroke, when the exhaust valve was opened, passing steam to the 'cold' cylinder, where it partly condensed, fully condensing when water was injected. By

using two cylinders, the working power was distributed equally through the entire up and down stroke, and these engines worked steadily (*Engineering*, 29 May 1868, 530). See also Rojas-Sola and Morena-De la Fuente 2019a, 2019b.
38. Trinder 1981, 79–80.
39. Barritt 2000, 37–40, 58, 64–9; *Lancaster Gazette* 9 January 1802, 25 August 1804. It was intended as a temporary measure but the canal link was never constructed, and it operated until 1879.
40. Skempton (1978–9, 96, 105) suggests that this was the engine ordered by the contractors Bough and Holmes from Boulton and Watt in 1802–3.
41. Smiles 1857, 29–34; Young 1923, 36–41.
42. Van Laun 2001b, 190–3; Hughes 1990, 316.
43. Mountford (2003) sets out the challenges of moving coal to navigable water and the use of inclined planes in the north-east of England, using the Lambton 'way' as a case study.
44. Mountford (2013, 25–6)
45. Dendy Marshall 1938, 19. Tomlinson (1915, 18) describes it as the first use of a fixed engine for drawing railway waggons, which is incorrect.
46. Bennett, Clavering and Rounding 1990, 12.
47. Uglow 2002, 126–32.
48. Fletcher 1891, 57.
49. Trevithick 1872, 155 (Trevithick to Giddy, 22 August 1802).
50. Trevithick 1872, 158 (Trevithick to Giddy, 2 May 1803).
51. A reference to a fatal explosion has led to a suggestion that it was the locomotive itself which blew up, which was not the case (Trinder 2003, 19). For the Greenwich accident, see *Newcastle Courant*, 24 September 1803. Trevithick, who was aware of the Coalbrookdale project, stated in 1831 'The first locomotive engine ever seen was one that I set to work in 1804, on a railroad at Merthyr Tydvil, in Wales …' – Select Committee of the House of Commons on Steam Carriages: *Report* (House of Commons, 1831), 78.
52. Letter from A.W. Pearce to W.G. Armstrong dated 21 June 1871: NRM: 2002-8348/100/8.
53. This section draws heavily on Guy et al., 2018.
54. The invention of the 'blast', whereby exhaust steam is released into the chimney along with gases from the fire to draw them through the tubes, became a matter of a controversy which rumbled on for well over a hundred years and was credited to George Stephenson and to Timothy Hackworth by their partisans. Trevithick claimed credit for the blast in his evidence to the Select Committee of the House of Commons on Steam Carriages: *Report* (House of Commons, 1831), 65. In the same *Report*, 50, John Farey Jr. credited Trevithick with its discovery and George Stephenson with its development.
55. Trevithick 1872, 162.
56. Guy et al. 2018.
57. It was picked up by the *Gloucester Journal* on 27 February; the *Bath Chronicle* on 1 March; the *London Courier and Evening Gazette* on 2 March; and the *Ipswich Journal* on 10 March.
58. *Universal Magazine*, March 1804, 323 ('Provincial Occurrences'); see also 286.
59. *Royal Cornwall Gazette*, 3 March 1804.
60. Cossons 1972, vol. 1, 327–8.
61. These are set out in Guy et al. 2018. They are *Retrospect of Philosophical, Mechanical, Chemical & Agricultural Discoveries* 1 (London, 1806), 344; *Bibliothèque Britannique ou Recueil: Sciences et Arts*, vol 31 (Geneva, 1806), 85–7; *L'esprit des journaux Français et étrangers par une Société de Gens de Lettres* 4 (Brussels, April 1806), 162–4; *Annalen der Physik* 22 (Halle, 1806), 403–4; *Efemeriden der Berg- und Hüttenkunde* vol 4 (Nuremberg, 1807), 82–3; *The Emporium of Arts & Sciences*, new series, 2 (Philadelphia, 1813), 374–5.

62. *Cambrian*, 10 March 1804. Martin was later consulting engineer for the Oystermouth plateway (Skempton, 433–4).
63. Rees and Guy 2006, 199–202.
64. Trevithick 1872, 186–7, 196.
65. *The Times*, 8 July 1808.
66. Liffen 2010.
67. Trevithick 1872, 196.
68. Bye 2003.
69. Andrieux 1815; this was the first detailed elevation of a locomotive. French practice in technical drawing was considerably in advance of British at this period (Cardwell 1994, 204–5; Bailey 2014, 23).
70. Bye 2014.
71. Winstanley 2014; Skempton 2002, 169–70.
72. Guy 2001, 122–3; Turnbull 2019, 100–5.
73. Lewis 2020, 51–5.
74. Oeynhausen and Dechen 1829, 43.
75. Bruyn 1987, 93 gives 1814, *Annales des travaux publics de Belgique*, XIX 1860–61, 227–8 n. 1 gives 1817. In 1815 John Cockerill sent over one Nicholas Dethier to England to gather information on different types of engines; Dethier was the maiden name of the mother of Joseph-Frédéric Braconier (1785–1858), the lawyer who opened the Murébure-Horloz pits. He may have received personally transmitted news about the Murray–Blenkinsop locomotives from a relative of his mother, and perhaps even before that had seen the drawings in the *Bulletin de la Société d'Encouragement pour l'Industrie Nationale* published in April 1815. I am grateful to Sheila Bye for this information. See also Caulier-Mathy (1971, 1980).
76. The Silesian foot measured 283 mm (11.14"), the Prussian *reichsfuß* 313.8536 mm (12.36").
77. Clarke 2001; Hills 2006, 250–1; Gomersall 2006. The Holbeck Foundry opened its doors in 1796, the year when Boulton and Watt invested in the Soho Foundry at Smethwick near Birmingham to create the first specialised engine-building works. The Haigh Foundry had been established around 1790, and after a shaky start acquired a reputation for excellent workmanship under Robert Daglish from 1804.
78. Bye 2003, 143.
79. Winstanley 2014, 125.
80. The patent is 3632 of 1812. See *The Repertory of Arts and Manufactures*, Series 2 vol. 24, February 1814, 139–42. These subassemblies could be in the form of either a trailing bogie or a double bogie. The patent was entitled *Facilitating and Reducing the Expense of Carriage on Railways and other Roads*. See Forward (1951–3) and Dendy Marshall (1953, 61–76).
81. Dendy Marshall 1953, 64–72.
82. *The Repertory of Arts and Manufactures*, Series 2 vol. 26, February 1815, 161–2 refers to the Lambton locomotive and to the chain system, but implies that the chain was only useful on gradients. The article was probably deliberately worded in such a way as to remind readers that it was an option for any future locomotives and that the patent was still valid. An article on the locomotive in the *Tyne Mercury; Northumberland and Durham and Cumberland Gazette* on 3 January 1815 is uninformative in this respect. See Dendy Marshall 1953, 64–72; Guy 2001, 123–5; Rees 2001, 155–6; Mulholland 1978, 177–9.
83. Lewis 2020, 56–8.
84. Sykes 1833, 92–3; Tomlinson 1915, 27; Guy 2001, 126–8; Hopkin 2014, 236–7.
85. Skempton 2002, 92–3.
86. Trevithick 1872, 188.
87. Young 1923, 41. Little is known of Waters; he may be the Thomas Waters whose death was recorded in the *Newcastle Courant* for 11 April 1829.

88. Young 1923, 47.
89. Young 1923, 48.
90. The uncertainty as to the precise dates of construction derives in part from the later and long-lasting quarrels about the precedence of Hedley and George Stephenson as to who had built the first successful adhesion locomotives and the claims that were made in later publications.
91. Crompton 2003; Guy 2001, 120. 'Puffing Billy' may reference William Hedley's asthma (Guy 2003, 66).
92. Young 1923, 48.
93. The sketch artist was either Archduke John of Austria, his brother Louis or one of their entourage, then on a goodwill mission to Britain. John had a particular interest in technology (Galignani 1819, 262; Sykes 1833, 95). It has not proved possible to identify copyright of the sketch.
94. This is the locomotive referred to as *My Lord* in some sources which Bailey (2019, 82) confirms is a later error.
95. Anonymous contributor to the *York Herald*, 27 September 1875, describing the Wylam plateway.

Chapter 5 War and peace 1814–1834

1. All the more remarkable in that Great Britain and then the United Kingdom had for many years sought to prevent skilled artificers and knowledge of innovative machinery from leaving the country. Until the end of the eighteenth century, it was widely accepted that high tariffs, prohibitive laws and isolationist ideals were economically advantageous for each country. Adam Smith's *Wealth of Nations* (published in 1776) formed the intellectual basis of the free-trade policies which dominated nineteenth-century economic thought (Harris 1998).
2. Gamst (2001) analyses the transfer of railway technology in the context of Britain and the USA in this period.
3. Murfitt 2017.
4. An example is the artist, architect and city planner Karl Friedrich Schinkel (1781–1841), better known for designing the Iron Cross medal and for the Romantic–classical buildings of central Berlin. He had made the familiar trip to Italy to study its architecture in 1824, then followed it up with a trip to Britain two years later where he had inspected railways (Gamst 1997, 800).
5. Baader (1832), a proposal for a railway from Munich to Starnberg is dedicated 'to all patriotically minded Bavarians and all capitalists who want to increase their wealth in a safe and advantageous, and at the same time charitable and glorious way'. Baader (1830), 'A word in time for a good cause', attempts to reassure public concern for safety in the aftermath of William Huskisson's death on the opening day of the Liverpool and Manchester Railway (chapter 11). See Eckert (2019) for von Baader's career.
6. Harkort 1833, 2.
7. 'Eisenbahnen, Schnellposten, Dampfschiffe und alle möglichen Fazilitäten der Kommunikation sind es, woraus die gebildete Welt ausgeht'; Goethe and Zelter 1915, 339, 6 June 1825, Johann Wolfgang von Goethe to Carl Friedrich Zelter. For the model, see Jessing, Lutz and Wild 2004, 97.
8. Boyes 2001, 203.
9. Chevalier 1832; Rieber 1990, 555–6.
10. For distrust of railways, see chapters 11 and 13. Curiously, a significant influence on Tsar Alexander was Franz von Baader (1765–1841) the brother of Joseph. Both men trained as mining engineers and resided for a while in England and Scotland, but took different trajectories thereafter. Franz's rejection of the empiricism he encountered in Britain led him to conclude that humankind cannot throw aside the presuppositions of

faith, church and tradition. He became Professor of Philosophy and Speculative Theology at the University of Munich and was instrumental in the revival of scholasticism in the Catholic Church. His memorandum *Sur le nécessité créée par la Révolution française d'une nouvelle et plus étroite union de la religion avec la politique* formed Tsar Alexander's ideas for a post-revolutionary and post-Napoleonic Holy Alliance among the Christian states of Europe. See Baader (1857).
11. Tredgold 1825; Nicholson 1825. Encyclopaedias such as Abraham Rees's multivolume publication (1801–20) served a wider public interest in general knowledge, including technology (Cossons 1972).
12. Hartley 2019.
13. A few examples will have to suffice. George Stephenson was very much a man of the north-east of England but worked on railways in Kent, Lancashire and Caernarvonshire. John Hodgkinson worked mainly in South Wales and the borders but was also involved in railway building in Gloucestershire and Derbyshire (Skempton 2002, 324–5, 655–7).
14. See Northover 2014, 147; Riden 1973.
15. Kirby 1993, 26–53.
16. Bailey 1978–9, 130.
17. Warren 1923, 32. Achard and Seguin (1926b) contains an itinerary.
18. These were Jonathan Knight, William Gibbs McNeill and George Washington Whistler (Dilts 1993, 85).
19. Pennsylvania Society for the Promotion of Internal Improvements in the Commonwealth 1826, 39–40; Carlson 1964.
20. Warren 1923, 135–7.
21. Allen 1953. See also Allen 1884.
22. For the Archdukes, see Galignani (1819, 262; Sykes 1833, 95). For the Grand Duke, see Warren (1923, 16); Young (1923, 276–7); Haywood 1969, 67.
23. Anon 1835.
24. Haywood 1969, 50–61, 112; Esper 1982, 601–2.
25. The visitor was Major George Washington Whistler (Dilts 1993, 86).
26. A devout Methodist, he nevertheless rode to hounds, was fond of dancing and sent his daughter to be educated by the Dominican nuns of Vilvorde. His biographer records an 'amiable weakness' for fine glassware and china, and some 'rather terrible oil paintings' (Young 1923, 331–55).
27. Hartley (2019, 33–6) discusses 'the coincidence of genius' in the north-east of England coalfield.
28. Guy 2001, 131.
29. Anon 1828, 298.
30. For George Stephenson's charisma, see Kemble 1879, 283–4.
31. Allen 1953, 113.
32. Cowburn 2001, 249.
33. Warren 1923, 180.
34. George Stephenson's brother Robert was more widely travelled, working on the Stratford and Moreton, the Nantlle and the Bolton and Leigh (Roper 2003).
35. Caldwell, Campbell and Brougham, 2014.
36. Reynolds 2003, 169–70; Cowburn 2001, 250; Thomas 1980, 231–2.
37. Rolt 2016, 333; Young 1923, 346; Elton 2003, 225–31. Tellingly, in 1830 George Stephenson took on as an apprentice the nephew of Lord Ravensworth, one of the 'Grand Allies' (Hartley 2019, 39).
38. Thompson 1942, 10–81 (1806–34); Gamst 1992, 78; Guy 2003, 74–5.
39. Haywood 1969, 65n–67n.
40. Anon 12 (18 July 1804).
41. Anon 1815.
42. The patent is 3959/1815, 15 November 1815 (Boyes 2001).

43. Baader 1822, unpaginated following viii.
44. A later revision (Wood 1831) was translated into French and published in both Paris and Brussels (Wood 1834a, 1834b).
45. Pennsylvania Society 1826, 19, 35; Strickland 1826; Rees 2003, 186–92. Strickland's original manuscript is in the Beamish Museum (acc. 2000–71).
46. Rubin 1961, 19–20.
47. For instance, Baader (1817, 50) 'Dampfmaschinen (loco-motive Engines) oder sogenannten Dampf-Pferde (Steamhorses)' or Gerstner (1831, 604) 'Bei allen neuern Eisenbahnen in England, Schottland und Frankreich werden ohne Ausnahme Schienen von gewalztem Eisen (*malleable iron rails*) verwendet'.
48. Gerstner 1813, 1827.
49. Gerstner 1831. Of the 1,900 advance orders published in the 1834 edition, 30% came from the construction and transportation sector, 23% from the book trade, 10% from the trade, factory, machine and iron and steel sector, 7% from libraries, 5% from the mining industry, 4% from research and education, 4% from the military, 3% from agriculture and forestry, and 14% from other sectors (Kurrer 2006, 1810).
50. Nicholson 1825, 1826.
51. Baader 1829, v–vi, 120, 137, 144.
52. Barker and Gerhold 1993; Guldi 2012.
53. Stevens (1812) may be said to have initiated the genre. See Anon 1828; Kilbourne 1828; Rubin 1961; Gibbons 1990.
54. Gamst 2001; Rees 2010.
55. Guy 2001, 136, referring to Durham Record Office: NCB1/JB/619-623; Hamel certainly received drawings.
56. Baader 1822, ii.
57. Turnbull 1928, 478.
58. Turnbull 1928, 474.
59. *Journal of the Franklin Institute and American Mechanics' Magazine*, vol. 1 (1826), 130–15. See also Rees 2003, 186–92.
60. Calkins 1867, 49; Baldwin Locomotive Works 1920, 9; Dawson 2004.
61. Renwick 1830.
62. This soon became the *American Railroad Journal*, later adding *and Advocate of Internal Improvements*, though it is remarkable how little of the material published in the first few volumes was either American or to do with railroads. A railwayman would have to read his way through pages that contained little of immediate interest to him.
63. Renwick and Lardner 1828. For Lardner, see Odlyzko (2019).
64. For Lamé, see Cotte (2009).
65. Articles on railways were few. Anon (1831) describes portable plateways in Sweden, and Destrem (1831) argues the case for canals served by ancillary railways.
66. Rieber 1990.
67. Minard 1834; Biot 1834.
68. The 'railroad university' phrase comes from *American Railroad Journal and Advocate of Internal Improvements* 4 (28 November 1835), 401.

Chapter 6 'Geometrical precision': wrought-iron rails 1808–1834

1. The phrase 'geometrical precision' comes from the *Dublin Penny Journal*'s description of the Dublin and Kingstown (1834, 67).
2. Cast-iron rails were 20% of the overall cost in the case of the Plymouth and Dartmoor plateway in 1823, wrought-iron rails 13% in the case of the Liverpool and Manchester in 1830 (Northover 2014, 153; Skempton 2002, 825).
3. Dow (2015) is an overview of British permanent way technology from 1804.
4. Elsas, 1960, xviii–xix.

5. Wood 1825, 60–1.
6. Wood 1825, 61; Gwyn 1999; Tomlinson 1915, 15; Bangor University Porth yr Aur mss 29599, 29619 and 29621.
7. Oeynhausen and Dechen 1829, 136–43.
8. Gwyn 2015, 134.
9. Dendy Marshall 1938, 136–7.
10. Stevenson 1878, 123. The Edinburgh Railway proposal of 1818 would have connected the Lothian collieries to the town (National Library of Scotland accession 10706, no 415; see also Robertson 1983, 34–7).
11. Birkinshaw 1821, 1822, 1824; Longridge 1822.
12. Tomlinson 1915, 77.
13. Ince 2001, 62–3.
14. *Newcastle Courant*, 18 December 1824. This was in the context of a proposed railway between Newcastle and Carlisle.
15. Bailey 1978–9, 122; Martin 1976, 1981; Boyd 1981, 18–19; Ferguson 2006, 183 – the dates are those of agreeing a contract or of a delivery, rather than the opening of the railway.
16. Thomas 1980, 59; Oeynhausen and Dechen 1829, 106.
17. Clinker 1954, 64.
18. Lewis 1968, 30.
19. Ferguson 2006, 182–5.
20. Allen 1953, 116; Thomas 1980, 59; Cotte 2003b, 22–3, figure 6.
21. *Aris's Birmingham Gazette*, 25 April 1831; Whishaw 1842, 60; Watkins 1891; Messenger 2012, 17, 24.
22. Tomlinson 1915, 239.
23. Vignoles 1889, 182.
24. Lewis 1968, 30.
25. Tomlinson 1915, 255.
26. Tomlinson 1915, 91.
27. Dilts 1993, 128–32.
28. Dilts 1993, 122–8; Thomas 1980, 59.
29. White 1976, 39–40.
30. Wayt 2016, 29.
31. Gamst 1997, 302.
32. Gamst 1997, 563–7, 579–80.
33. Vignoles 1889, 84, 137.
34. Wilson and Roberts 1879, 37. Watkins (1891, 661) states that they arrived on the Allegheny Portage Railroad in 1832, which may reflect a confusion with the laying of the strap rails; Franz Anton von Gerstner suggests that on the Philadelphia and Columbia, at least, the Harford Davis rails arrived after five years' operation (Gamst 1997, 564).
35. Watkins 1891, 664, 671.
36. Watkins 1891, 666–70; Watkins 1892.
37. Trautwine 1835.
38. Hartley 2019, 35 suggests that George Stephenson had become aware of this benefit by 1821.
39. Wood 1838, 163.
40. When they do, it is evident from the distinctive squealing sound they make going round a curve, as they interact with the rails
41. Thomas 1980, 59.
42. Dilts 1993, 129.
43. Allen 1953, 111, 126, 135–8.
44. These lasted as long as they did because they were better suited to their work. It would be hard to imagine a more rudimentary railway waggon than the flat wooden trolleys

with loose double-flanged wheels on loose axles used in Welsh slate mines until around 1998, yet they were ideal for moving long slabs on sharp curves underground since the loose wheels meant that they could be slewed from side to side.
45. The process by which this entered formal scientific and engineering practice is obscure. Credit was claimed by Brunel's awkward colleague William Gravatt (1806–66) and by William Froude (1810–79), who worked together on the South Devon Railway from 1836, Gravatt at least claiming that his views were based on earlier experience (Greenfield 2011, 70–5). Rankine (1883, 651–3) states that Gravatt produced a paper on the subject in 1828, which is not otherwise attested.
46. White (1976, 40), Hartley (2010, 159–61, 2014, 267–8) and Lewis (2020, 131–62) detail the extraordinary complexity of the conversion process on plateways.
47. Carter and Carter 2014.

Chapter 7 'Most suitable for hilly countries': rope and chain haulage 1815–1834

1. Walker et al. 1831, 53.
2. Rope is often referred to as 'cable', a thick, heavy form of rope. One definition of a cable is that it is made of nine strands and is 9 inches (23 cm) and upwards in circumference.
3. Edgeworth 1802, 223.
4. Thompson 1847; Skempton 2002, 701–2.
5. Hazard 1827, 276.
6. Skempton 2002, 701–2.
7. An example of Thompson's controversial style is to be seen in the correspondence between him and Nicholas Wood in the *Birmingham Chronicle* on 14 October and 18 November 1824.
8. Oeynhausen and Dechen 1829, 46–9; Warren 1923, 130–4.
9. Gamst 1992, 78; Stuart 1871, 123.
10. Wayt 2016, 26; Gamst 1997, 710–15.
11. Gamst 1997, 129–43 (Mohawk and Hudson), 213 (Rochester), 236–44 (Ithaca and Owego), 561–78 (Philadelphia and Columbia), 686–7 (Chesterfield), 732–3 (Tuscambia, Courtland and Decatur).
12. Gamst 1997, 625–6. The spectacular inclined planes on the Mauch Chunk railroad at Mt Pisgah and Mt Jefferson date from the construction of the 'back-track' return route for empties in 1844–5 (see *New York Times*, 21 September 1865).
13. Wilson 1883; Gamst 1997, 514.
14. Gueneau 1931b, 229.
15. Cowburn 2001, 239.
16. The following section draws on Warren 1923, 165–70 and Bailey 1980–81.
17. For James Walker, John Urpeth Rastrick and William Cubitt, see Skempton 2002. For Philip Taylor see *Dictionary of National Biography*.
18. Tomlinson 1915, 116.
19. Walker and Rastrick 1829, 4.
20. Walker and Rastrick 1829, 5.
21. Walker and Rastrick 1829, 14.
22. Walker and Rastrick 1829, 11, 55.
23. Stephenson and Locke 1830; Chrimes 2019, 68. Locke was another son of the north-east of England, and went on to build much of the British network as well as railways in France and Holland. These documents were published together as Walker, Stephenson, Lock and Booth 1831 in Philadelphia.
24. Wood 1838, 271.
25. Thomas 1980, 108–14. It is not entirely clear how this operated. Whishaw's description is unenlightening and may not reflect the original design. He states that 'When the

train reaches the top of the Crown Street tunnel Incline, the end of the rope is brought back by a small four-wheeled carriage, called a pilot, drawn by one horse; an operation which occupies about five minutes' (Whishaw 1842, 197).
26. Osborne 1921, 245.
27. He had taken his mathematics degree at the age of 17 and already had several years' experience of canal work behind before he went to England. Stapleton (1978, 65) states that he went to England independently; Roberts (1872, 582) states that he was already an employee of the Commonwealth of Pennsylvania.
28. Roberts 1872, 582.
29. Jones 1835.
30. Mountford 2013, 94.
31. Poussin 1836, 102.
32. Wild 1840, Plate 4.
33. Wayt 2016, 26.
34. Wayt 2016, 25–6.
35. Dilts 1993, 146–9.
36. Gamst 1997, 625.
37. Wilson 1883.
38. Most of the goods moved on the Allegheny Portage Railroad were transported on canal barges on railed flats.
39. Gamst 1997, 213.
40. Tomlinson 1915, 243–5. For Stephenson's involvement in the Stanhope and Tyne's design, see Bailey 2003a, 65–7.
41. Baader 1822, 138–9, tafel XV; Hughes 1990, 109.
42. Gamst 1997, 239–40; Lee 2008, 11; Gamst 1997, 733.
43. Monroe 1914, 221.
44. Smith 1888, 14.
45. *Durham County Advertiser*, 30 December 1826.
46. Bailey 1978–9, 128–9; Bailey 1980–81, 173–4. Up until 1828, the value of fixed engines produced by Robert Stephenson & Co. had been twice that of its locomotive production; not all of these were for railway operation (Hills 2006, 249).
47. Mountford 2013, 94–100 (Cromford and High Peak), 91–2, 150 (Lime Street), 145–7 (Stanhope and Tyne), 106–10 (Leicester and Swannington); Young 1923, 130–2 (Brusselton); Hughes 1990, 118–19 (Ynysgedwyn).
48. Thomas 1980, 108–110.
49. Hills 1989, 142; Mountford 2013, 91–2, 150.
50. Mountford 2013, 147–9.
51. Welch 1835; Toogood 1973, 167–9; Poussin 1836, 148.
52. United States Department of the Treasury 1838, 9–10.
53. On 20 May 1833, the Board approved contracts for two engines each from Linton Rogers, Smith and Minis, Boyle and Johnston, McClurg Pratt and Wade, and Stackhouse Tomlinson & Co (Toogood 1973, 11).
54. Roberts 1872, 582 (Allegheny); United States Department of the Treasury 1838, 169 (Philadelphia and Columbia), 263 (Charleston and Hamburg); Monroe 1914, 221, 227 (Mohawk and Hudson); Dilts 1993, 147 (Baltimore and Ohio).
55. Bousson 1863, 333.
56. The following descriptions derive from Mountford 2013, 139–50.
57. Whishaw 1842, 51; Fellows 1930, 67.
58. Mountford 2013, 145–50.
59. Skempton 2002, 275–6.
60. Whishaw 1842, 43.
61. The 1822 patent is 4699 of 16 April. For the Épinac inclined plane, see Minard 1834, 55–6.

62. Roebling 1843, 322. These had been acquired from 'four establishments of reputation', Durfee & May of Hudson, New York, Tucker & Carter of New York City, Jacob Dunton and William Ker, both of Philadelphia (Toogood 1973, 12).
63. Gamst 1997, 572.
64. Mountford 2013, 41; Tomlinson 1915, 378.
65. An endless iron chain was apparently used on an underground incline plane at a colliery near Dukinfield in Cheshire in 1822 (*Morning Chronicle*, 11 November 1822).
66. Harland 2013.
67. *Derby Mercury*, 23 December 1829.
68. 'La remorque s'effectue au moyen d'un câble en fer sans fin' (Poussin 1836, 102).
69. Lehigh Coal and Navigation Company 1839, 31.
70. Mountford 2013, 150.
71. Thomas 1980, 199–201.
72. Dilts 1993, 146–9.
73. After it had been operating for over a year on the short summit section, the Bolton and Leigh's *Lancashire Witch* was driven up the 1/30 Chequerbent incline, three-quarters of a mile long, in December 1829 (*Worcester Journal*, 24 December 1829). *Aris's Birmingham Gazette*'s description of the opening of the first section records that the locomotive operated on the section between Pendlebury and the top of the Daubhill plane (11 August 1828). The plane continued to be wound by a fixed engine for some time to come (Whishaw 1842, 43). A Norris locomotive was driven up the 1/15 Belmont inclined plane at Philadelphia in 1836 (White 1997, 71–2). See also Norris (1838, 17) for a later demonstration on the Belmont plane.
74. Young 1923, 177–8.
75. Fixed engine haulage to and from Lime Street lasted until 1870 (Thomas 1980, 116–19, 235; Reed 1996, 86).
76. Bailey 2003b, 193–6.
77. Bennett, Clavering and Rounding 1990, 12. The last major standard gauge incline plane in Britain was constructed in 1955, at a chemical works in Whitehaven in Cumberland (Mountford 2013, 314–15). An inclined plane was constructed as late as 1970 in a slate quarry in the Corris area of Wales (personal knowledge).

Chapter 8 'That truly astonishing machine': locomotives 1815–1834

1. Abbé Jean-Antoine Dubois (1765–1848), Catholic missionary and Hindu *sannyasi* (Simmons 1991, 14).
2. Bailey and Glithero 2000, 9; Vignoles 1889, 185.
3. Rees 2001.
4. Guy 2001, 138.
5. Wood 1838, 358–9.
6. Bailey 2019. See also Paxton 2001, 94; Reynolds 1980, 1999, 2003; Rees 2003, 177.
7. Warren 1923, 53–9; Bailey 2003a, 11–21. See also Bailey 1978–9, 111–12; Rolt 2016, 89–119.
8. Rees 2003.
9. Ahrons 1927, 1–3 and personal communication Dr Michael Bailey.
10. Ferguson 2006, 179–82; Bailey 2014, 33–5; Martin 2021. The locomotives built c. 1848 were for Hetton Colliery. One operated until 1912 when it was preserved.
11. Tomlinson 1915, 118–19; Achard and Seguin 1926b, 66–7 and plate XI.
12. Tomlinson 1915, 143–8; Warren 1923, 116. A contemporary model survives in the Science Museum: object number 1985–8741.
13. Achard 1926, 80 corrects Warren 1923, 135; see also Forward 1943–4; McNair 2019.
14. Bailey 2021 charts the history of the independent locomotive manufacturing industry in Britain, which until mid-century constructed the great majority of the locomotives used on British, Irish and continental European railways as well as many for the USA.

15. For the 1828–9 experimental locomotive, see Kobayashi and Cotte 1997, which references Archives départementales de l'Ardèche 41J 268-1, December 1828 and 267, January 1829. For the others, see Achard and Seguin 1926a, 1926b; Payen 1988; Kobayashi and Cotte 1997. The identity of the locomotive shown as plate XIII in Achard 1926 has never been confirmed.
16. *Rocket* was trialled at Killingworth from 3 to 5 September 1829 (Bailey and Glithero 2000, 18).
17. Achard and Seguin 1926a, 1926b.
18. Smiles 1857, 285, 287; Booth 1980, 83. It is hard to imagine the scrupulous Henry Booth making an unwarranted claim but he may not have been aware of the extent of discussion between Stephenson and Seguin.
19. The precise sequence is debated. See Fort 1989a, 1989b. Raybould (1968) provides the context for the Earl's mineral estate and its development.
20. *The Agenoria* has other innovative features as well, including a spring safety valve, wheel-mounted balance weights and a mechanical lubricator. It may have been built with a fusible plug.
21. Allen 1884, 14–15, 16–20.
22. *Morning Courier and New York Inquirer*, 12 June 1829. The West Point Foundry was based in Cold Springs but had a fitting and a machine shop in New York City.
23. 'Light engine' means without a train.
24. Withuhn 2006; Allen 1884, 20–2; White 1997, 239–47.
25. Bailey 1980–81.
26. Bailey and Glithero 2001, 173.
27. For the 'ordeal', see Wood 1838, 301–33; Thomas 1980, 63–75.
28. Thomas 1980, 68–9; Fletcher 1891, 68–74; Dendy Marshall 1953, 187–92; Abbott 1989, 30–2; *Mechanics' Magazine* 4, 15 October 1825, 12, 10 October 1829, 114–16, 31 October 1829, 163.
29. Warren 1923, 205–8 (quote at 208); Ahrons 1927, 13–16. Young (1923, 180–6, 195–209) is more sympathetic to Hackworth.
30. Dendy Marshall 1953, 181–5; Thomas 1980, 155–6; Harkort 1833, plate 4.
31. Calkins 1867, 49.
32. Ahrons 1927, 128–32.
33. Bailey and Glithero 2000.
34. Bailey 2003b.
35. Oeynhausen and Dechen 1829, 40–2.
36. Warren 1923, 120–6.
37. Warren 1923, 140–9.
38. Withuhn 2006, 157–62.
39. Warren 1923, 152–9.
40. Lewis 2020, 65–73, 120. Thirteen were built; the last was *Bedwellty* in 1853, which worked until about 1882.
41. Bailey and Davidson 2020.
42. Thomas 1980, 152.
43. Kemble (1879, 278–84) quotes a letter she wrote the day after her trip on the footplate. The locomotive was equipped with a sight glass, suggesting that it was not the *Rocket* (Kemble 1879, 281); Bailey and Glithero 2000, 85.
44. This was the *Dreadnought*, placed on the Liverpool and Manchester Railway on 13 March 1830 for trials on ballasting duties before being sold to the Bolton and Leigh. It is said to have had six wheels and horizontal cylinders, and to have been 'a clumsy machine with a chain to an accelerating wheel' (Warren 1923, 245; Dendy Marshall 1953, 203; Thomas 1980, 56).
45. Warren 1923, 245, 256–7, 260; Ahrons 1927, 18–19. The derailment and fatalities were reported in the press (*Lancaster Gazette* 30 July 1831; *Derby Mercury* 3 August 1831;

Sheffield Independent 6 August 1831) in terms which leave no doubt as to the identity of the locomotive. Thomas (1980, 179) states that it was returned to service. *The Engineer* for 3 November 1898 claims that it was sold to the Petersburg Railroad in Virginia where it was at work in May 1831. The Petersburg Railroad did indeed have a Bury locomotive of this name but it was built in 1833 (United States Department of the Treasury 1838, 237).
46. Tomlinson 1915, 187–8.
47. Young 1923, 233–8. See Hopkin (2010) for the history of Hackworth's business after 1830.
48. The *Morning Post* for 26 September 1834 and the *Morning Advertiser* for 29 September 1834 recount the struggle that *Nelson* had to move the inaugural train on the Leeds and Selby, when it 'sent forth groans resembling those of an elephant'.
49. Virginskiï 1971; Bailey 2014, 75–6. A description of the boilers of *Rocket* and *Novelty* translated from German had appeared in *Gornyi Zhurnal* in 1832 (4, 137–53). It is possible that Demidov had obtained plans or drawings at some stage.
50. Vignoles 1889, 185.
51. Ahrons 1927, 29–30; another locomotive to this design went to the Liverpool and Manchester.
52. Ahrons 1927, 27–9.
53. Ahrons 1927, 33–47.
54. Young 1923, 247–84.
55. *Herepath's Railway Journal* (*The Railway Magazine and Steam Navigation Journal*), 6 (1839), 275; Ahrons 1927, 48–9. Ferguson (2006, 189) shows *Trotter*, a locomotive with a partial return flue boiler and with vertical cylinders, operating a jackshaft drive to a single driving axle on a 2-2-2-0 wheel arrangements built by the Dundee Foundry for the Dundee and Newtyle in 1834 but Whishaw (1842, 85) states that all three of the railway's locomotives were the 'modified Planet' type referred to above.
56. Ince 1990, 2001. Lewis (2019b) discusses an early, probably never built, Neath Abbey locomotive design; Lewis (2020) describes the use of Neath Abbey locomotives in South Wales.
57. Coste and Perdonnet (1830, 162) confirms Price's attendance at Rainhill. See also *Cambrian* 31 July 1830.
58. Thomas 1980, 151.
59. West Glamorgan Record Office: D/D NAI L/36/1-3, L/38/7.
60. Lewis (2020, 77–8) provides the evidence for these two locomotives.
61. Rattenbury and Lewis 2004, 63–9.
62. Rattenbury and Lewis 2004, 64; Lewis 2020, 97–9.
63. Rattenbury and Lewis 2004, 69–71.
64. Messenger (2012, 146) gives the cost of the first Bodmin and Wadebridge Neath Abbey locomotive as £720; Rattenbury and Lewis (2004, 73) state that Neath Abbey's *Dowlais* cost £604 19s 1d; Dendy Marshall (1953, 160) gives £800 as the price of a 'Planet', £1,000 for a 'Samson'.
65. Pease 1907, 390; Ince 1990, Ince 2001, 63–6; Rattenbury and Lewis 2004, 63–79; Lewis 2020, 73–130.
66. Vertical boiler locomotives were built for industrial use until 1957 in the United Kingdom. Vertical boilers were also commonly used on railway breakdown cranes, which in some cases remained in use well after the last steam locomotives were retired.
67. Thomas 1980, 56–7, 65; a plateway locomotive similar to *Twin Sisters* was designed for Bute Ironworks in South Wales but built with a horizontal boiler (Lewis 2020, 109–10).
68. Dendy Marshall 1953, 214–16; Abbott 1989, 113–14, 75–77.
69. Bailey 2014, 77–8.
70. Dilts 1993, 92–4.
71. Brown 1871, 146–63; White 1997, 512–13.

72. Dilts 1993, 98.
73. Dilts 1993, 196–201; White 1997, 71, 95.
74. Dilts 1993, 207.
75. Dendy Marshall 1953, 145; Bailey 1978–79, 127; White 1997, 66.
76. Baldwin Locomotive Works 1920, 10; White 1997, 8, 24, 249, 538. From 1835 George Washington Whistler and his English-born mechanic James Francis began to copy the Boston and Lowell's 'Planets' at the Locks and Canals Machine Shop in Lowell, Massachusetts (White 1997, 545).
77. White 1997, 7–12.
78. White 1997, 33–57.
79. White 1997, 21, 66, 86, 176; for David Matthew, see White (1997, 543).
80. Bailey 2014, 86–8; White 1997, 21, 86, 176.
81. White 1997, 34–7. Having been thus 'Americanised' it was ironic that it was renamed *John Bull*. In 1845 it was again rebuilt as a 4-4-0 and renamed *Rochester* (White 1997, 45).
82. Colburn 1871, 96; Warren 1923, 305. Gwyn (2004) explores the possibility that two engineers from Tredegar Ironworks, which had been running bogie waggons since 1821, formed part of the discussion.
83. White 1997, 248–68; Treese 2006, 24; Alvarez 1974, 88.
84. White 1997, 537–8.
85. Bailey 2014, 69–72.
86. White 1997, 151–2, 167–75. The *Herald* of the Baltimore and Susquehanna was another Stephenson 0-4-0 rebuilt to suit American circumstances, in this case as a 4-2-0, in 1832 (see Snowden Bell 1927; White 1997, 168).
87. White 1997, 167. Jervis clearly believed the subassemblies were swivelling bogies.
88. White 1997, 33–4.
89. Baldwin Locomotive Works 1920, 14–21; White 1997, 33–46, 167–9. For Jervis, see Finch (1930–1). For the *Experiment* see Jarvis Main Collection, document 100, Jervis Public Library, Rome, New York. For the Charleston and Hamburg's *E.L. Miller* see Baldwin Locomotive Works (1920, 15) and White (1979, 270); this was only Baldwin's second full-size locomotive and the first to utilise their patented 'half crank' in which the wheel formed an arm of the driving crank by the use of an offset extension of the axle fastened to a wheel spoke.
90. White 1997, 7–12.
91. White 1997, 83–90.
92. White 1994, 80–3; Allen 1884, 26–8; White 1997, 509–10.
93. Thomas 1980, 64.
94. Wood 1838, 448–9; for Brandreth, see *Dictionary of National Biography*.
95. Young 1923, 187–9.
96. Baader 1817, 50.
97. Baader 1822, 145–61, 173–7, table 16, figs 3 and 4.
98. Baader 1822, 157.
99. Thomas 1980, 65.
100. Young 1923, 93–4.
101. Young 1923, 92.
102. Badnall 1833.
103. *Mechanics Magazine* 1834, 9–14, 20–5, 36–7, 69–77, 116–18, 259–65; Meredith 1998. Badnall's papers concerning the 'undulating railway' are preserved in the University of Salford Archives and Special Collections: GB 427 RBP.
104. Anon 1826; Tomlinson 1915, 118; Bailey 2017.
105. Thomas 1980, 172, 174; Emerson 1909, vol. 3, 184, 187, 189, 191; White 1997, 142.
106. Dilts 1993, 98; White 1997, 118, 541.
107. Dilts 1993, 99; *Railroad Journal and Advocate of Internal Improvements* 1, 43, 20 October 1832.

108. White (1967, 3–4) states that this locomotive, the *George W. Johnston*, was completed in 1831 and won the $1,000 second prize at the trials. Dilts (1993, 99) says that it was built in 1832, in the aftermath of the trials, and was a rebuild.
109. Dilts 1993, 99.
110. Wilkins 1888, 187; Dendy Marshall 1953, 217–22; Rattenbury and Lewis 2004, 60–1.
111. For Admiral Thomas Cochrane, Tenth Earl of Dundonald, Marquess of Maranhão, GCB, ODM (Chile) (1775–1860), senior British naval flag officer and radical politician, see *Dictionary of National Biography*. For the use of Dundonald's rotary engine on the *Rocket*, see Bailey and Glithero (2000, 35–8).
112. Bailey 1978–9, 113–15; Hills 2006, 255.
113. Pennington 1894, 21–38; Williams-Ellis 2004, 61.
114. Dawson 2004, 14–29.
115. White 1997, 20.
116. White 1984a; White 1997, 13–24.
117. *Lancaster Gazette* 21 August 1830.
118. The Prussian Field Marshall Gebhard Leberecht von Blücher had forced Napoleon on to the defensive after the Battle of Leipzig in 1813; when the locomotive was built, his crucial participation at Waterloo lay in the future. The British press picked up on his soldiers' nickname for him, 'Marshall Forwards'.
119. Bailey and Glithero 2000, 21.
120. Thomas 1980, 151.
121. Augustine of Hippo, *De Civitate Dei*, iv, 11, 16. Agenoria weds Labour ('work') and earns the praise of Politia ('civilisation') in a Latin fable by Pandolfo Collenuccio in 1497.
122. Bailey and Glithero 2000, 103; National Museums Liverpool: Archives Department B/VF/4/8.
123. White 1994, 74.
124. Brown 1871, 179–81.
125. Bick 1987, 68–9.

Chapter 9 Coal carriers 1815–34

1. Davies and Moorhouse 2003, 221; Muhle 2015, 179. Dr Karsten published a memorandum in August 1816 entitled *Über der Versorgung der Hauptund Residenz-städte Berlin und Bresslau mit steinkohlen* ('Concerning the Supply of the Capital and Royal Cities of Berlin and Breslau with Hard Coal'). It has not proved possible to locate a copy of this document.
2. Robertson 1983, 34–7; Skempton 2002, 661–9. His more ambitious coal-carrying schemes included an Edinburgh Railway (1818) from the Lothian collieries, a railway from Brechin to the sea at Montrose (1819), a railway or canal from collieries near Kirkaldy to Aberbrothock, Montrose and Aberdeen, and from Midlothian to Berwick-on-Tweed (1821) (National Library of Scotland accession 10706, nos 405, 415, 439, 458 and 441). For the Newton Colliery railway, see National Library of Scotland accession 10706, no 446.
3. Lewis 1970, 296; Banham 2001.
4. Hartley 2019; Bailey 2019.
5. Bailey 2019, 102.
6. Oeynhausen and Dechen 1829, 74–96. See also Mountford 2006, 2013, 74–80.
7. Mountford 1976, 2001, 2013, 289–92.
8. Skempton 2002, 701.
9. Oeynhausen and Dechen 1829, 10–74.
10. Stokes 2001.
11. Kirby 1993, 1–8, 26–53.

12. Tomlinson 1915, 80–1.
13. Tomlinson 1915, 112; Historic England list entry 1475481. The elegantly curving wing walls were added in 1829 and the bridge has seen much subsequent alteration.
14. Tomlinson 1915, 107.
15. Pease 1907, 91.
16. *Durham County Advertiser* and the *Newcastle Courant* 1 October 1825. The sketch is in NRM archive GB 756 1943-238 2005-7137. See also Tomlinson 1915, 109.
17. Tomlinson 1915, 106–15.
18. Tomlinson 1915, 153–6.
19. Moorsom 1975; Kirby 1993, 73–4.
20. It operated for about 15 months until it was relegated to serving as a workmen's cabin at the foot of the Brusselton Incline before being accidentally burnt down (Tomlinson 1915, 129; Young 1923, 122–4).
21. Tomlinson 1915, 122–30; Young 1923, 122–7. Although the *Globe* locomotive (chapter 8) was built in 1830 for passenger service, it was not for several more years that locomotives were regularly used to haul trains of coaches on the Stockton and Darlington. The surviving manager's reports for 1830 to 1833 are full of the difficulties of running a railway that mixed horse and locomotive haulage (NRM: GRA/1/39).
22. Skempton 2002, 85–8; Tomlinson 1915, 182–3, 187; see Jones (2019) for the difficulties of railway suspension bridges.
23. Stokes 2001, 2003, 2010.
24. Priestley (1831, 81) refers to 'the conveyance of coal, slate, stone, and other commodities, from the interior of the country to the port of Liverpool, by the Leeds and Liverpool Canal from Leigh; and the return of corn, iron, lime, and merchandize from the above port, and from Warrington and other places, to Bolton, Bury, and their populous environs'.
25. Dendy Marshall 1953, 214–16.
26. *Manchester Mercury*, 5 August 1828.
27. Pennington 1894, 1–10; Williams-Ellis 2004, 61.
28. Priestley 1831, 553–7.
29. For Scotland as a whole, see Robertson (1983, 44–96). See Brotchie and Jack (2007) and Ferguson (2006, 177–9) for the Fife coalfield. For Lanarkshire, see Ferguson (2006, 179–82) and Martin (1976).
30. *Glasgow Herald*, 19 May 1826; *London Courier and Evening Gazette*, 11 February 1828.
31. *The Scotsman*, 14 May 1831.
32. *Caledonian Mercury*, 6 February 1832.
33. This section draws on Cowburn (2001).
34. Loi du 21 avril 1810 concernant les mines, les minières et les carriers.
35. De Gallois 1818.
36. Napoleonic legislation also continued to govern the construction of public works (Loi du 8 Mars 1810, la procédure d'expropriation et les indemnités prévues).
37. The engravings in Deroy and Motte (1836) show the railway passing through a suitably challenging landscape.
38. Mellet and Henry 1828. Mellet (1828) had also undertaken a translation of Tredgold (1825).
39. Seguin, Seguin and Biot (1826) is their outline of the proposed works submitted to the Director-General of the Ponts-et-Chaussées et des Mines.
40. Combe, Escudié and Payen 1991, 11.
41. Lamming 2003.
42. Cotte 2003a.
43. Bouvier 1945, 100–3.
44. It was produced by 'Engelmann père et fils', a company set up in 1833.
45. Cotte 2003b.

46. Gueneau 1930, 1931a, 1931b.
47. Gueneau 1931b, 235.
48. Lewis 1970, 338–9; Mayberg, Blackarts and Andriessen 1845.
49. Harkort 1833, plate 1. Dr Egen, a *gymnasium* teacher and economist, became a member of the Leipzig–Dresden Railway Committee and prepared drawings for the Nuremberg–Fürth railway (https://www.deutsche-biographie.de/sfz12545.html), accessed 12 November 2019.
50. Lewis 1970, 338–9.
51. Chambers 1876, 9.
52. Chambers 1876, 17; Heydinger 1962.
53. Danville and Pottsville Railroad Company 1831; Gamst 1997, 622–5.
54. Allen 1884, 14–15, 16–22, 1953; White 1976, 36–8; Withuhn 2006.
55. Hazard 1827.
56. A description appears in *Executive Documents of the House of Representatives at the Second Session of the Twenty-First Congress, begun and held at the City of Washington December 6, 1830. Vol 3*, document 53, 82-85. See also Gamst 1997, 611–12.
57. *Niles Weekly Register*, 13 September 1828, 40–1.
58. Caldwell, Campbell and Brougham, 2014.
59. The shipping of coal from rail-served harbours remains important in the USA, Australia, India and China to this day. Railways retain an advantage over roads as coal carriers, particularly where long overland distances are involved.

Chapter 10 Internal communications 1815–1834

1. Telford 1810.
2. Henriet 1912, 108. These would have served Genoa, Bordeaux, Nantes, Le Havre, Calais, Ghent and Mainz.
3. Phillips 1817, 75–6.
4. Kirby 1993, 39–40.
5. Gray 1821 and subsequent reissues: Gray 1823, 1825.
6. Cumming 1824, 6, 33.
7. English 1827.
8. See e.g. Reynolds 2006.
9. Saunders 1949; Urquijo 2006, 12–23.
10. Nimmo 1826.
11. McNair (2007) is a detailed biography of William James.
12. McNair 2010. James may have seen one of Trevithick's locomotives at work, and had certainly witnessed the Middleton engines in operation in 1816 (McNair 2007, 33, 52, 58).
13. James 1823.
14. Norris 1987; Roper 2003, 27–30.
15. Fellows 1930. It nevertheless lasted until 1953. Its demise is referenced in the film *The Titfield Thunderbolt*: 'Perhaps there were not men of sufficient faith in Canterbury'.
16. Others, patented by his son, but in which he may have had a hand, were more ingenious but only became practical propositions many years later, and were mainly to do with the problems encountered by road and rail vehicles negotiating curves.
17. The account of the Budweis–Linz railway here draws on Weidmann (1842), Huyer (1892-3), Weidmann, Fobbe and Prechtler (1962), Freudenburger (1997), Gamst (1997, 3–15) and Sima (2008).
18. Rimmer 1985.
19. Oeynhausen and Dechen 1827, 109–14.
20. *Biographisches Lexikon des Kaisertums Österreich*.
21. For the rate of exchange, see Gerstner (1831, 10).

22. Taylor-Cockayne 2020.
23. Skempton 2002, 325, 362–4, 796.
24. Rimmer 1985, 7.
25. See Riden (1973) for the role of the Butterley Company in railway construction.
26. Gerstner 1831, vii.
27. This followed the publication of a book by the younger von Gerstner on the advantages and costs of the proposed railway (Gerstner 1824). See Kaiserlich Königlich Priv. Erste Eisenbahn-Gesellschaft 1830, 13.
28. Turnock 2004, 84.
29. Gamst 1997, 12.
30. Weidmann 1842, 61–4.
31. http://biography.hiu.cas.cz/Personal/index.php/BO%C5%BDEK_Josef_Jan_28.2.1782-21.10.1835, accessed 8 February 2019.
32. Gerstner 1831, 640.
33. Gerstner 1831, 37.
34. *Linzer Zeitung*, 23 July 1832.
35. Weidmann, Fobbe and Prechtler 1962, 116 ff; Sima 2008, 112–15.
36. Kaiserlich Königlich Priv. Erste Eisenbahn-Gesellschaft 1830, 6–7.
37. *Derby Mercury*, 3 February 1841.
38. *Derby Mercury*, 29 May 1833, 19 June 1833; *Derbyshire Courier*, 15 June 1833; Pigot 1818, 65, 433; 1828, 121, 148, 493, 709, 718, 899, 900, 1024, 1025, 1101, 1123; 1837, 32.
39. Stolberg-Wernigerode 1982.
40. Weidmann 1842, 20–47; Sima 2008; Steingruber 2021.
41. Sima 2008, 7–8. Zolla, also known as François Zola, was the father of the novelist Émile Zola.
42. It was on this section that steam locomotives were ultimately employed, from 1854 onwards (Ransom 1996, 49–50).
43. Freundl 2016, 56; Ardeleanu 2009.
44. Hodgkins 2003.
45. Railway & Canal Historical Society Chronology.
46. Guldi 2012, 199.
47. Obituary, *Minutes of the Proceedings of the Institution of Civil Engineers* 16, issue 1857, 1857, 150–4.
48. Haywood 1969.
49. Baranyi 1968. 377.
50. Geršlová 2003.

Chapter 11 The first main lines 1824–1834

1. Technical changes included structural engineering, power generation and transmission, machine tools, gas lighting, developments in the chemical industry and elevators (Cardwell 1994, 228).
2. Booth 1830, 89–90. Packet ships were so called because they were contracted by governments to carry mail as well as passengers, goods and time-limited items such as newspapers. A related development was the growing use of steamships on ocean voyages. The first to cross the Atlantic was the American *SS Savannah* in 1819, though the engine was only used on part of the voyage. Claimants to the title of the first ship to make the transatlantic trip substantially under steam are the British-built but Dutch-owned *Curaçao*, from Hellevoetsluis to Surinam in 1827, and the Canadian *SS Royal William* in 1833. Steamships also plied Asian waters from 1819 (Gibson-Hill 1954). Booth's (1830) account was published in Philadelphia the following year (Booth 1831).
3. Richards 1972; Booth 1980; Thomas 1980, 11–32.

4. Fitzgerald 1980, 1–13. The 1824 proposal would have led from Princes Dock at the northern end of Liverpool by means of an inclined plane at Everton leading ultimately to a terminus at Quay Street in Manchester.
5. Abolition of the Atlantic slave trade had not brought to an end the coastal traffic in enslaved people from one state of the USA to another.
6. Grinde 1976, 84–5. As early as 1805 a Georgia newspaper had carried an account of the Surrey Iron Railway (*Augusta Chronicle and Gazette of the State*, 19 October).
7. Terry Sharrer 1982.
8. See Brown (1909) and Kent (1925) for the Brown family of Ballymena (Co. Antrim), Liverpool, Baltimore, Philadelphia, New York and Boston. The Liverpool representatives were British agents for the transatlantic packet line inaugurated in 1822 by Thomas Cope as Philadelphia's response to the New York–Liverpool Black Ball line.
9. Dilts 1993, 128, 131.
10. Dilts 1993, 128.
11. In 1830 £1 was worth $4.56. The Liverpool and Manchester cost £600,0000, the Baltimore and Ohio $4,000,000, the Charleston and Hamburg a mere $951,140, though still considerably in excess of the original estimate of $600,000 (Grinde 1976, 91). Only eight other engineering projects in the United Kingdom had cost more than the Liverpool and Manchester: the Royal Canal in Ireland, the Worcester and Birmingham, the Grand Junction, the Birmingham and Liverpool Junction and the Caledonian canals, Plymouth Breakwater, Sheerness Dockyard and Kingstown Harbour (Skempton 2002, 834–6).
12. https://www.ucl.ac.uk/lbs/person/view/8961, accessed 18 April 2018.
13. Booth 1980, 12, 51; Allen 1953, 117–18. The argument was forgotten over cards, but Allen was shocked to see clergymen playing.
14. Thomas 1980, 29; Skempton 2002, 690. Reynolds (2011) describes how Manchester investors became reluctant to put money into railway schemes after the economic collapse following the boom of 1824–5.
15. Stephenson miscalculated on the Liverpool and Manchester but was clearly an able surveyor and was mostly supportive of his colleagues, as Hartley (2016b, 2017a, 2017b) makes clear.
16. Dilts 1993, 140–4.
17. For the Irish in England in this period, see Thompson (1980, 469–85).
18. Thomas (1980, 55), quoting from the papers of the Armenian engineer and archaeologist, Joseph Hekeyan Bey (1807–75).
19. Dilts 1993, 62–80, 132–3, 176–81.
20. Dilts 1993, 62, 162–3.
21. Grinde 1976, 88.
22. *The Georgia Constitutionalist*, 9 August 1833.
23. Martineau 1837, vol. 2, 9.
24. Martineau 1837, vol. 2, 11.
25. Crockett 1835, 15–16. A 'tizzick' is a consumptive cough.
26. Alvarez 1974, 74–5. *The Georgia Constitutionalist* for 9 August 1833 claimed that this had been anticipated from the start.
27. Gamst 1997, 715. The enslaved men owned by the Company were entered as assets in published inventories (see e.g. *American Railroad Journal* 4, 1835, 6.
28. Thomas 1980, 108–11; Dawson 2021.
29. Thomas 1980, 199–200. Banking or 'help-up' locomotives were used instead.
30. Dilts 1993, 146–9.
31. Thomas 1980, 244–5.
32. Dilts 1993, 97–9.
33. Wayt 2016, 35.

NOTES to pp. 245–252

34. *Charleston Courier*, February 1830. See also White 1984b, 1997, 518. Miller took out patents for a vertical boiler on 21 June 1830 and for improving the adhesion of locomotive driving wheels on 19 June 1834 (United States Patent Office 1847, 153, 213).
35. In the words of the *Georgia Constitutionalist* (9 August 1832) the eight-wheelers were 'more liable for derangement'.
36. For goods rolling stock on American systems, see White (1993).
37. Dilts 1993, 80.
38. Bury 1831, plate 7; Thomas 1980, 180–2; Donaghy 1972, 59–63.
39. Brown 1871, opp. 146.
40. Thomas 1980, 183.
41. 'The carriage in which the Duke of Wellington and other distinguished strangers will travel on the railway will be truly magnificent. Messrs Edmonson of Liverpool, are preparing it. The floor is 32 feet long by 8 wide, and is supported from 8 large iron wheels. The sides are beautifully ornamented, superb Grecian scrolls and balustrades richly gilt, supporting a massy hand-rail all around the carriage, along the whole centre of which an ottoman will be the seat for the company. A grand canopy 24 feet long is placed aloft on gilded pillars, and is so continued as to be lowered for passing through the tunnel. The drapery is of rich crimson cloth, and the whole is surmounted by the ducal coronet' (Manchester Central Reference Library: BR Q 942 72 L44).
42. As chapter 3 notes, in 1826 James Renwick of Columbia University had proposed independent subassemblies for railed vehicles on the Morris Canal in New Jersey, and the Quincy granite railway was carrying large blocks of stone on a timber frame connecting two flats from 1828.
43. Dilts 1993, 70.
44. Thomas 1980, 69–70.
45. White 1993, 160–2, 166; Whiting 1853, 112–17, 128.
46. Dilts 1993, unpaginated illustrations between pp. 158–9.
47. Brown 1871, 102–3.
48. Dilts 1993, 148–9, 183–4, unpaginated illustrations between 158–9; White 1978, 10–11.
49. Thomas 1980, 108, 136; Dickinson 1969, 60–4.
50. Dilts 1993, 197–9, 207.
51. Snowden Bell 1912, 22.
52. Kirby 1993, 50–3, 79–80.
53. Jarvis 1991, 1–16.
54. Booth 1980, 46–51.
55. Thomas 1980, 139.
56. Thomas 1980, 186–207.
57. Thomas 1980, 148, Donaghy 1972, 138.
58. White 1997, flyleaf.
59. Alvarez 1974, illustrations following 67. No date is given but the engineer's name is given as 'H. Allen', suggesting 1834 at the latest.
60. Samson and Previtts 1999.
61. Dilts 1993, 142. Though Richards and MacKenzie (1986, 39) date the surviving octagonal Mont Clare building to 1830, Harwood (1994, 54, 60) dates it to *c.* 1850–1.
62. E. Weber, *Ellicott's Mills drawn from natur* [sic], *& lithographed by Ed. Weber*, Baltimore, *c.* 1836: print; colour; 44 x 55 cm; G. Harvey, *Scene of the Baltimore & Ohio Railroad and the Chesapeake & Ohio Canal at Harpers Ferry, Virginia*, *c.* 1837–40, oil on panel, 18¼" x 24" (46.36 × 60.96 cm).
63. Library of Congress Prints and Photographs Division, LC-USZ62-52635 (b&w film copy neg.).
64. Topham 1925, 179. In May 1850, the Washington Board of Aldermen and Common Council approved a resolution calling for the relocation of the Baltimore and Ohio depot to a more suitable location east of New Jersey Avenue.

65. *Georgia Constitutionalist* 9 August 1832; South Carolina Archives: Plate of Hamburg, South Carolina.
66. Aitken and Summerville were the first two towns in the USA to be founded by a railroad company (Wayt 2016, 20).
67. Wosk (2013, 30–104) and Livesey (2016, 56–88) discusses 'the March of Intellect'.
68. Grant 1938–41, 127.
69. Bury 1831; Gwyn 2019b. The sketches 'Railway Office, Liverpool', 'View on the Line of the Liverpool and Manchester Railway', 'View on the Liverpool and Manchester Railway with the Locomotive "Twin Sisters" in a Siding', 'Bridge over the Irwell and Entrance into Manchester', 'Chat Moss' and 'The Viaduct over the Sankey Canal' are now in the Yale Center for British Art, New Haven, CT. Later sketches by Shaw which have not survived were turned into hand-coloured aquatints by S. G. Hughes; these are the well-known side views of *Jupiter*, *North Star*, *Liverpool* and *Fury* with their trains, *Travelling on the Liverpool & Manchester Railway: A Train of the First Class Carriages, with the Mail and A Train of the Second Class for Outside Passengers*, and *Travelling on the Liverpool & Manchester Railway: A Train of Waggons with Goods and A Train of Carriages with Cattle*.
70. Hollinghurst 2009.
71. Fitzgerald 1980, 11, 56–7.
72. Fitzgerald 1980, 29–49.
73. Triumphal arches and city gates began to replace town walls and defensive towers in many European cities in the eighteenth century. Buildings were laid out on wide thoroughfares. Government-sponsored roads served to impress the might of central power on potentially fractious provinces.
74. Walker 1830, 19.
75. Bacon 1620; Webster 1975, 12; Gwyn 2019b.
76. Dilts 1993, 10–11.
77. Dilts (1993, between pp. 80 and 81). *Maestoso con moto ben marcato* appropriately translates as 'in a majestic manner with accentuated movement'.
78. Dilts 1993, 100–1.
79. King William and Queen Adelaide could not have accepted an invitation so shortly after the death of King George IV on 26 June.
80. For István Széchenyi, 'the greatest of the Magyars', see Baranyi (1968, 273–4, 338–9).
81. Thomas 1980, 89.
82. Elton 1963.
83. Duffy 1982–3, 56; Fitzgerald 1980, 58
84. It is hard not to draw an analogy with the Moscow Metro, opened by Stalin with his customary sinister charm in 1935, another railway built to show what a new social order could accomplish.
85. Dilts 1993, 128–9, 136.
86. Latrobe 1868, 14; Snowden Bell 1912, 3–4.
87. Green 1963, 214. On 25 August 1835 he watched the formal opening of the Washington branch from a nearby hill (Dilts 1993, 184).

Chapter 12 Coming of age: the public railway 1830–1834

1. In 1833, the Tsar promulgated a general plan for building a network of major roads, with little result, due to a lack of funds, engineers and labour. See Haywood (1969, 24).
2. It is possible that Belgian enthusiasm for railways owed something to Thomas Gray, the English advocate of iron railways (Gray 1821, 1823, 1825). By 1815 he was living in Brussels where he was soon advocating a general European railway system (Wilson 1845, 14–20). Anon (1866) summarises Gray's career.
3. Weill 1896.

4. Simons, Nicolas and de Ridder 1833.
5. Messenger 2012.
6. Gamst 1997, 611-626; Gueneau 1930, 1931a, 1931b; Stokes 2003, 2006; Clinker 1954; Tomlinson 1915, 240–9.
7. Robertson 1983, 44–69.
8. The lithographs were published as Hill (David Octavius), *Opening of the Glasgow and Garnkirk Railway View near Proven Mill Bridge looking West*, *Opening of the Glasgow and Garnkirk Railway View at St. Rollox looking South-East*, *Opening of the Glasgow and Garnkirk Railway View of the Germiston Embankment looking West* [and] *Glasgow and Garnkirk Railway View of the Depot looking South*, Alexander Hill, Edinburgh, 2 January 1832.
9. Tomlinson 1915, 202–5; 253–60; Dawson 2020b.
10. Crompton 2014.
11. Bury 1831, additional plate A. The Braithwaite and Ericsson locomotive appears to be either *William IV* or *Queen Adelaide*, both of which are believed to have worked on the Liverpool & Manchester and not the St Helens and Runcorn Gap.
12. Roberts 1836.
13. Martineau 1837, 7–8.
14. Gamst 1997, 133.
15. Gamst 1997, 146.
16. 'Nothing but Passengers and their Luggage, and Parcels are taken nor is any other Branch of Traffic meditated.' By 1836 it was carrying a million passengers a year (The National Archives: RAIL 384/25)9.
17. Hardy 1834.
18. Gamst 1997, 245.
19. Gamst 1997, 252.
20. Baldwin 2001, 326–8; see also Tomlinson (1915, 355) for other 'wayleave' lines.
21. Cowburn 2001, 247.
22. Stokes 2003, 81.
23. *Moniteur belge / Belgisch Staatsblad* 1 May 1834, 25–8, Législation et jurisprudence des chemins de fer belges. Chemins de fer de l'Etat. 1834. Loi du 1er mai 1834.
24. Murray 1938, 24; Clinker 1954, 72.
25. Gamst 1997, 248–9.
26. Gamst 1997, 785–9.
27. Lewis 1960, 11; Gueneau 1931a, 1931b. The Épinac was funded directly by its owner, the banker Jacob-Samuel Blum.
28. 'Oh, Molesworth, great reformer, thou who promisest Utopia' (Trollope 1855, 153; Messenger 2012, 7–19, 51–2). Messenger (2014) discusses sources of finance within a peripheral, though mineral-rich, part of the United Kingdom.
29. Clinker 1954, 113; Donaghy 1972, 99–103.
30. Chandler 1954, 254.
31. Chandler 1954.
32. Boyes 1978.
33. Stokes 2003, 2010; Skempton 2002, 690–1.
34. Gamst 1997, 537.
35. Callender 1902; the quotation comes from p. 146. Between 1835 and 1837 states issued over 60 million dollars in debt for infrastructure. Indiana legislators adopted a 'Mammoth Internal Improvement' bill in 1836, which authorised the borrowing of $10,000,000 to give the state a network of canals, turnpikes and railroads. Maryland borrowed $15,000,000 for canals and railroads, and Pennsylvania $27,000,000.
36. Stoskopf 2000.
37. Dunham 1941.
38. Lewis 1968, 17–25, 1996.

39. 3 & 4 William IV c. 73.
40. Watkins 1892, 7 and plate opposite; Treese 2006, 22.
41. Trautwine 1835; Watkins 1891, 661, 696.
42. Gamst 1997, 144, 151–2, 179, 189, 220, 246, 274, 279, 300, 453, 500, 521, 529, 600, 604.
43. Gamst 1997, 530–1.
44. Gamst 1997, 151.
45. Library Company of Philadelphia, James McClees Philadelphia Photograph Collection (3)1322.F.123a.
46. Wilson and Roberts 1879, 71–3; Gamst 1997, 578–5; Harper 2002.
47. Gamst 1997, 138.
48. Gamst 1997, 255.
49. Gamst 1997, 245, 152.
50. Gamst 1997, 569.
51. Among overland railways built before 1834, the one which made longest use of horse traction was the Nantlle in North Wales, where 'Prince' and 'Corwen' were hauling runs on its short remaining stub until they were replaced by a road tractor shortly before its closure in 1963.
52. Tomlinson 1915, 158.
53. Haywood 1969, 164–5, 171.
54. Edgerton (2006, 33) points out that major twentieth-century railway companies owned almost as many horses as locomotives, mainly for road haulage but also for shunting. The most extensive horse-drawn railway systems were not installed until the twentieth century; these were prefabricated 'Decauvilles' on the henequen plantations in the Yucatán which ran to thousands of kilometres, as extensive as they were flimsy, and became the region's mass transit system. Some remain in use.
55. Bergin 1835.
56. Constantine and Frame 2012, 181–2; White 1978, 9; *American (Boston) Traveller* 14 April 1829; separate broadsheet. Curiously, though the elevation suggests that it ran on two double-axle bogies, the plan and the accompanying text suggests that each axle was independent and that the whole was kept on the rails by vertical guides, since the wheels had no flanges. Morgan patented a railroad car of which nothing is known in 1827.
57. Other than on plateways in south-east Wales and in the Forest of Dean and the short-lived carriages on the Stockton and Darlington and the Liverpool and Manchester, bogie vehicles were not used in the United Kingdom until 1846, when Richard Boyes Osborne (1815–99), an Irishman who had been engineer of the Philadelphia and Reading, used them on the Waterford and Limerick (Murray 1995, 56, drawing on National Library of Ireland mss 7888-7895; also MS 9606).
58. White 1978, 9, 12–13, 18.
59. Tomlinson 1915, 256.
60. Murray 1938, 35–6.
61. Alvarez 1974, 126–49; Marrs 2009, 150–5.
62. John Quincy Adams diary 39, 1 December 1832- 31 May 1835, 178; http://www.masshist.org/jqadiaries/php/doc?id=jqad39_178, accessed 4 February 2020.
63. Crockett 1835, 37; the Colonel, still no doubt mindful of his recent trip on the Charleston and Hamburg, performed this experiment on the Camden and Amboy while making his way to New York.
64. The plans in Tomlinson (1915, 258) derive from those accompanying Brees (1847), and it cannot be certain that they reflect arrangements of 1834. However, a tourist guide prepared shortly after the opening states that at Leeds the 'warehouses are admirably arranged, they are vastly extensive, and afford every possible convenience for the reception and transmission of passengers and goods', and the description of arrangements at Selby, with the six tracks entering a terminal building, and four projecting further to jetties on the Ouse, agrees with the plan published in Tomlinson (Parsons 1835, 60, 82–3).

65. National Library of Ireland. Kirkwood J *c.* 1834. *View of a map of the route of the railway from Westland Row to Killiney, Co. Dublin*; *The Opening of the Dublin & Kingstown Railway at the Rere [sic] of Entrance Westland Row Shewing Cumberland St. & arch, part of All Hallows Chapel with Brunswick Street in the distance.*
66. Martin 1981, 34–41.
67. Munsell 1875, 14–15.
68. Monroe 1914, 219–20.
69. Gamst 1997, 249.
70. Free Library of Philadelphia A917.481 P536 v.10; a marginalium notes '1832'.
71. https://www.philadelphiabuildings.org/pab/app/ar_display_projects.cfm/51793, accessed 13 March 2019.
72. *American Railroad Journal* 1833, 50.
73. View of New Orleans in 1836. Drawing by George Washington Sully. Sully panorama of New Orleans, 1836, Louisiana Research Collection, Tulane Special Collections, Howard-Tilton Memorial Library, Tulane University, New Orleans, LA, 70118.
74. Elysian Fields Avenue is the setting for Tennessee Williams's play *A Streetcar Named Desire*, but by 1947 'Smoky Mary', as the Pontchartrain was known, was no more.
75. Richards and MacKenzie 1986, 39
76. Fiero 1984.
77. Hills (2006, 255–6) records purchases of machinery by the Liverpool and Manchester.
78. Clinker 1954, 72, 78.
79. It soon established a more extensive 'Railway Coach Factory' in a former distillery by the Grand Canal Dock near the terminus (Whishaw 1842, 70). This is shown on the Irish 6 inch ordnance survey of 1837, where the Serpentine Road facilities are no more than 'old engine house'. See also Hardy (1834, 13–14).
80. *American Railroad Journal* 1834, 385.
81. Gamst 1997, 522–3.
82. A Londoner anxious to see a locomotive in operation in 1834 would have had to make a pilgrimage to Canterbury to see *Invicta* at work. Otherwise, the choice lay between Kingswinford in Staffordshire and Newport, Monmouthshire.
83. *Spectator* May 1834.
84. Anon, 1834. Toy trains made of sheet tin were being manufactured in France from 1835 (Freeman 1999, 212).
85. *Athenaeum*, 5 July 1834, 509.

Chapter 13 'The new avenues of iron road' 1834–1850

1. Emerson 1844, 481.
2. O'Rourke and Williamson 2002; Freeman 1999.
3. Bailey (2016) provides a context for technological development of mainline railways in the period 1830 to 1850 and suggests further areas for research. Gomersall and Guy (2010) also set out research questions for the early railway period.
4. Gamst 1997, 39.
5. Marrs 2009, 5.
6. Carver 1847.
7. Whitney 1848.
8. White 1994, 72; provincial statutes of Canada, cap. 29, 'An Act to provide for affording the Guarantee of the Province to the bonds of Rail-way Companies', 30 May 1849. See also MacDonald, 2006, 2010; MacDonald and Tennant 2016.
9. Chrimes 2019.
10. Hartley 2016b, 2017a, 2017b.
11. Buchanan 2002. Marc Brunel (1769–1849) initially made his way to the USA, where he became the chief engineer of New York City. In 1799 he came to England and set

up his block-making machines at the Royal Navy's Portsmouth dockyard. He designed and built the world's first underwater tunnel through soft ground, between Wapping and Rotherhithe from 1825 to 1843, for which he was knighted.
12. Boyes 2016, 30.
13. Cox and O'Dwyer 2016, 62–3.
14. The medal's motto *'Dant ignotas marti novasque mercurio alas'* may be translated as 'They [railways] give hitherto unknown wings to Mars and Mercury', emphasising both defence and communications in the national interest.
15. De Block 2011.
16. King 1991.
17. Specifically, Le Havre, Dieppe, Boulogne, Calais, Dunkirk, Ostend, Amsterdam, Bremen, Hamburg, Rostock and Stettin.
18. Haywood 1969, 24, 136.
19. Young 1923, 276–7. For his trip to Russia, young John Wesley Hackworth was prudent enough to change his middle name to William to avoid any suggestion of disrespect to established religion. National Railway Museum: HACK/1/3/5. See Guy 2003, 69–72.
20. Rieber 1990: Haywood 1969, 134–46.
21. See e.g. Destrem 1831.
22. Haywood 1969, 146–237.
23. White 1994, 67–9.
24. Oostindie 1984, 1988; Brehony 2013.
25. Saunders 1949.
26. Kerr 2016; Mukherjee 2016; Gwyn 2019a; Raman and Balakrishnan 2020. The first Indian railway ran 3.5 miles from Red Hill (Sengundram) to Chintadrapettah Bridge in Madras, carrying granite for road building to a waterway (*Asiatic Journal* November 1836 and August 1837; *Madras Gazette* 4 May 1836) using rails cast at Parangipettai (*The Standard* 23 August 1836).
27. Longworth and Rickard 2014, 2019.
28. Wang 2010.
29. Katsumasa 1979. A Japanese translation was made of van der Burg (1844).
30. Hawks 1856, vol. 1, 248, 464.
31. Lewis 2020, 161–2.
32. Bailey 2003a, 76–7, 193–6.
33. Bailey 2003a, 146.
34. Buchanan 2002, 90; *The Glamorgan Monmouth and Brecon Gazette and Merthyr Guardian* 3 October 1840, 6 March 1841; *Monmouthshire Merlin* 24 April 1841.
35. Ghega 1853, 72; Tusch 2019.
36. Robert 1843, 110–44.
37. Webster 1972, 110.
38. Elsas 1960, xviii–xix.
39. Stapleton 1978, 38–40.
40. Anon 1836, 4; Serlo 1878, 15.
41. Norris 1838 is a trilingual (English, French and German) publicity brochure.
42. This was a tiny 33-ton Baldwin loco for the United Railways of Yucatan (White 1997, 57).
43. Ghega 1853, 94–140.
44. Bailey 2014, 125.
45. A tank locomotive carrying its water and fuel on the one frame can work with equal facility in either direction, and any locomotive which works in a shunting yard has to do so. Locomotives can operate tender-first on mainline duties if necessary but tenders will vary in weight in the course of operations, with the result that they ride poorly, and a high load of fuel can impair visibility.
46. Reed 1982, 7–51; Cattell and Falconer 1995; Whishaw 1842, 234–8.

47. The National Archive: RAIL 384/68, 74, 76, 77, RAIL 1110/260; White 1997, 543. The first reference to a *rotonde* in the *Journal de Chemins de Fer* comes in 1847, vol. 6, 459.
48. Atmore 2004.
49. Ghega 1853, 72.
50. Cavicchi 2016; *American Polytechnic Journal* October–November 1854, 257–64 contains an engraving of the locomotive and an account of the trip.
51. Skempton 1996.
52. Haywood 1998, 107, 213.
53. Rosenberg and Vincenti 1978.
54. Chrimes 2003, 252–7; Buchanan 2002, 73.
55. Ghega 1854, 19
56. Buchanan 2002.
57. Haywood 1969, 229–30.
58. Veenendaal 1995, 7–9; Hilton 1990, 35; *Cincinnati Commercial* 4 January 1879.
59. Mountford, Young and Burdsall 2019; Buchanan 2002, 69, 90, 168.
60. Gamst 1997, 20.
61. Wolmar 2010, 18.
62. Haywood 1969, 232.
63. White 1993, 28–9.
64. Haywood 1969, 214–19, 234.
65. Brooke 1989, 38.
66. Skempton 1996, 34; Haywood 1998, 214.
67. Coleman 2000, 204.
68. *L'Illustration* 1 avril 1843.
69. Dilts 1993, 132–3.
70. They would be driving the American transcontinental Union Pacific through the Rocky Mountains in the 1860s (Brehony 2013; Oostindie 1988, 30).
71. Haywood 1998, 131–40, 196–7, 250.
72. *The Monthly Railway Record*, July 1847, 314.
73. Fördös 2017, 45. Saar (1874) is a novella which tells the story of the relationship between Georg Huber and Tertschka, two Semmeringbahn stone-breakers.
74. Marrs 2009, 55–83.
75. Dilts 1993, 138.
76. For an Irish–Welsh punch-up on the Chester and Holyhead, see *The North Wales Chronicle* for 26 May 1846. For an Irish–native 'melée' in Georgia, see Marrs (2009, 69).
77. Marrs 2009, 69.
78. Dawson, forthcoming.
79. Alvarez 1974, 141.
80. Tarsaidze 1950, 291.
81. Alvarez 1974, 40; Marrs 2009, 57–66.
82. Public Act, 7 and 8 Victoria, ch. 85. See Boyes 2016.
83. Marrs 2009, 178. A German immigrant considered a southern depot as little better than the pigsties in his native land.
84. Hartley 2016a.
85. Bourne and Britton 1839; Bourne 1846; Viau 1847.
86. Matsumoto-Best 2003, 32–4.
87. The first known instance was on 19 September 1841, during the pontificate of Gregory XVI, when Mgr André Raess pronounced benediction on the locomotives at Mulhouse in the presence of a largely Protestant gathering, an interfaith event which gave great joy to Michel Chevalier (Droulers 1983, 120, 124).
88. El Sayed and Gwyn 2016; Gwyn 2019a.
89. Simmons 1991, 318.

90. Haywood 1969, 178, 214–15, 217, 232n, 224.
91. Rudé 1981, 186, 169.
92. Chevalier 1836, 164.
93. Thomas Cook's 1841 temperance campaigners' train from Leicester to Loughborough was the first venture of what became a prosperous business but was not the first excursion, however defined. The Bodmin and Wadebridge ran special trains to enable people to watch public executions, the first in 1835 (Messenger 2012, 34).
94. Robertson 1983, 252–7; Simmons 1991, 282–93, Marrs 2009, 164–9.
95. Simmons 1991, 332–4; Alvarez 1974, 132, 138–40; Marrs 2009, 101, 151, 156–7, 177.
96. Gwyn 2003.
97. Marrs 2009, 153. The traveller was Frederick Law Olmstead, the landscape architect, best known for his work on New York's Central Park.
98. Marrs 2009, 110–12, 153, 155; Gamst 1997, 694. Frederick Douglass, who had escaped from enslavement by boarding a train at Baltimore, staged an early anti-segregation transport protest on the Eastern Railroad (MA) in 1841 (Douglass 1857, 346, 419–20).
99. The mechanisation of paper making in the nineteenth century promoted letter writing and made newspapers more easily affordable. As the Introduction points out, the iron railway developed in almost exactly the same time frame as the electric telegraph.
100. In Britain the mails had first been transported by train on the Liverpool and Manchester in 1830. In 1839 Parliament required railway companies to convey the Royal Mail throughout the UK at reasonable rates, and, thereafter, the expansion of postal services and railway networks went hand in hand. Gamst (1997, 148, 164, 166, 178, 185, 694, 701, 713) describes carriage of mails on American railways, some of them already making use of dedicated sorting carriages. See also Marrs (2009, 169–71). Tsar Nicholas I appreciated that railways could carry post as well as speedily communicate orders in the absence of a telegraph system (Haywood 1969, 232).
101. Marrs (2009, 198) discusses the embrace of the railway by the antebellum southern USA as the 'pursuit of modernity in the service of its own demands'.

Chapter 14 'You can't hinder the railroad'

1. Booth 1830, 89–90.
2. Hopkin 2001.
3. Bourne and Britton 1839; Bourne 1846.
4. The Rev. Dr Thomas Arnold, the reforming headmaster of Rugby School, watching a train from a bridge over the London and Birmingham, famously observed 'I rejoice to see it, and think that feudality is gone for ever' (Stanley 1860, 353). Anne Broome, recalling her enslaved childhood in South Carolina, described risking a beating to watch the train, and admitted that for years she had accepted her parents' story that she had been delivered to them on the locomotive's cow-catcher and not by a stork, still less by any other means (Marrs 172; Library of Congress: Federal Writers' Project: Slave Narrative Project, Vol. 14, South Carolina, Part 1, Abrams-Durant).
5. Thompson 2015.
6. Livesey (2016) suggests that road-coach travel in the writings of these authors offered an antidote to the displacements of modernity by weaving together a British nation 'out of strongly rendered, disjointed localities'.
7. Wordsworth 1845; *Morning Post* 16 and 24 October, 20 December 1844.
8. Wallach 2002. The Canajoharie and Catskill was an unsuccessful concern and had already closed following a bridge collapse by the time the painting was completed.
9. Booth 1830, 89–90.
10. Booth 1830, 90.
11. Schivelbusch (2014, 33–44) considers the theme of the annihilation of time and space by railway travel.

12. *Contre Vaudois: Journal de la Suisse Romande*, 16 July 1892, 1–2.
13. Emerson 1910, 305–6, 482.
14. Emerson 1910, 190–1.
15. Emerson 1910, 305
16. Cronkhite 1951, 308.
17. Thoreau 1854, 130.
18. Thoreau 1854, 126.
19. Thoreau 1854, 128.
20. Ruskin 1905, 60–1.
21. Ruskin 1905, 61–2.
22. Lee 1973. Newdigate is better known for having instigated a prize at the University of Oxford for composition in English verse.
23. Livesey 2016, 188–9; Bailey 2003a, 33–6.
24. The preliminary surveys of 1831 agree well with George Eliot's narrative preceding the Reform Bill of the following year.
25. Eliot 1872, 235–42. For the often haphazard recruitment of surveyors in this period, see Conder (1868, 12–44).

A Note on Sources and Terminology

1. Dendy Marshall 1938; Baxter 1966.
2. Cossons 2001.
3. Divall 2003.
4. Tomlinson 1915, 15.
5. Bangor University mss: Porth yr Aur 29599, 29619 and 29621.
6. Gwyn 2001b, 2002.

BIBLIOGRAPHY

Abbott, R.A.S., 1989. *Vertical Boiler Locomotives and Railmotors Built in Great Britain*. Oxford: Oakwood.
Achard, F., 1926. 'The first British locomotives of the St. Etienne-Lyon Railway', *Transactions of the Newcomen Society* 7, 1, 68–80.
Achard, F. and L. Seguin, 1926a. 'Marc Seguin and the invention of the tubular boiler', *Transactions of the Newcomen Society* 7, 1, 97–116.
Achard, F. and L. Seguin, 1926b. 'British railways of 1825 as seen by Marc Seguin', *Transactions of the Newcomen Society* 7, 63–7.
Ahrons, E.L., 1927 (1987). *The British Steam Railway Locomotive 1825–1925*. London: Ian Allen.
Alfrey, J. and C. Clark, 1993. *The Landscape of Industry: Patterns of Change in the Ironbridge Gorge*. London and New York: Routledge.
Allen, H., 1884. *The Railroad Era: First Five Years of its Development*. New York: privately printed.
Allen, H., 1953. 'Diary of Horatio Allen 1828 (England)', *The Railway and Locomotive Historical Society Bulletin*, 89 (November), 97–138.
Allison, W., S. Murphy and R. Smith, 2010. 'An early railway in the German mines of Caldbeck', in *Early Railways 4*. Sudbury: Six Martlets, 52–69.
Alvarez, E., 1974. *Travel on Southern Antebellum Railroads, 1828–1860*. Tuscaloosa: University of Alabama.
Andrieux, 1815. 'Description d'un chariot à vapeur [steam carriage] imaginé par M. Blenkinsop, pour le transport du charbon de terre', *Bulletin de la Société d'encouragement pour l'industrie nationale* (avril), 80–6, 121.
Angerstein, R.R., 2001. *R.R. Angerstein's Illustrated Travel Diary, 1753–1755*. London: Science Museum.
Anon, 1804 (*an* XII). 'Description des deux nouvelles espèces de chemins de fer pour les grosses voitures, et pour les voitures de voyages', *Annales des Arts et Manufactures* (30 Messidor), 89–96.
Anon, 1805 (*an* XIII). 'Notice sur diverses Méthodes nouvelles pour faciliter la Navigation intérieure dans les lieux où la pente est considérable', *Annales des arts et manufactures* 22 (30 Fructidor), 306–24.
Anon, 1809. 'Description d'un sas mobile, descendant et montant alternativement le long d'un plan incline', *Annales des arts et manufactures* 32, 94, 42–82.

BIBLIOGRAPHY

Anon, 1815. 'Chariot à vapeur destine au transport des charbons de terre, imagine par M. Blenkinsop', *Archives des découvertes et des inventions nouvelles*, 257–61.
Anon, 1819. 'Notes sur les *rail-ways* ou chemins de fer', *Annales des arts et manufactures* 2, 5, 205–10.
Anon, 1826. 'On locomotive steam carriages', *Newton's London Journal of Arts and Sciences* 2, 11, 142–6 and plate VII, fig. 4.
Anon, 1828. 'Rail roads in England &c., public documents concerning the Ohio Canals which are to connect Lake Erie with the Ohio River', *Civil Engineer and Herald of Internal Improvements* 1, 19, 298–9.
Anon, 1831. 'Description des chemins de fer mobiles employés en Suède dans les divers travaux de déblai, de construction et d'exploitation', *Journal des Voies de Communication* (1 Janvier), 54–62 and plates.
Anon, 1834. *Descriptive Catalogue of the Padorama of the Manchester and Liverpool Rail-Road.* London: Colyer.
Anon, 1835. 'Izvestiye o sukhoputnom parokhode ustroyennom v Uralskikh zavodah v 1833', *Gornyĭ zhurnal* 5, 445–8.
Anon, 1836. *Die Eisenbahnen. Eine faßliche Beschreibung der Bestandtheile, Darstellung des Nutzens und Geschichte der Eisenbahnen.* Leipzig: Ludwig Schred.
Anon, 1866. 'Not quite a man for a straight-jacket', *All the Year Round* 16 (28 July), 62–5.
Ardeleanu, C., 2009. 'From Vienna to Constantinople on board the vessels of the Austrian Danube Steam-Navigation Company (1834–1842)', *Romanian Academy History Institute Historical Yearbook* 6, 187–202.
Atmore, H., 2004. 'Railway interests and the "rope of air", 1840–8', *British Journal for the History of Science* 37, 3, 245–79.
Ayris, I., J. Nolan and A. Durkin, 1998. 'The archaeological excavation of wooden waggonway remains at Lambton D Pit, Sunderland', *Industrial Archaeology Review* 20, 1, 5–22.
Baader, F. von, 1857. *Sämmtliche Werke* 2, 5. Leipzig: H. Bethmann.
Baader, J. von, 1817. *Uber ein neues System der fortschaffenden Mechanik.* Munich: EA Fleischmann.
Baader, J. von, 1822. *Neues System der Fortschaffenden Mechanik.* Munich: Verfassers.
Baader, J. von, 1829. *Sur l'avantage de substituer des chemins de fer d'une construction améliorée à plusieurs canaux navigables projetés en France.* Paris: Bachelier.
Baader, J. von, 1830. *Huskisson und die Eisenbahnen: Noch ein Wort zu seiner Zeit für eine gute Sache.* Munich: F.G. Franckh.
Baader, J. von, 1832. *Vorschlag zur Herstellung einer Eisenbahn zwischen München und Starnberg in Verbindung mit einer Dampf-Schifffahrt auf dem Würmsee.* Munich: Franz.
Bacon, F., 1620. *Novum Organum.* London: John Bill.
Badnall, R., 1833. *A Treatise on Railway Improvements.* London: Sherwood, Gilbert and Piper.
Bailey, M.R., 1978–9. 'Robert Stephenson and Company, 1823–1829', *Transactions of the Newcomen Society* 50, 109–38.
Bailey, M.R., 1980–81. 'George Stephenson, locomotive advocate: the background to the Rainhill Trials', *Newcomen Society Transactions* 52, 171–9.
Bailey, M.R. (ed.), 2003a. *Robert Stephenson: The Eminent Engineer.* Aldershot: Ashgate.
Bailey, M.R., 2003b. 'The mechanical business', in *Robert Stephenson: The Eminent Engineer*, ed. M.R. Bailey, 163–210.
Bailey, M.R., 2014. *Locomotion: The World's Oldest Steam Locomotives.* Stroud: History Press.
Bailey, M.R., 2016. 'Technology on the move: engineering development on early main line railways', in *Main Line Railways.* Clare: Six Martlets, 17–29.
Bailey, M.R., 2017. 'Brunel's fan: his locomotive draught experiments of 1840/41', *International Journal for the History of Engineering & Technology* 87, 20–41.

BIBLIOGRAPHY

Bailey, M.R., 2019. '*Blücher* and after: a re-assessment of George Stephenson's first locomotives', in *Early Railways 6*. Milton Keynes: Six Martlets, 79–102.

Bailey, M.R., 2021. *Built in Britain: The Independent Locomotive Manufacturing Industry in the Nineteenth Century*. Market Drayton: Railway and Canal Historical Society.

Bailey, M.R. and P.H. Davidson, 2020. *Canterbury & Whitstable Railway* Invicta *Locomotive*. Whitstable: Whitstable Community Museum and Gallery.

Bailey, M.R. and J.P. Glithero, 2000. *The Engineering and History of* Rocket. London and York: National Railway Museum.

Bailey, M.R. and J.P. Glithero, 2001. 'Learning through restoration: the Samson locomotive project', in *Early Railways*. London: Newcomen Society, 278–93.

Baldwin, J.H., 2001. 'The Stanhope & Tyne Railway: a study in business failure', in *Early Railways*. London: Newcomen Society, 325–41.

Baldwin Locomotive Works, 1920. *History of the Baldwin Locomotive Works, 1831–1920*. Philadelphia: Martino-Pflieger Co.

Banham, J., 2001. 'Coal, banks and railways', in *Early Railways*. London: Newcomen Society, 297–310.

Baranyi, G., 1968. *Stephen Széchenyi and the Awakening of Hungarian Nationalism, 1791–1841*. Princeton, NJ: Princeton University Press.

Barker, T. and D. Gerhold, 1993. *The Rise and Rise of Road Transport 1700–1900*. Cambridge: Cambridge University Press.

Barritt, S. 2000. *Preston to Walton Summit: The Old Tram Road*. Lancaster: Carnegie.

Baxter, B., 1966. *Stone Blocks and Iron Rails*. Newton Abbot: David and Charles.

Beatty, F.M., 1955. 'The scientific work of the Third Earl Stanhope', *Notes and Records of the Royal Society of London* 11, 2, 202–21.

Beazley, E., 1967. *Madocks and the Wonder of Wales*. London.

Bennett G., E. Clavering and A. Rounding, 1990. *A Fighting Trade*. Gateshead: Portcullis Press.

Bergin, T.F., 1835. 'Description of the new buffer, invented and applied to railway coaches by Mr. T.F. Bergin 1835', *Dublin Penny Journal*, 4, 182 (26 December), 201–3.

Betancourt y Molina, A., 1807. *Mémoire sur un nouveau système de navigation intérieure*. Paris: Institut National de France.

Bick, D., 1987. *The Gloucester & Cheltenham Tramroad*. Oxford: Oakwood Press.

Biot, E., 1834. *Manuel du constructeur de chemins de fer*. Paris: de Roret.

Birkinshaw, J., 1821, 1822, 1824. *Specification of John Birkenshaw's Patent for an Improvement in the Construction of Malleable Iron Rails*. Newcastle: E. Walker.

Booth, H., 1830. *An Account of the Liverpool and Manchester Railway*. Liverpool: Wales and Baines.

Booth, H., 1831. *An Account of the Liverpool and Manchester Railway*. Philadelphia: Carey & Lea.

Booth, H., 1980. *Henry Booth: Inventor, Partner in the Rocket and Father of Railway Management*. Ilfracombe: Stockwell.

Borsay, P. 2001. 'London 1600–1800: a distinctive culture?', *Proceedings of the British Academy* 107, 167–84.

Bourne, J.C., 1846. *The History and Description of the Great Western Railway*. London: David Bogue.

Bourne, J.C. and J. Britton, 1839. *Drawings of the London and Birmingham Railway*. London: Bourne.

Bousson, M., 1863. 'Sur les resultants pratiques des different modes de traction et d'exploitation sucessivement employés sur les anciennes lignes de Rhône et Loire', in *Mémoires et documents relatifs à l'art des constructions et au service de l'ingénieur Tome 5, 4 ième série, 1ier semestre. Annales des Ponts et Chausses 1863: Tome 5*. Paris: Dunod, 314–82.

Bouvier, J.-A., 1945. 'La gare de Perrache. Ses conditions de site et les conséquences de son établissement', *Les Études rhodaniennes* 20, 1–2, 97–111.

BIBLIOGRAPHY

Boyd, J.I.C., 1981. *Narrow Gauge Railways in North Caernarvonshire. Volume 1. The West.* Oxford: Oakwood Press.

Boyd, J.I.C., 1991. *The Wrexham, Mold & Connah's Quay Railway Including the Buckley Railway.* Oxford: Oakwood Press.

Boyes, G., 1978. 'The Exchequer Bill Loan Commissioners as a source of canal and railway finance, 1817–42', *Journal of the Railway & Canal Historical Society* 24, 85–92.

Boyes, G., 2001. 'An alternative railway technology; early monorail systems', in *Early Railways*. London: Newcomen Society, 192–207.

Boyes, G., 2014. 'Working the Peak Forest Railway: some revised interpretations', in *Early Railways 5*. Clare: Six Martlets, 250—9.

Boyes, G., 2016. 'Early progress towards common standards for Britain's railways', in *Early Main Line Railways*. Clare: Six Martlets, 30–47.

Boyes, G. and B. Lamb, 2012. *The Peak Forest Canal and Railway: An Engineering and Business History*. Derby: Railway and Canal Historical Society.

Bradley, S.A., 2016. *The Railways: Network, Nation and People*. London: Profile.

Brees, S.C., 1847. *Railway Practice: A Collection of Working Plans and Practical Details of Construction in the Public Works of the Most Celebrated Engineers . . . on the Several Railways, Canals, and Other Public Works Throughout the Kingdom. Third Series.* London: Williams & Co.

Brehony, M., 2013. 'Free labour and "whitening" the nation: Irish migrants in colonial Cuba', *Saothar* 38, 7–18.

Brooke, D., 1989. 'The railway navvy: a reassessment', *Construction History* 5, 35–45.

Brotchie, A.W. and H. Jack, 2007. *Early Railways of West Fife: An Industrial and Social Commentary*. Catrine: Stenlake.

Brown, J.C., 1909. *A Hundred Years of Merchant Banking*. New York: privately printed.

Brown, W.H., 1871. *The History of the First Locomotives in America*. New York: Appleton.

Brunton, J., 1939. *John Brunton's Book*. Cambridge: Cambridge University Press.

Bruton, J.F., 1955–6. 'Some industrial steam locomotives and railways', *Transactions of the Newcomen Society* 30, 77–92.

Bruton, R.N., 2015. 'The Shropshire Enlightenment: A Regional Study of Intellectual Activity in the Late Eighteenth and Early Nineteenth Centuries'. PhD thesis, University of Birmingham.

Bruyn, A. de, 1987. *Histoire des Rues et des Lieux-Dits de la Commune de Saint-Nicolas*. Liège: Dricot.

Buchanan, A., 2002. *Brunel: The Life and Times of Isambard Kingdom Brunel*. London and New York: Hambledon and London.

Buchanan, R., 1811. *Report Relative to the Proposed Rail-way from Dumfries to Sanquhar*. Dumfries: Geo. Johnstone and Co.

Bury, T.T., 1831. *Six Coloured Views on the Liverpool and Manchester Railway*. London: Ackermann.

Butterfield, H., 1931. *The Whig Interpretation of History*. London: G. Bell & Sons.

Bye, S., 2003. 'John Blenkinsop and the patent steam carriages', in *Early Railways 2*. London: Newcomen Society, 134–48.

Bye, S., 2010. 'Regarding old rails: a Middleton Railway miscellany', in *Early Railways 4*. Sudbury: Six Martlets, 141–50.

Bye, S., 2014. 'As others saw us: some views of Salamanca', in *Early Railways 5*. Clare: Six Martlets, 243–9.

Caldwell, R., D. Campbell and J. Brougham, 2014. 'Australia's first railway: an entrepreneurial adaptation of skill and technology from leading international collieries', in *Early Railways 5*. Clare: Six Martlets, 54–73.

Calkins, W., 1867. *Memorial of Matthias W. Baldwin*. Philadelphia: Collins.

Callender, G.S., 1902. 'The early transportation and banking enterprises of the States in relation to the growth of corporations', *Quarterly Journal of Economics* 17, 1, 111–62.

BIBLIOGRAPHY

Cantle, G.S., 1978–9. 'The steel spring suspensions of horse-drawn carriages (*circa* 1760 to 1900)', *Transactions of the Newcomen Society* 50, 1, 25–36.

Cardwell, D.S.L., 1994. *The Fontana History of Technology*. London: Fontana.

Carlson, R.E., 1964. 'The Pennsylvania Improvement Society and its promotion of canals and railroads, 1824–1826', *Pennsylvania History: A Journal of Mid-Atlantic Studies* 31, 3, 295–310.

Carlton, R., L. Turnbull and A. Williams, 2019. 'Discovering the Willington Waggonway', in *Early Railways 6*. Milton Keynes: Six Martlets, 1–10.

Carter, I. and E. Carter, 2014. 'Space, time and early railways: oddities from the other end of the world', in *Early Railways 5*. Clare: Six Martlets, 74–96.

Carver, H., 1847. *Proposal for a Charter to Build a Railroad from Lake Michigan to the Pacific Ocean*. Washington: Gideon.

Cattell, J. and K. Falconer, 1995. *Swindon: The Legacy of a Railway Town*. Swindon: English Heritage.

Caulier-Mathy, N., 1971. *La modernisation des charbonnages liégeois pendant la première moitié du XIX*. Paris: Société d'Édition 'Les Belles Lettres'.

Caulier-Mathy, N., 1980. 'Industrie et politique au Pays de Liège Frederic Braconier (1826–1912)', *Belgisch Tijdschrift voor Nieuwste Geschiedenis* 1–2, 3–83.

Cavicchi, E., 2016. 'Dream trains, electromagnetic possibilities and trial runs: early explorations in electromagnetic traction by rail', in *Early Main Line Railways*. Clare: Six Martlets, 164–87.

Chambers, G., 1876. *Historical Sketch of Pottsville, Schuylkill County, Pa*. Pottsville: Standard Publishing Company.

Chandler, A.D., 1954. 'Patterns of American railroad finance, 1830–50', *Business History Review* 28, 3, 248–63.

Chapman W., 1797. *Observations on the Various Systems of Canal Navigation*. London: I and J Taylor.

Chapman, W., 1813. *Report of William Chapman, Civil Engineer, on Various Projected Lines of Navigation from Sheffield*. Sheffield: James Montgomery.

Chevalier, M., 1832. *Religion Saint-Simonienne. Politique Industrielle. Système De La Méditerranée*. Paris: Au Bureau du Globe.

Chevalier, M., 1836. *Lettres sur l'Amérique du Nord, Tome Deuxième*. Paris: Gosselin.

Chrimes, M., 2003. 'Building the London & Birmingham Railway', in *Robert Stephenson: The Eminent Engineer*, ed. M.R. Bailey. Aldershot: Ashgate, 241–61.

Chrimes, M., 2019. 'A safe pair of hands: Joseph Locke and the construction of main-line railways 1823–1840', *Early Main Line Railways 2*. Sponsors of Second International Early Main Line Railways Conference, 59–94.

Clark, G. and D. Jacks, 2007. 'Coal and the Industrial Revolution, 1700–1869', *European Review of Economic History* 11, 39–72.

Clarke, M., 2001. 'The first steam locomotives on the European mainland', *Early Railways*. London: Newcomen Society, 219–32.

Clinker, C.R., 1954. *The Leicester & Swannington Railway*. Leicester: Leicestershire Archaeological Society.

Clinker, C.R. and C. Hadfield, 1958. 'The Ashby-de-la-Zouch Canal and its railways', *Transactions of the Leicestershire Archaeological and Historical Society* 34, 53–76.

Colburn, Z., 1871. *Locomotive Engineering, and the Mechanism of Railways, Volume 1*. London and Glasgow: Collins.

Coleman, D.C., 1992. *Myth, History and the Industrial Revolution*. London: Hambledon.

Coleman, T., 2000. *The Railway Navvies*. London: Pimlico.

Collins, P. (ed.), 1989. *Stourbridge and its Historic Locomotives*. Dudley: Dudley Leisure Services.

Combe, J.-M., B. Escudié and J. Payen, 1991. *Vapeurs sur le Rhône: histoire scientifique et technique de la navigation à vapeur de Lyon à la mer*. Lyon: Presses Universitaires de Lyon.

BIBLIOGRAPHY

Conder, F.R., 1868. *Personal Recollections of English Engineers and of the Introduction of the Railway System into the United Kingdom.* London: Hodder and Stoughton.

Constantine, M.-A. and P. Frame (eds), 2012. *Travels in Revolutionary France and a Journey Across America: George Cadogan Morgan and Richard Price Morgan.* Cardiff: University of Wales Press.

Cook, R.A. and C.R. Clinker, 1984. *Early Railways Between Abergavenny and Hereford.* Oakham: Railway and Canal Historical Society.

Cort, R., 1834. *Rail-road Impositions Detected.* London: W Lake.

Cossons, N. (ed.), 1972. *Rees' Manufacturing Industry 1819–1820: A Selection from The Cyclopaedia, or Universal Dictionary of Arts, Sciences and Literature.* London: Longman, Hurst, Rees, Orme and Brown.

Cossons, N., 2001. Keynote address. *Early Railways.* London: Newcomen Society, 1–6.

Coste, L. and A. Perdonnet, 1830. *Mémoire sur les chemins à ornières.* Paris: Huzard.

Cotte, M., 2003a. 'Les débuts de la ligne ferroviaire de Saint-Étienne à Lyon et les événements de 1830', *Revue d'histoire des chemins de fer* 26, 218–28.

Cotte, M., 2003b. 'Définition de la voie ferrée moderne: la synthèse du Saint-Étienne-Lyon (1825–1835)', *Revue d'histoire des chemins de fer* 27, 7–26.

Cotte, M., 2009. 'Gabriel Lamé, itinéraire d'un jeune ingénieur en France et en Europe (1820–1832)', *Bulletin de la SABIX* 44, 1–16.

Cowburn, I., 2001. 'The origins of the St-Etienne Rail Roads, 1816–38: French industrial espionage and British technology transfer', in *Early Railways.* London: Newcomen Society, 233–50.

Cox, R. and D. O'Dwyer, 2016. 'Construction of the Dublin–Galway main line, 1845–1851', in *Early Main Line Railways.* Clare: Six Martlets, 62–75.

Crockett, D., 1835. *An Account of Col. Crockett's Tour to the North and down East.* Philadelphia and Baltimore: Carey, Hart, and Co.

Crompton, J., 2003. 'The Hedley mysteries', in *Early Railways 2.* London: Newcomen Society, 149–64.

Crompton, J., 2014. 'A Lancashire colliery railway in transition: the Haydock Colliery Railways 1757–1835', in *Early Railways 5.* Clare: Six Martlets, 155–70.

Cronkhite, G.F., 1951. 'The Transcendental Railroad', *The New England Quarterly* 24, 3, 306–28.

Cumming, T.G., 1824. *Illustrations of the Origin and Progress of Rail and Tram Roads.* Denbigh: printed for the author.

Curr, J., 1797. *The Coal-Viewer and Engine Builder's Practical Companion.* Sheffield: Taylor.

Damus, S., 2016. 'The Central Argentine Railway, from inception to maturity', in *Early Main Line Railways.* Clare: Six Martlets, 264–76.

Danville and Pottsville Railroad Company, 1831. *Reports of the Engineers of the Danville and Pottsville Railroad Company.* Philadelphia: Clark and Raser.

Darsley, R., 2006, 'Some considerations on the origins of the chaldron wagon in the northeast of England', in *Early Railways 3.* Sudbury: Six Martlets, 221–41.

Darsley, R., 2010. 'The Durban Bluff Railway and the early railway scene in South Africa', in *Early Railways 4.* Sudbury: Six Martlets, 223–31.

Davies, A.S., 1943–5. 'Early railways of the Ellesmere and of the Montgomeryshire Canals, 1794–1914', *Transactions of the Newcomen Society* 24, 141–6.

Davies, N. and R. Moorhouse, 2003. *Microcosm: Portrait of a Central European City.* London: Jonathan Cape.

Davies, W.L., 1992. *Bridges of Merthyr Tydfil.* Cardiff: Glamorgan Record Office.

Dawson, A., 2004. *Lives of the Philadelphia Engineers: Capital, Class and Revolution, 1830–1890.* Aldershot: Ashgate.

Dawson, A., 2020a. 'The Lake Lock Rail Road: the first public railway?', *Journal of the Railway and Canal Historical Society* 239, 134–43.

BIBLIOGRAPHY

Dawson, A., 2020b. *Yorkshire's First Main Line: The Leeds & Selby Railway*. Market Drayton: Railway and Canal Historical Society.

Dawson, A., 2021. 'The tunnels on the Liverpool & Manchester Railway', *Journal of the Railway and Canal Historical Society* 241, 271–86.

Dawson, A., forthcoming. 'Understanding railway uniforms in the early mainline period'.

De Block, G., 2011. 'Designing the nation: the Belgian Railway Project, 1830–1837', *Technology and Culture* 52, 4, 703–32.

De Gallois, L., 1818. 'Des Chemins de Fer en Angleterre, Notamment à Newcastle, dans le Northumberland', *Annales des mines*, Mai, 129–44.

Dendy Marshall, C.F., 1938. *A History of British Railways Down to the Year 1830*. Oxford: Oxford University Press.

Dendy Marshall, C.F., 1953. *A History of British Railway Locomotives Down to the Year 1833*. London: The Locomotive Publishing Company.

Deroy, A.V. and C. Motte, 1836. *Les rives de la Loire dessinées d'après nature et lithographiées*. Paris: Motte.

Destrem, M.G., 1831. 'Considérations générale sur les avantages relatives des canaux et des chemins a ornières, et application du mode de transport le plus avantageux pour la Russie', *Journal des Voies de Communication* 21, 1–77.

Dickinson, F., 1969. 'The Mellings of Rainhill, 1830–70', *Transactions of the Historical Society of Lancashire and Cheshire* 121, 59–75.

Dickinson, H.W., 1913. *Robert Fulton: Engineer and Artist*. London: John Lane.

Dilts, J.D., 1993. *The Great Road*. Stanford, CA: Stanford University Press.

Divall, C., 2003. 'Beyond the history of early railways', in *Early Railways 2*. London: Newcomen Society, 1–9.

Donaghy, T.J., 1972. *Liverpool & Manchester Railway Operations 1831–1845*. Newton Abbot: David and Charles.

Douglass, F., 1857. *My Bondage and my Freedom: Part I. Life as a Slave; Part II. Life as a Freeman*. New York: Auburn.

Dow, A., 2015. *The Railway: British Track since 1804*. Barnsley: Pen & Sword.

Droulers, P., 1983. 'Christianisme et innovation technologique: les premiers chemins de fer', *Histoire, économie & société* 2, 1, 119–32.

Dublin Penny Journal, 1834. 'The Dublin and Kingstown Railway Source', 3, 113 (30 August), 65–8; 3 121 (5 October), 132–3.

Duffy, M.C., 1982–3. 'Technomorphology and the Stephenson traction system', *Transactions of the Newcomen Society* 54, 55–78.

Dunham, A.L., 1941. 'How the first French Railways were planned', *Journal of Economic History* 1, 1, 12–25.

Dunn, M., 1844. *An Historical, Geological, and Descriptive View of the Coal Trade of the North of England*. Newcastle upon Tyne: Pattison and Ross.

Dunn, M. 1848. *A Treatise on the Winning and Working of Collieries*. Newcastle: printed for and published by the author.

Durnford, M., 1863. *Family Recollections of Lieut. General Elias Walker Durnford*. Montreal: John Lovell.

Dutens, J., 1819. *Memoires sur les travaux publics de l'Angleterre*. Paris: Imprimerie Royale.

Eckert, M., 2019. 'Inspired by British inventions: Joseph von Baader (1763–1835) – a Bavarian engineer fighting a losing battle', *International Journal for the History of Engineering & Technology* 89, 1–2, 216–37.

Edgerton, D., 2006. *The Shock of the Old*. New York: Oxford.

Edgeworth, R.L., 1802. 'On the practicability and advantages of a general system of rail roads', *Journal of Natural Philosophy, Chemistry & the Arts*, 221–3.

Edgeworth, R.L., 1820. *Memoirs of Richard Lovell Edgeworth, Esq*. London: R. Hunter.

Egerton, F.H., 1802. 'Account of the under-ground inclined plane, executed at Walkden Moor, in Lancashire', *Repertory of Arts and Manufactures* 16 (first series edn), 153–64.

BIBLIOGRAPHY

Egerton, F.H., 1812. *Description du plan incliné souterrain*. Paris: Chaignieau Ainé.

Eliot, G., 1872. *Middlemarch, A Study of Provincial Life*. Edinburgh: Blackwood.

Ellison, F.B., 1936–38. 'The history of the Hay Railway 1810–1864', *Transactions of the Woolhope Naturalists' Field Club*, 76–87.

Ellison, F.B., 1937. 'The Hay Railway 1810–1863', *Transactions of the Newcomen Society* 18, 29–42.

Elsas, M., 1960. *Iron in the Making: Dowlais Iron Company Letters, 1782–1860*. Cardiff: Glamorgan County Records Committee.

El Sayed, A. and D. Gwyn, 2016. 'Early main line railways in Egypt: the tainted gift', in *Early Main Line Railways*. Clare: Six Martlets, 76–88.

Elton, A., 1963. 'The pre-history of railways with particular reference to the early quarry railways of North Somerset', *Proceedings of the Somersetshire Archaeological & Natural History Society* 107, 31–59.

Elton, J., 2003. 'Robert Stephenson in society', in *Robert Stephenson: The Eminent Engineer*, ed. M.R. Bailey. Aldershot: Ashgate, 211–37.

Emerson, R.W., 1844. *Essays: Second Series*. Boston: James Munroe and Company.

Emerson, R.W., 1910. *Journals of Ralph Waldo Emerson, volume 3, 1833–1835*, ed. E.W. Forbes and W.E. Forbes. London, Boston and New York: Houghton Mifflin.

English, H., 1827. *A Complete View of the Joint Stock Companies Formed During the Years 1824 and 1825*. London: Boosey and Sons.

Erskine May, T., 1844. *A Treatise upon the Law, Privileges, Proceedings and Usage of Parliament*. London: Charles Knight.

Esper, T., 1982. 'Industrial serfdom and metallurgical technology in 19th-century Russia', *Technology and Culture* 23, 4, 583–608.

Ewans, M.C., 1977. *The Haytor Granite Tramway and Stover Canal*. Newton Abbot: David and Charles.

Executive Documents of the House of Representatives at the Second Session of the Twenty-First Congress, begun and held at the City of Washington December 6, 1830. Volume 3, document 53, 1–108.

Farey, J., 1817. *General View of the Agriculture and Minerals of Derbyshire*. London: printed for Board of Agriculture.

Faujas-de-St-Fond, B., 1797. *Voyage en Angleterre, en Écosse et aux iles Hébrides. Tome 1*. Paris: Jansen.

Faujas-de-St-Fond, B., 1799. *Travels in England, Scotland, and the Hebrides*. London: Ridgway.

Fellows, R.B., 1930. *History of the Canterbury and Whitstable Railway*. Canterbury: J. A. Jennings.

Ferguson, N., 2006. 'Anglo-Scottish transfer of railway technology in the 1830s', in *Early Railways 3*. Sudbury: Six Martlets, 176–90.

Fiero, K.W., 1984. *The Lemon House: Allegheny Portage Railroad National Historic Site, Pennsylvania*. Washington, DC: US Dept. of the Interior, National Park Service.

Finch, J.K., 1930–1. 'John B. Jervis, civil engineer', *Transactions of the Newcomen Society* 11, 108–19.

Fitzgerald, R.S., 1980. *Liverpool Road Station, Manchester: An Historical and Architectural Survey*. Manchester: Manchester University Press.

Fletcher, W., 1891. *The History and Development of Steam Locomotion on Common Roads*. London, New York and Ipswich: Spon and Cowell.

Flinn, M.W., 1984. *The History of the British Coal Industry. Vol. 2 1700–1830: The Industrial Revolution*. Oxford: Clarendon Press.

Fördös, L., 2017. *Die Wahrnehmung des Semmering-Rax-Gebietes in der Literatur im Wandel der Zeit*. Magistra der Philosophie thesis, University of Vienna.

Fort, D., 1989a. 'The "Agenoria" and the Shutt End Railway', in *Stourbridge and its Historic Locomotives*, ed. P. Collins. Dudley: Dudley Leisure Services, 41–7.

BIBLIOGRAPHY

Fort, D., 1989b. 'The Stourbridge Lion, Delaware and Hudson', in *Stourbridge and its Historic Locomotives*, ed. P. Collins. Dudley: Dudley Leisure Services, 30–40.

Forward, E.A., 1943–4, 'Stephenson locomotives for the St. Etienne-Lyons railway', 1828, *Transactions of the Newcomen Society* 24, 89–98.

Forward, E.A., 1951–3. 'Chapman's Locomotives, 1812–1815: some facts and some speculations', *Transactions of the Newcomen Society* 28, 1–19.

Freeman, M., 1999. *Railways and the Victorian Imagination*. New Haven, CT: Yale University Press.

Freudenberger, H., 1997. 'The Linz–Budweis Railways: technology, science and the economy', *Journal of European Economic History* 26, 2, 239–68.

Freundl, S., 2016. ‚Die Dampfschiffahrt auf der bayerischen Donau von den Anfängen bis zum Beginn des Ersten Weltkrieges: Ein Rückblickl', *Verhandlungen des Historischen Vereins für Oberpfalz und Regensburg*, 51–117.

Fulton, R., 1796. *A Treatise on the Improvement of Canal Navigation*. London: I. and J. Taylor.

Fulton, R., 1799. *Recherches sur les moyens de perfectionner les canaux de navigation, et sur les nombreux avantages de petits canaux*. Paris: Dupain-Triel et Bernard.

Galignani, J.A., 1819. *Galignani's Repertory or Literary Gazette and Journal of the Belles Lettres 6*. Paris.

Gamst, F.C., 1992. 'The context and significance of America's first railroad, on Boston's Beacon Hill', *Technology and Culture* 33, 1, 66–100.

Gamst, F.C., 1997. *Early American Railroads: Franz Anton Ritter von Gerstner's 'Die innern Communication' (1842–1843)*. Stanford, CA: Stanford University Press.

Gamst, F.C., 2001. 'The transfer of pioneering British railroad technology to North America', in *Early Railways*. London: Newcomen Society, 251–65.

Geraghty, P.J., 2010. 'Promotion of road steam transport at the dawn of the railway age', *Journal of the Railway and Canal Historical Society* 36, 8, 88–105.

Gerhold, D., 2010. 'The rise and fall of the Surrey Iron Railway, 1802–46', *Surrey Archaeological Collections* 95, 193–210.

Geršlová, J., 2003. 'Riepl, Franz', in *Neue Deutsche Biographie 21*. Berlin: Duncker & Humblot, 602–3.

Gerstner, F.A. von, 1824. *Über die Vortheile der Anlage einer Eisenbahn zwischen der Moldau und Donau*. Vienna: Tendler and von Manstein.

Gerstner, F.J. von, 1813. *Zwey Abhandlungen über Frachtwägen und Strassen und über die Frage, ob, und in welchen Fällen der Bau schiffbarer Kanäle, Eisenwege, oder gemachter Strassen vorzuziehen sei*. Prague: Gottlieb Haase.

Gerstner, F.J. von, 1827. *Mémoire sur les grandes routes, les chemins de fer et les canaux de navigation, traduit de l'allemand de M. F. de Gerstner*. Paris: Bachelier.

Gerstner, F.J. von, 1831. *Handbuch der Mechanik Mechanik fester Körper*. Prague: Johann Spurny.

Ghega, K. von, 1853. *Uebersicht der Hauptfortschritte des Eisenbahnwesens in dem Jahrzehnde 1840–1850 und die Ergebnisse der Probefahrten auf einer Strecke der Staatsbahn ueber den Semmering in Österreich: Band 1*. Vienna: Sollinger.

Ghega, K. von, 1854. *Malerischer Atlas der Eisenbahn über den Semmering*. Vienna: Carl Gerold.

Gibbons, E.J., 1990. 'The building of the Schuylkill Navigation System, 1815–1828', *Pennsylvania History* 57, 1, 13–43.

Gibson-Hill, C.A., 1954. 'The steamers employed in Asian Waters, 1819–39', *Journal of the Malayan Branch of the Royal Asiatic Society* 27, 1, 120–62.

Goethe, J.W., von and C.F. Zelter, 1915. *Der Briefwechsel zwischen Goethe und Zelter: 1819–1827*. Leipzig: Insel-Verlag.

Gomersall, H., 2006. 'The Round Foundry of Leeds', in *Early Railways 3*. Sudbury: Six Martlets, 260–9.

BIBLIOGRAPHY

Gomersall, H. and A. Guy, 2010. 'A research agenda for the early British railway', in *Early Railways 4*, Sudbury: Six Martlets, 327–51.

Goodbody, R., 2010. *The Metals: From Dalkey to Dún Laoghaire*. Dún Laoghaire: Dún Laoghaire Rathdown County Council.

Goodchild, J., 2006. 'The Lake Lock Railroad', in *Early Railways 3*. Sudbury: Six Martlets, 40–50.

Gordon, A., 1832. *An Historical and Practical Treatise upon Elemental Locomotion*. London: Steuart and Blackwood.

Grant, M.H., 1938–41. 'British medals since 1760', *British Numismatic Journal* 23, 119–52.

Gray, T., 1821. *Observations on a General Iron Rail-Way: Showing its Great Superiority Over All the Present Methods of Conveyance*. London: Baldwin Cradock and Joy (and subsequent editions of 1822 and 1823).

Gray, T., 1823. 'Proposition for a general iron rail-way, with steam-engines', *Mechanic's Magazine* 19 (3 January), 290–1.

Gray, T., 1825. *Observations on a General Iron Rail-way, Or Land Steam-conveyance*. London: Baldwin, Cradock and Joy.

Green, F.M., 1963. 'On tour with President Andrew Jackson', *New England Quarterly* 36, 2, 209–28.

Greenfield, D.J., 2011. 'I.K. Brunel and William Gravatt, 1826–1841: Their Professional and Personal Relationship'. PhD thesis, University of Portsmouth.

Grinde, D.A., 1976. 'Building the South Carolina Railroad,' *South Carolina Historical Magazine* 77, 2, 84–96.

Grudgings, S., 2019. 'Early railways in the Bristol coalfield', in *Early Railways 6*. Milton Keynes: Six Martlets, 11–24.

Gueneau, L., 1930. 'Les houillères d'Épinac vers 1830', *Annales de Bourgogne*, 2, 159–70.

Gueneau, L., 1931a. 'Le Chemin de Fer d'Épinac à Pont d'Ouche: la concession de la ligne', *Annales de Bourgogne*, 3, 38–65.

Gueneau, L., 1931b. 'Le Chemin de Fer d'Épinac à Pont d'Ouche: construction et aménagement de la ligne', *Annales de Bourgogne*, 3, 224–52.

Guldi, J., 2012. *Roads to Power: Britain Invents the Infrastructure State*. Cambridge, MA and London: Harvard University Press.

Guy, A., 2001. 'North eastern locomotive pioneers 1805 to 1827: a reassessment', in *Early Railways*. London: Newcomen Society, 117–44.

Guy, A., 2003. 'Early railways: some curiosities and conundrums', in *Early Railways 2*. London: Newcomen Society, 64–78.

Guy, A., 2014. 'Missing links: some atypical early railways in Britain', in *Early Railways 5*. Clare: Six Martlets, 171–98.

Guy, A. and J. Rees, 2011. *Early Railways 1569–1830*. Oxford: Shire.

Guy, A., M. Bailey, D. Gwyn, R. Protheroe Jones, M.J.T. Lewis, J. Liffen and J Rees, 2018. 'Penydarren re-examined', in *Early Railways 6*. Milton Keynes: Six Martlets, 147–93.

Gwyn, D., 1999. 'From blacksmith to engineer: artisan technology in North Wales in the early nineteenth century', *Llafur* 7, 51–65.

Gwyn, D., 2001a. 'Transitional technology: the Nantlle Railway', in *Early Railways*. London: Newcomen Society, 46–62.

Gwyn, D., 2001b. '"Ignorant of all science": technology transfer and peripheral culture, the case of Gwynedd, 1750–1850', in *From Industrial Revolution to Consumer Revolution: International Perspectives on the Archaeology of Industrialisation/De la Révolution industrielle à la Revolution de la Comsommation: perspectives internationales sur l'archéologie de l'industrialisation*, ed. M. Palmer and P. Neaverson. Leeds: Association for Industrial Archaeology, 39–45.

Gwyn, D., 2002a. 'An early high-pressure steam engine at Cloddfa'r Coed', *Transactions of the Caernarvonshire Historical Society* 63, 26–43.

Gwyn, D., 2003. 'Artists, Chartists, railways and riot', in *Early Railways 2*, London: Newcomen Society, 37–51.

Gwyn, D., 2004. 'Tredegar, Newcastle, Baltimore: the swivel truck as paradigm of technology transfer', *Technology & Culture* 45, 778–94.

Gwyn, D., 2008. '"Best adapted to the general carriage": railways of the Llangollen canal, their history and archaeology', *Patrimoine de l'Industrie: resources, pratiques* 18, 61–70.

Gwyn, D., 2010. '"What passes and endures": the early railway in Wales', in *Early Railways 4*. Sudbury: Six Martlets, 125–40.

Gwyn, D., 2014. 'Railways in Gwynedd 1759–1848', in *Early Railways 5*. Clare: Six Martlets, 97–111.

Gwyn, D., 2015. *Welsh Slate: History and Archaeology of an Industry*. Aberystwyth: Royal Commission on the Ancient and Historical Monuments of Wales.

Gwyn, D., 2019a. 'The first railways in Africa', in *Early Railways 6*. Milton Keynes: Six Martlets, 238–43.

Gwyn, D., 2019b. '"Many shall run to and fro, and knowledge shall be increased": presenting the Liverpool and Manchester Railway', *Early Main Line Railways 2*. Sponsors of Second International Early Main Line Railways Conference, 309–26.

Hair, P.E.H., 1988. *Coal on Rails or The Reason of my Wrighting: The Autobiography of Anthony Errington*. Liverpool: Liverpool University Press.

Hair, T.H., 1987. *A Series of Views of the Collieries in the Counties of Northumberland and Durham*. Newcastle upon Tyne: Davis (reproduction of 1844 original).

Hardy, P.D., 1834. *Thirteen Views of the Dublin and Kingstown Railway*. Dublin: P. Dixon Hardy.

Harkort, F.H., 1833. *Die Eisenbahn von Minden nach Cöln*. Hagen: Brune.

Harland, J.H., 2013. 'The transition from hemp to chain cable: innovations and innovators', *The Mariner's Mirror*, 99, 1, 72–85.

Harper, J.A., 2002. *The History and Geology of the Allegheny Portage Railroad*. Pittsburgh: Pittsburgh Geological Society.

Harris, J.R., 1998. *Industrial Espionage and Technology Transfer: Britain and France in the Eighteenth Century*. Aldershot and Brookfield: Ashgate.

Hartley, R.F., 1994. 'Tudor miners of Coleorton, Leicestershire', in *Mining Before Powder*, ed. T.D. Ford and L.M. Willies. Matlock Bath: Peak District Mines Historical Society, 91–101.

Hartley, R.F., 2010. 'The Coleorton Railway', in *Early Railways 4*. Sudbury: Six Martlets, 151–66.

Hartley, R.F., 2014. 'Early iron railways and plateways in Leicestershire', in *Early Railways 5*. Clare: Six Martlets, 260–69.

Hartley, R.F., 2016a. 'The architecture of early main lines in the British Isles (1825–1850): heritage under pressure', in *Early Main Line Railways*. Clare: Six Martlets, 246–63.

Hartley, R.F., 2016b. 'George Stephenson: the railway surveys. Part 1, 1819–1832', *Journal of the Railway and Canal Historical Society* 38, 9, 550–60.

Hartley, R.F., 2017a. 'George Stephenson: the railway surveys. Part 2, 1832–1848', *Journal of the Railway and Canal Historical Society* 39, 1, 2–12.

Hartley, R.F., 2017b. 'George Stephenson: the railway surveys. Part 3', *Journal of the Railway and Canal Historical Society* 39, 2, 83–7.

Hartley, R.F., 2019. 'Why Killingworth?' in *Early Railways 6*. Milton Keynes: Six Martlets, 25–40.

Harwood, H.H., 1994. 'The early Baltimore & Ohio Railroad and its physical remains', in R. Shorland-Ball (ed.), *Common Roots – Separate Branches: Railway History and Preservation*. York and London: Science Museum, 48–62.

Hawks, F.L., 1856. *Narrative of the Expedition of an American Squadron to the China Seas and Japan: Performed in the Years 1852, 1853, and 1854*. Washington: Beverley Tucker.

BIBLIOGRAPHY

Haywood, R.M., 1969. *The Beginnings of Railway Development in Russia in the Reign of Nicholas I, 1835–1842.* Durham, NC: Duke University Press.

Haywood, R.M., 1998. *Russia Enters the Railway Age, 1842–1855.* New York: Columbia University Press.

Hazard, E., 1827. 'Observations on rail-roads', *Journal of the Franklin Institute* 3, 275–7.

Henaux, F., 1843. 'De la découverte de la houille dans le Pays de Liège', *Messager des sciences historiques de Belgique*, 258–80.

Henriet, J., 1912. 'Les transports par voie ferrée', *Association française pour l'avancement des sciences* 20, 1, 107–17.

Heydinger, E.J., 1962. 'Schuylkill County Railroad activity', *Railroad and Locomotive Historical Society Bulletin* 106, 37–9.

Hills, R., 1989. *Power from Steam: A History of the Stationary Steam Engine.* Cambridge and New York: Cambridge University Press.

Hills, R., 2006. 'The development of machine tools in the early railway era', in *Early Railways 3.* Sudbury: Six Martlets, 242–83.

Hilton, G.W., 1990. *American Narrow Gauge Railways.* Stanford, CA: Stanford University Press.

Hiskey, C.E., 1978. *John Buddle (1773–1843): Agent and Entrepreneur in the North-East Coal Trade.* MLitt thesis, Durham University.

HMSO, 1856. *English Patents of Inventions, Specifications.* London: HMSO.

Hodgkins, D., 2003. 'Success and failure in making the transition to a modern railway: the Liverpool & Manchester and Cromford & High Peak', in *Early Railways 2.* London: Newcomen Society, 52–63.

Hoffmann, K. and N. Zrnić, 2012. 'A contribution on the history of ropeways', in *Explorations in the History of Machines and Mechanisms*, ed. T. Koetsier and M. Ceccarelli, 381–94. Springer: Dordrecht.

Hollinghurst, H., 2009. *John Foster and Sons: Kings of Georgian Liverpool.* Liverpool: Liverpool History Society.

Hopkin, D., 2001. 'Reflections on the iconography of early railways', in *Early Railways.* London: Newcomen Society, 342–54.

Hopkin, D., 2010. 'Timothy Hackworth and the Soho Works, circa 1830–1850', in *Early Railways 4.* Sudbury: Six Martlets, 280–301.

Hopkin, D., 2014. 'William Brunton's walking engines and the Crich rail-road', in *Early Railways 5.* Clare: Six Martlets, 221–42.

Hughes H.D., 1866. *Hynafiaethau Llandegai a Llanllechid.* Bethesda: privately published.

Hughes, S., 1990. *Brecon Forest Tramroad: The Archaeology of an Early Railway System.* Aberystwyth: Royal Commission on the Ancient and Historical Monuments of Wales.

Hughes, S., 2000. *Copperopolis: Landscapes of the Early Industrial Period in Swansea.* Aberystwyth: Royal Commission on the Ancient and Historical Monuments of Wales.

Hughes, S., 2010. 'The emergence of the public railway in Wales', in *Early Railways 4.* Sudbury: Six Martlets, 107–24.

Hughes, S., forthcoming. *The Swansea Canal and its Early Railways.* Aberystwyth and Swansea: Royal Commission on the Ancient and Historical Monuments of Wales and Swansea Canal Society.

Huyer, R., 1892–3. 'Die Budweis-Linzer Pferdeeisenbahn', *Mitteilungen des Vereines für Geschichte der Deutschen in Böhmen*, 77–88.

Hyde Hall, E., 1952. *A Description of Caernarvonshire (1809–1811).* Caernarfon: Caernarvonshire Historical Society.

Ince, L., 1990. 'Locomotives of the Neath Abbey Iron Company', *Industrial Railway Record* 191, 120–31.

Ince, L., 2001. *Neath Abbey and the Industrial Revolution.* Stroud: Tempus.

James, W., 1823. *Report, or Essay to Illustrate the Advantages of Direct Inland Communication.* London: J. and A. Arch.

BIBLIOGRAPHY

Jars, G., 1774–81. *Voyages Métallurgiques*. Lyon: Gabriel Regnault.

Jarvis, A., 1991. *Liverpool Central Docks*. Stroud: Sutton and National Museums and Galleries on Merseyside.

Jessing, B., B. Lutz and I. Wild, 2004. *Goethe Lexikon: Personen-Sachen-Begriffe*, Ausgabe 2. Stuttgart and Weimar: Metzler.

Jones, N. and R. Hankinson, 2004. 'Brinore Tramroad, Talybont on Usk', *Archaeology in Wales* 44, 196–7.

Jones, N.W., 1987. 'A wooden wagon way at Bedlam Furnace, Ironbridge', *Post-medieval Archaeology* 21, 259–62.

Jones, S., 1835. 'Allegheny Portage Rail-Way', *Hazard's Register of Pennsylvania* 15, 321–7.

Jones, S.K., 2019. 'Rise and fall: steam and the suspension bridge', *Early Main Line Railways* 2. Sponsors of Second International Early Main Line Railways Conference, 375–98.

Kaiserlich Königlich Priv. Erste Eisenbahn-Gesellschaft, 1830. *Directions-Bericht an die P.T. Herren Actionäre der t. t. priv. ersten Eisenbahn-Gesellschaft über den Stand der Eisenbahn zwischen der Moldau und Donau*. Vienna.

Karlsson, L.O., 2001. 'A rediscovered early rail wagon', in *Early Railways*. London: Newcomen Society, 20–3.

Katsumasa, H., 1979. *Japan's Discovery, Import and Technical Mastery of Railways*. Tokyo: United Nations University.

Kemble, F.A., 1879. *Record of a Girlhood*. New York: H. Holt.

Kent, F., 1925. *The Story of Alexander Brown & Sons*. Baltimore: privately published.

Kerr, I.J., 2016. 'The early main line railway as concept and period, and its utility for the history of railways in 19th-century India', in *Early Main Line Railways*. Clare: Six Martlets, 89–101.

Kilbourne, J., 1828. *Public Documents Concerning the Ohio Canals, which are to connect Lake Erie with The Ohio Rivers, Comprising a Complete Official History of These Great Works Of Internal Improvement*. Columbus: Whiting.

King, D.J.S., 1991. 'The ideology behind a business activity: the case of the Nuremburg-Fürth railway', *Business and Economic History* 20, 162–70.

Kirby, M.W., 1993. *The Origins of Railway Enterprise: The Stockton and Darlington Railway 1821–1863*. Cambridge: Cambridge University Press.

Kobayashi, I. and M. Cotte, 1997. 'The locomotives of the Saint-Etienne & Lyon Railway: design, construction and first uses (1825–1835)', *Historical Studies in Civil Engineering* 17, 101–10.

Kurrer, K.-E., 2006. 'On the relationship between construction engineering and strength of materials in Gerstner's "Handbook of Mechanics"', *Proceedings of the Second International Congress on Construction History* 2, 1809–28.

Lamming, C., 2003. 'Retour aux origines et aux années 1820: de l'atelier de charronnage primitif anglais aux premiers dépôts organisés en France par Marc Seguin', *Revue d'histoire des chemins de fer* 28–9, 257–74.

Latrobe, J.H.B., 1868. *The Baltimore and Ohio Railroad: Personal Recollections*. Baltimore: Sun Printing.

Le Neve Foster, P. 1875. 'Report on stone tramways in Italy', *Journal of the Society of Arts* (25 June), 709–13.

Lee, C.E., 1954. *The Swansea & Mumbles Railway*. South Godstone: Oakwood.

Lee, H.C., 2008 (revised and expanded by W.F. Rossiter and J. Marcham). *A History of Railroads in Tompkins County*. Ithaca, NY: The History Center in Tompkins County.

Lee, M.J., 1973. 'Griff Collieries', *Industrial Railway Record* 47, 1–19.

Lehigh Coal and Navigation Company, 1839. *Report of the Board of Managers of the Lehigh Coal and Navigation Company*. Philadelphia: James Kay.

Lewis, D.C., 1911. *Hanes Plwyf Defynnog*. Merthyr Tydfil: HW Southey a'i Feibion.

Lewis, M.J.T., 1960, *The Pentewan Railway 1829–1918*. Truro: D.B. Barton.

BIBLIOGRAPHY

Lewis, M.J.T., 1968. *How Ffestiniog Got its Railway*. Caterham: Railway and Canal Historical Society.

Lewis, M.J.T., 1970. *Early Wooden Railways*. London: Routledge and Kegan Paul.

Lewis, M.J.T., 1996. 'Archery and Spoonerisms: the creators of the Festiniog Railway', *Journal of the Merionethshire Historical and Record Society* 13, 3, 263–76.

Lewis, M.J.T., 2001. 'Railways in the Greek and Roman World', *Early Railways*. London: Newcomen Society, 8–19.

Lewis, M.J.T., 2003. 'Bar to fish-belly: the evolution of the cast-iron edge rail', in *Early Railways 2*. London: Newcomen Society, 102–17.

Lewis, M.J.T., 2005–7. 'George Overton on tramroads and railways', *Journal of the Railway and Canal Historical Society* 35, 322–36.

Lewis, M.J.T., 2006. 'Reflections on 1604', in *Early Railways 2*. London: Newcomen Society, 8–22.

Lewis, M.J.T., 2010. 'Construction and temporary railways', in *Early Railways 4*. Sudbury: Six Martlets, 177–92.

Lewis, M.J.T., 2014. 'Early passenger carriage by rail', in *Early Railways 5*. Clare: Six Martlets, 199–220.

Lewis, M.J.T., 2019a. 'A new locomotive drawing from the 1820s', *Journal of the Railway and Canal Historical Society* 235, 476–86.

Lewis, M.J.T., 2019b. 'Point work to 1830', in *Early Railways 6*. Milton Keynes: Six Martlets, 41–57.

Lewis, M.J.T., 2020. *Steam on the Sirhowy Tramroad and its Neighbours*. Market Drayton: Railway and Canal Historical Society.

Liffen, J., 2006. 'The iconography of the Wylam Waggonway', in *Early Railways 3*. Sudbury: Six Martlets, 51–75.

Liffen, J., 2010. 'Searching for Trevithick's London railway of 1808', in *Early Railways 4*. Sudbury: Six Martlets, 1–29.

Livesey, R., 2016. *Writing the Stagecoach Nation: Locality on the Move in Nineteenth-Century British Literature*. Oxford: Oxford University Press.

Longridge, M., 1822. *Specification of John Birkinshaw's Patent: For an Improvement in the Construction of Malleable Iron Rails, to be Used in Rail Roads, With Remarks on the Comparative Merits of Cast Metal and Malleable Iron Rail-ways*. Newcastle: Walker.

Longworth, J. and P. Rickard, 2014. 'Early Australian railed-ways: 1788–1855', in *Early Railways 5*. Clare: Six Martlets, 36–53.

Longworth, J. and P. Rickard, 2019. 'Plateways, iron and steel road rails, and rutways in Australia', in *Early Railways 6*. Milton Keynes: Six Martlets, 221–37.

Lord, P., 1998. *Industrial Society: The Visual Culture of Wales*. Cardiff: University of Wales Press.

McCarthy, D., 2014. 'Dalkey Quarry Tramway 1815–c1855", in *Early Railways 5*. Clare: Six Martlets, 330–43.

McCutcheon, W.A., 1980. *The Industrial Archaeology of Northern Ireland*. Belfast: HMSO.

MacDonald, H., 2006. 'The Cape Breton waggonways of the General Mining Association', in *Early Railways 3*. Sudbury: Six Martlets, 96–125.

MacDonald, H., 2010. 'The Rideau Waggonway that never was: rail vs canal in the defence of Canada, 1814–1825', in *Early Railways 4*. Sudbury: Six Martlets, 206–22.

MacDonald, H. and R. Tennant, 2016. 'The inter-colonial railway idea in British North America: 1835–1867', in *Early Main Line Railways*. Clare: Six Martlets, 204–26.

Mackenzie, E. and M. Ross, 1834. *An Historical, Topographical, and Descriptive View of the County Palatine of Durham*. Newcastle upon Tyne: Mackenzie and Dent.

McKeown, J., 2006. 'Inland waterways', in *Engineering Ireland*, ed. R. Cox. Cork: Collins Press, 102–15.

McNair, M., 2007. *William James (1771–1837): The Man Who Discovered George Stephenson*. Oxford: Railway and Canal Historical Society.

BIBLIOGRAPHY

McNair, M., 2010. 'The Central Junction Railway of 1820: a study in historical perspective', in *Early Railways 4*. Sudbury: Six Martlets, 167–76.

McNair, M., 2019. Early locomotives of the St Etienne-Lyon Railway, in *Early Railways 6*. Milton Keynes: Six Martlets, 71–8.

Madison, N.V., 2015. *Tredegar Iron Works: Richmond's Foundry on the James*. Charleston, SC: The History Press.

Magennis, E. 2014. 'The Irish Parliament and the regulatory impulse, 1692–1800: the case of the coal trade', *Parliamentary History Special Issue* 33, 1, 54–72.

Maillard, S., 1784. *Théorie des machines mués par la force de la vapeur de l'eau*. Vienna and Strasbourg: Gay, Cellot, Jombert and Jombert.

Marrs, A.W., 2009. *Railroads in the Old South*. Baltimore: Johns Hopkins University Press.

Martin, D., 1976. *The Monkland & Kirkintilloch Railway*. Strathkelvin: Strathkelvin District Libraries and Museums.

Martin, D., 1981. *The Garnkirk & Glasgow Railway*. Strathkelvin: Strathkelvin District Libraries and Museums.

Martin, D., 2021. 'The resilience of locomotives with "vertical in-line" cylinders: a Scottish perspective', *Railway and Canal Historical Society Journal* 242 (November), 369–76.

Martineau, H., 1837. *Society in America*, 2 vols. New York and London: Saunders and Otley.

Matsumoto-Best, S., 2003. *Britain and the Papacy in the Age of Revolution 1846–1851*. Rochester, NY: The Boydell Press for the Royal Historical Society.

Mayberg, F., C. Blanckarts and F. Andriessen, 1845. *Bericht an die Aktionäre der Prinz-Wilhelm-Eisenbahn von Steele nach Vohwinkel*. Düsseldorf: Wolf.

Mazzoni, M., 1836. *The Traveller's Guide of Milan*. Milan: Senzogno.

Mellet, F.N., 1828. *Traité des machines à vapeur et de leur application à la navigation, aux mines, aux manufactures, etc. traduit de l'anglais de Th. Tredgold*. Paris and Brussels: Bachelier.

Mellet, F.N. and C.J. Henry, 1828. *Mémoire sur le chemin de fer de la Loire, d'Andrezieux à Roanne*. Paris: Huzard-Courcier.

Meredith, J.C., 1998. 'A wild ride on Badnall's famous undulating railway', *Railroad History* 178, 107–19.

Messenger, M., 2012. *The Bodmin & Wadebridge Railway, 1834–1983*. Truro: Twelveheads Press.

Messenger, M., 2014. 'Sources of finance for early Cornish railways', in *Early Railways 5*. Clare: Six Martlets, 132–9.

Millin, A.-L., 1807. *Voyage dans les départemens du Midi de la France; Tome Premier*. Paris: Tourneisen.

Minard, C.J., 1834. *Leçons faites sur les chemins de fer à l'Ecole des ponts et chaussées en 1833–1834*. Paris: Carillan-Gœury, Libraire Des Corps Royaux des Ponts et Chaussées et des Mines.

Monroe, J.H., 1914. *Schenectady, Ancient and Modern*. Geneva, NY: W.F. Humphrey.

Moorsom, N., 1975. *Stockton and Darlington Railway: The Foundation of Middlesbrough*. Middlesbrough: J.G. Peckston.

Morris, E., 1841. 'On cast iron for railways', *Journal of the Franklin Institute* 3, 2, 304–18.

Mountford, C., 1976. *The Bowes Railway*. London: Industrial Railway Society and Tyne and Wear Industrial Monuments Trust.

Mountford, C., 2001. 'Rope haulage: the forgotten element of railway history', in *Early Railways*. London: Newcomen Society, 171–91.

Mountford, C., 2003. 'Researching rope-haulage – a case study: the Lambton Railway, 1800–1835', in *Early Railways 2*. London: Newcomen Society, 118–33.

Mountford, C., 2006. 'The Hetton Railway: Stephenson's original design and its evolution', in *Early Railways 3*. Sudbury: Six Martlets, 76–95.

Mountford, C., 2013. *Rope and Chain Haulage*. Melton Mowbray: Industrial Railway Society.

Mountford, C., M. Young and B. Burdsall, 2019. Interpreting sources for the Durham & Sunderland Railway, 1836–60, in *Early Railways 6*. Milton Keynes: Six Martlets, 103–23.
Muhle, E., 2015. *Breslau: Geschichte einer europäischen Metropole*. Cologne: Böhlau Verlag.
Mukherjee, E., 2016. 'Managing technology transfer: land acquisition for the East Indian Railway, 1850–1854', in *Early Main Line Railways*. Clare: Six Martlets, 102–14.
Mulholland, P., 1978. 'The first locomotive in Whitehaven', *Industrial Railway Record* 75, 177–9.
Mullineux, F., 1971. *The Duke of Bridgewater's Underground Canal at Worseley*. Manchester: Lancashire and Cheshire Antiquarian Society.
Munsell, J., 1875. *The Origin, Progress and Vicissitudes of the Mohawk and Hudson Rail Road*. Albany, NY: privately published.
Murfitt, S.E., 2017. 'The English patent system and early railway technology 1800–1852'. PhD thesis, University of York.
Murray, K., 1938. 'Dublin's First Railway'. *Dublin Historical Record* 1, 1, 19–26.
Murray, K.A., 1995. 'Richard Osborne at Limerick', *The Old Limerick Journal* 32, 50–7.
Nef, J.U., 1966. *The Rise of the British Coal Industry*. London: Cass.
Nehru, J. 1936. *An Autobiography*. London: Bodley Head.
New, J., 2019. 'Why displace the horse?', in *Early Railways 6*. Milton Keynes: Six Martlets, 58–70.
Nicholson, J., 1825. *The Operative Mechanic and British Machinist*. London: Knight and Lacey.
Nicholson, J., 1826. *Description des machines à vapeur . . . Traduit de l'Anglais, par T. Duverne*. Paris: Bachelier.
Nimmo, A., 1826. *The Report of Alexander Nimmo, Civil Engineer, M.R.I.A., F.R.S.E., &c: on the Proposed Railway Between Limerick and Waterford*. Dublin: Thom and Johnston.
Norris, J., 1987. *The Stratford and Moreton Tramway*. Guildford: Railway and Canal Historical Society.
Norris, W., 1838. *Locomotive Steam Engine of William Norris*. Philadelphia: Kiderlen and Stollmeyer.
Northover, P., 2014, 'Buying iron: the case of the Plymouth and Dartmoor railway', in *Early Railways 5*. Clare: Six Martlets, 140–54.
O'Rourke, K. and J.G. Williamson, 2002. 'When did globalisation begin?' *European Review of Economic History* 6, 23–50.
Odlyzko, A., 2019. 'Dionysius Lardner: the denigrated sage of early railways', *Early Main Line Railways 2*, Sponsors of Second International Early Main Line Railways Conference, 39–58.
Oeynhausen, C. von and H. von Dechen, 1829. *Ueber Die Schienenwege in England: Bemerkungen Gesammelt Auf Einer Reise in den Jahren 1826 und 1827*. Berlin: G Reimer.
Oostindie, G.J., 1984. 'La burguesía cubana y sus caminos de hierro, 1830–1868', *Boletín de Estudios Latinoamericanos y del Caribe* 37, 99–115.
Oostindie, G.J., 1988. 'Cuban railroads, 1830–1868: origins and effects of "progressive entrepreneurialism"', *Caribbean Studies* 20, 3–4, 24–45.
Osborne, R.B., 1921. 'Professional biography of Moncure Robinson', *The William and Mary Quarterly* 1, 237–60.
Outram, B., 1801. 'Minutes to be observed in the construction of railways', *Recreations in Agriculture* 4, 473–7.
Owen, D., 2001. *South Wales Collieries, vol. 1*. Stroud: Tempus.
Palmer, M. and P. Neaverson, 2002. 'Pre-locomotive railways of Leicestershire and South Derbyshire', in *The Impact of the Railway on Society in Britain. Essays in Honour of Jack Simmons*, ed. A.K.B. Evans and J.V. Gough. Aldershot: Ashgate, 11–32.
Parliamentary Papers, 1808. *First Report from the Committee on the Highways of the Kingdom*.

BIBLIOGRAPHY

Parsons, E., 1835. *The Tourist's Companion, Or, The History of the Scenes and Places on the Route by the Rail-Road and Steam-Packet from Leeds and Selby to Hull.* London: Whittaker; Manchester: Bancks; Selby: Galpine.

Parthasarathi, P., 2011. *Why Europe Grew Rich and Asia Did Not: Global Economic Divergence, 1600–1850.* Cambridge: Cambridge University Press.

Patel, R., 2018. 'The early development of the Outram-pattern plateway 1793–1796', *Journal of the Railway and Canal Historical Society* 233, 326–37.

Patel, R., 2020. 'The Permanent Way of the 1805 Congleton Railway: new evidence from fieldwork', *Industrial Archaeology Review* 42, 1, 62–78.

Paxton, R., 2001. 'An engineering assessment of the Kilmarnock & Troon railway (1807–1846)', in *Early Railways*. London: Newcomen Society, 82–102.

Paxton, R., 2014. 'The Bell Rock Lighthouse Railway', in *Early Railways 5*. Clare: Six Martlets, 312–29.

Payen, J., 1988. 'Seguin, Stephenson et la naissance de la locomotive à chaudière tubulaire (1828–1829)', *History and Technology* 6, 145–71.

Pease, A.E., 1907. *The Diaries of Edward Pease, the Father of English Railways.* London: Headley Bros.

Pennington, M., 1894. *Railways and Other Ways: Being Reminiscences of Canal and Railway Life During a Period of Sixty-seven Years.* Toronto: Williamson and Co.

Pennsylvania Society for the Promotion of Internal Improvements in the Commonwealth, 1826. *The First Annual Report of the Acting Committee.* Philadelphia: Pennsylvania Society.

Pevsner, N. and E. Williamson, 1979. *Nottinghamshire (Pevsner Architectural Guides: Buildings of England).* New Haven, CT and London: Yale University Press.

Phillips, J., 1803. *The General History of Inland Navigation.* London: C and R Baldwin,

Phillips, J.R.S., 1972. 'The earliest passenger carrying railway vehicle? A note', *Transport History* 5, 152–4.

Phillips, R., 1817. *A Morning's Walk from London to Kew.* London: J. Adlard.

Pigot J., 1818. *The Commercial Directory for 1818–19–20.* Manchester: Pigot.

Pigot J., 1828. *Pigot and Co.'s National Commercial Directory for 1828–9.* London: Pigot.

Pigot J., 1837. *Pigot and Co.'s National Commercial Directory for the Whole of Scotland and of the Isle of Man . . . to which are Added, Classified Directories of . . . Manchester, Liverpool, Leeds, etc.* London: Pigot.

Poussin, G.T., 1836. *Chemins de fer americains.* Paris: L. Mathias.

Powell, T., 2000, *Staith to Conveyor.* Houghton-le-Spring: Chiltern Ironworks.

Price, M.R.C., 1982. *Industrial Saundersfoot.* Llandysul: Gomer.

Priestley, J., 1831. *Historical Account of the Navigable Rivers, Canals, and Railways, of Great Britain.* London: Longman, Rees, Orme, Brown and Green.

Raman, A. and V. Balakrishnan, 2020. 'The Vadavār Railroad in Tanjāvûr district, Madras Presidency, reported in 1836', *Current Science* 119, 7, 1216–21.

Rankine, W.J.M., 1883. *A Manual of Civil Engineering.* London.

Ransom, J.P.G., 1996. *Narrow Gauge Steam: Its Origins and World-wide Development.* Sparkford: Oxford Publishing.

Rattenbury, P.G., 1964–5. 'Survivals of the Brinore Tramroad in Brecknockshire', *Journal of Industrial Archaology* 1, 173–83.

Rattenbury, P.G. and M.J.T. Lewis, 2004. *Merthyr Tydfil Tramroads and Their Locomotives.* Oxford: Railway and Canal Historical Society.

Rattenbury, P.G. and R. Cook, 1996. *The Hay and Kington Railways.* Gwernymynydd: Railway and Canal Historical Society.

Raybould, T.J., 1968. 'The development and organization of Lord Dudley's mineral estates, 1774–1845', *Economic History Review*, New Series 21, 3, 529–44.

Reed, B., 1982. *Crewe Locomotive Works and its Men.* Newton Abbot: David and Charles.

Reed, M., 1996. *The London & North Western Railway.* Penryn: Atlantic Transport.

Rees, J., 2001. 'The strange story of the steam elephant', in *Early Railways*. London: Newcomen Society, 145–70.
Rees, J., 2003. 'The Stephenson standard locomotive (1814–1825): a fresh appraisal', in *Early Railways 2*. London: Newcomen Society, 177–201.
Rees, J., 2010. 'The *Sans Pareil* model: its purpose and possible origins', in *Early Railways 4*. Sudbury: Six Martlets, 232–58.
Rees, J. and A. Guy, 2006. 'Richard Trevithick and pioneer locomotives', in *Early Railways 3*. Sudbury: Six Martlets, 191–220.
Reid, D., 1983. 'The origins of industrial labor management in France: the case of the Decazeville Ironworks during the July Monarchy', *Business History Review* 57, 1, 1–19.
Renwick, J., 1830. *Treatise on the Steam Engine*. New York: G and C and H Carvill.
Renwick, J. and D. Lardner, 1828. *Popular Lectures on The Steam Engine: In Which its Construction and Operation are Familiarly Explained*. London: J. Taylor.
Reynolds, P., 2011. 'Railway investment in Manchester in the 1820s', *Journal of the Railway and Canal Historical Society* 211, 38–48.
Reynolds, P.R., 1980. 'Scott's tramroad, Llansamlet', *Journal of the Railway and Canal Historical Society* 26, 85–95.
Reynolds P.R., 1999. 'Scott's Pit: new evidence from the north of England', *South West Wales Industrial Archaeology Society Bulletin* 74, 11–19.
Reynolds, P.R., 2003. 'George Stephenson's 1819 Llansamlet locomotive', in *Early Railways 2*. London: Newcomen Society, 165–76.
Reynolds, P.R., 2006. 'The London & South Wales Railway scheme of 1824/25', *South West Wales Industrial Archaeology Society Bulletin* 95, 3–7.
Richards, E.S., 1972. 'The finances of the Liverpool and Manchester Railway again', *Economic History Review* 25, 2, 284–92.
Richards, J. and J.M. MacKenzie, 1986. *The Railway Station: A Social History*. Oxford and New York: Oxford University Press.
Richardson, M.A., 1843. *The Local Historian's Table Book, of Remarkable Occurrences 3*. Newcastle: Richardson.
Riden, P.J., 1972. 'Outram's "Minutes to be observed in the construction of railways"', *Journal of the Railway and Canal Historical Society* 18, 61–4.
Riden, P.J., 1973. 'The Butterley Company and railway construction, 1790–1830', *Transport History* 6, 30–52.
Rieber, A.J., 1990. 'The rise of engineers in Russia', *Cahiers du monde russe et soviétique* 31, 4, 539–68.
Rimmer, A., 1985. *The Cromford & High Peak Railway*. Oxford: Oakwood.
Ripley, D., 1993. *The Little Eaton Gangway and Derby Canal*. Oxford: Oakwood.
Robert, L.E., 1843. *Histoire et description naturelle de la commune de Meudon*. Paris: Paulin.
Roberts, S.W., 1836. *An Account of the Portage Rail Road over the Allegheny Mountain, in Pennsylvania*. Philadelphia: Nathan Kite.
Roberts, S.W., 1872. Obituary notice of Edward Miller, civil engineer, *Proceedings of the American Philosophical Society* 12, 581–6.
Robertson, C.A., 1983. *The Origins of the Scottish Railway System 1722–1844*. Edinburgh: John Donald.
Roebling, J.A., 1843. 'American manufacture of wire rope', *American Railroad Journal* 16, 321–4.
Rojas-Sola, J.I. and E. De la Morena-De la Fuente 2019a. 'The Hay Inclined Plane in Coalbrookdale (Shropshire, England): analysis through computer-aided engineering', *Applied Sciences* 9, 16, 1–22.
Rojas-Sola, J.I. and E. De la Morena-De la Fuente 2019b. 'The Hay Inclined Plane in Coalbrookdale (Shropshire, England): geometric modeling and virtual reconstruction', *Symmetry* 11, 4, 589.
Rolt, L.T.C., 2005. *Landscape Trilogy, vol. 2: Landscape with Canals*. Stroud: Sutton.

Rolt, L.T.C., 2016. *George and Robert Stephenson: The Railway Revolution*. Stroud: Amberley.
Roper, R.S., 2003. 'Robert Stephenson, Senior, 1788–1837', in *Early Railways 2*. London: Newcomen Society, 26–36.
Rosenberg, N. and W.G. Vincenti, 1978. *The Britannia Bridge: The Generation and Diffusion of Technical Knowledge*. Monograph series/Society for the History of Technology 10. Cambridge, MA: MIT Press.
Rubin, J., 1961. 'Canal or railroad? Imitation and innovation in the response to the Erie Canal in Philadelphia, Baltimore, and Boston', *Transactions of the American Philosophical Society*, New Series, 51, 7, 1–106.
Rudé, G., 1981. *The Crowd in History 1730–1848*. London: Laurence and Wishart.
Ruskin, J., 1905. *The Cestus of Aglaia and the Queen of the Air with other Papers and Lectures on Art and Literature 1860–1870; Complete Works of John Ruskin vol. 19*. London: G. Allen.
Russeger, J., 1837. 'Einige Höhen in den Thälern Gastein und Rauris im Herzogthume Salzburg', *Zeitschrift für Physik und Mathematik*, 193–235.
Rynne, C. 2006. *Industrial Ireland 1750–1930: An Archaeology*. Cork: Collins Press.
Saar, F. von, 1874. *Die Steinklopfer: Eine Geschichte*. Heidelberg: Georg Weiss.
Samson, W.D. and G.J. Previtts, 1999. 'Reporting for success: the Baltimore and Ohio Railroad and management information, 1827–1856', *Business and Economic History* 28, 2, 235–54.
Saunders, A., 1949. 'Short history of the Panama Railroad', *Railway and Locomotive Historical Society Bulletin* 78, 8–44.
Schivelbusch, W., 2014. *The Railway Journey: The Industrialization of Time and Space in the Nineteenth Century*. Oakland: University of California Press.
Schofield, R.B., 2000. *Benjamin Outram 1764–1805*. London: Merton Priory Press.
Seguin, M., C. Seguin and E. Biot, 1826. *Mémoire sur le chemin de fer de St.-Étienne à Lyon*. Paris: Firmin Didot.
Serlo, A.L., 1878. *Leitfaden zur Bergbaukunde* 2. Berlin: Springer.
Shorland-Ball, R., 1994. *Common Roots – Separate Branches: Railway History and Preservation*. York and London: Science Museum.
Sima, J., 2008. 'Die Pferdeeisenbahn Budweis-Linz-Gmunden'. Doctor of Technology thesis, Technical University of Vienna.
Simmons, J., 1991. *The Victorian Railway*. London: Thames and Hudson.
Simons, P., G. Nicolas and J. de Ridder, 1833. *Mémoire a l'appui du projet d'un chemin à ornières de fer, a établir entre Anvers, Bruxelles, Liege et Verviers*, Brussels: de Vroom.
Skempton, A.W., 1953–5. 'The engineers of the English river navigations, 1620–1760', *Transactions of the Newcomen Society* 29, 25–55.
Skempton, A.W., 1973–4. 'William Chapman, 1749–1832, civil engineer (the 11th Dickinson memorial lecture)', *Transactions of the Newcomen Society*, 46 1, 45–82.
Skempton, A.W., 1978–9. 'Engineering in the Port of London, 1789–1808', *Transactions of the Newcomen Society* 50, 87–105.
Skempton, A.W., 1996. 'Embankments and cuttings on the early railways', *Construction History* 11, 330–49.
Skempton, A.W. (ed.), 2002. *Biographical Dictionary of Civil Engineers Volume 1 – 1560–1830*. London: Thomas Telford and Institution of Civil Engineers.
Skempton, A.W. and A. Andrews, 1976–7. 'Cast-iron edge-rails at Walker Colliery, 1798', *Transactions of the Newcomen Society* 48, 119–22.
Skempton, A. and C. Hadfield, 1979. *William Jessop, Engineer*. Newton Abbot: David & Charles.
Smiles, S., 1857. *Life of George Stephenson, Railway Engineer*. London: John Murray.
Smith, J.F., 1888. *Frederick Swanwick: A Sketch*. Edinburgh: privately published.
Smith, R.S., 1957. 'Huntingdon Beaumont, adventurer in coalmines', *Renaissance & Modern Studies* 1, 115–53.

BIBLIOGRAPHY

Smith, R.S., 1960. 'England's first rails: a reconsideration', *Renaissance and Modern Studies* 4, 119–34.
Smith, T., 2014, 'Stoneways', in *Early Railways 5*. Clare: Six Martlets, 295–311.
Snowden Bell, J., 1912. *Early Motive Power of the Baltimore & Ohio Railroad*. New York: Sinclair.
Snowden Bell, J., 1927. 'The genesis of the locomotive truck', *The Railway and Locomotive Historical Society Bulletin* 15, 72–8.
Stanley, A.P., 1860. *The Life and Correspondence of Thomas Arnold, D.D.* Boston: Ticknor and Fields.
Stapleton, D.H., 1978. 'The origin of American railroad technology, 1825–1840', *Railroad History* 139, 65–77.
Steingruber, C.K., 2021. *Pferdeeisenbahn Budweis-Linz-Gmunden. Dokumentation Der Als Bodendenkmal Erhaltenen Streckenabschnitte Mitsamt Kunstbauten*. St Gotthard.
Stephenson, R. and J. Locke, 1830. *Observations on the Comparative Merits of Locomotive and Fixed Engines, as Applied to Railways*. Liverpool: Wales and Baines.
Stevens, J. 1812. *Documents Tending to Prove the Superior Advantages of Rail-Ways and Steam Carriages over Canal Navigation*. New York: T and J Swords.
Stevenson, D., 1878. *Life of Robert Stevenson, Civil Engineer*. Edinburgh, London and New York: Black and Spon.
Stevenson, R., 1824. *Railway, Communicated to the Editor of The Edinburgh Encyclopaedia*. Edinburgh.
Stokes, W., 2001. ''Early railways and regional identity', in *Early Railways*. London: Newcomen Society, 311–24.
Stokes, W., 2003. 'Who ran the early railways? The case of the Clarence', in *Early Railways 2*. London: Newcomen Society, 79–92.
Stokes, W., 2006. 'The importance of the Northeast viewers in the development of early railways and locomotives', in *Early Railways 3*. Sudbury: Six Martlets, 165–75.
Stokes, W., 2010. 'A City Job? The London shareholders in the Clarence Railway', in *Early Railways 4*. Sudbury: Six Martlets, 192–205.
Stolberg-Wernigerode, O. zu, 1982. 'Lanna, Adalbert, Grossindustrieller', in *Neue deutsche Biographie 13*. Berlin, West: Duncker & Humblot, 618–19.
Stoskopf, N., 2000. 'Qu'est-ce que la haute banque parisienne au XIXe siècle?', *Journée d'études sur l'histoire de la haute banque* (16 novembre), Fondation pour l'histoire de la haute banque.
Strickland, W., 1826. *Reports on Canals Railways Roads & Other Subjects Made 1826*. Philadelphia: H.C. Carey and I. Lea.
Stuart, C.B., 1871. *Lives and Works of Civil and Military Engineers of America*. New York: D Van Nostrand.
Sykes, J., 1833. *Local Records; or, Historical Register of Remarkable Events: Which Have Occurred in Northumberland and Durham, Newcastle upon Tyne, and Berwick upon Tweed*. Newcastle: privately published.
Tarsaidze, A., 1950. 'American pioneers in Russian railroad building', *Russian Review* 9, 4, 286–95.
Taylor-Cockayne, M., 2020. 'Josias Jessop, civil engineer to railway engineer', *Journal of the Railway and Canal Historical Society* 238, 117–21.
Telford, T., 1810. *Report By Mr Telford, Relative to the Proposed Railway from Glasgow to Berwick-Upon-Tweed; with Mr Jessop's Opinion Thereon. And Minutes of a Meeting of the General and Sub-Committees, held 4th April 1810*. Edinburgh: Neill.
Terry Sharrer, G., 1982. 'The merchant-millers: Baltimore's flour milling industry, 1783–1860', *Agricultural History* 56, 1, Symposium on the History of Agricultural Trade and Marketing, 138–50.
Tew, D.H., 1951–53. 'Canal lifts and inclines with particular reference to those in the British Isles', *Transactions of the Newcomen Society* 28, 35–58.

BIBLIOGRAPHY

Thomas, R.H.G., 1980. *The Liverpool & Manchester Railway*. London: Batsford.

Thompson, B., 1847. *Inventions, Improvements, and Practice of Benjamin Thompson, in the Combined Character of Colliery Engineer, and General Manager*. Newcastle: Lambert.

Thompson, C., 2008. *Harnessed: Colliery Horses in Wales*. Cardiff: National Museums and Galleries of Wales.

Thompson, E.P., 1980. *The Making of the English Working Class*. London: Penguin.

Thompson, M., 2015. *The Picturesque Railway: The Lithographs of John Cooke Bourne*. Stroud: The History Press.

Thomson, T.R., 1942. *Check List of Publications on American Railroads before 1841*. New York: The New York Public Library.

Thoreau, H.D., 1854. *Walden; or, Life in the Woods*. Boston: Ticknor and Fields.

Tomlinson, W.W., 1915. *The North Eastern Railway: Its Rise and Development*. Newcastle-upon-Tyne and London: Andrew Reid and Longmans, Green.

Toogood, A.C., 1973. *Historic Resource Study Allegheny Portage Railroad*. Denver: United States Department of the Interior.

Topham, W., 1925. 'First railroad into Washington and its three depots', *Records of the Columbia Historical Society* 27, 175–247.

Trautwine, J.C., 1835. 'Sections of rails used in the United States', *Journal of the Franklin Institute* new series 16, 303–4.

Tredgold, T., 1825. *A Practical Treatise on Railroads*. London: Taylor.

Treese, L., 2006. *Railroads of New Jersey: Fragments of the Past in the Garden State Landscape*. Mechanicsburg, PA: Stackpole Books.

Trevithick, F. 1872. *Life of Richard Trevithick*. London: Spon.

Trinder, B., 1981. *The Industrial Revolution in Shropshire*. London and Chichester: Phillimore.

Trinder, B., 1996. *The Industrial Archaeology of Shropshire*. London and Chichester: Phillimore.

Trinder, B., 2003. 'Recent research on early Shropshire Railways', in *Early Railways 2*. London: Newcomen Society, 10–25.

Trinder, B., 2005. *'The Most Extraordinary District in the World': Ironbridge and Coalbrookdale*. Chichester: Phillimore.

Trinder, B., 2008. 'William Reynolds: polymath – a biographical strand through the Industrial Revolution", *Industrial Archaeology Review* 30, 1, 17–32.

Trinder, B., 2013. *Britain's Industrial Revolution: The Making of a Manufacturing People, 1780–1870*. Lancaster: Carnegie.

Trollope, A., 1855. *The Warden*. London: Longman, Brown, Green and Longmans.

Tucker, E.G., 1981.'The Coalbrookdale Railway, 1767–68', *Journal of the Railway and Canal Historical Society* 27, 2–6.

Turnbull, A.D., 1928. *John Stevens, an American Record*. New York and London: The American Society of Mechanical Engineers.

Turnbull, L., 2019. *The Railway Revolution*. Newcastle: North of England Institute of Mining and Mechanical Engineers and The Newcastle upon Tyne Centre of the Stephenson Locomotive Society.

Turnock, D., 2004. *The Economy of East Central Europe, 1815–1989: Stages of Transformation in a Peripheral Region*. London: Routledge.

Tusch, R., 2019. 'Railway architecture: the authenticity and identity of the Semmering Railway', *Early Main Line Railways 2*. Sponsors of Second International Early Main Line Railways Conference, 353–74.

Uglow, J., 2002. *The Lunar Men*. London: Faber and Faber.

United States Department of the Treasury, 1838. *Steam Engines: Letter from the Secretary of the Treasury, Transmitting, in Obedience to a Resolution of the House, of the 29th of June Last, Information in Relation to Steam Engines, &c*. Washington: Thomas Allen.

United States Patent Office, 1847. *List of Patents for Inventions and Designs: Issued by the United States, from 1790 to 1847, with the Patent Laws and Notes of Decisions of the Courts of the United States for the Same Period.* Washington: J. and G.S. Gideon.

Urquijo, J.I., 2006. 'Los Primeros ferrocarriles de Venezuela: proyectos, tanteos, suenos e ilusiones', *Revista sobre relaciones industriales y laborales* 42, 9–58.

Van der Burg, P., 1844. *Eerste grondbeginselen der natuurkunde.* Gouda: van Goor.

Van Laun, J., 1976. 'The Hay Railway: the passing of an early railway act', *Journal of the Railway and Canal Historical Society* 22, 2–10.

Van Laun, J., 1979. 'Hills Pits, Blaenavon', *Industrial Archaeology Review* 3, 3, 257–75.

Van Laun, J., 1996–8. 'A tramroad company seal', *Journal of the Railway and Canal Historical Society* 32, 132–4.

Van Laun, J., 2001a. 'Pre-1840s trackways in South Wales', in *Early Railways.* London: Newcomen Society, 27–45.

Van Laun, J., 2001b. *Early Limestone Railways.* London: The Newcomen Society.

Van Laun, J., 2003. 'In search of the first all-iron edge-rail', in *Early Railways 2.* London: Newcomen Society, 93–101.

Van Laun, J. and D. Bick, 2000. 'South Wales plateways 1788–1860', *Antiquaries Journal* 80, 321–31.

Vanags, J., 2001. *The Mansfield and Pinxton Railway.* Mansfield: Old Mansfield Society.

Vaxell, L. de, 1805. 'Opisanie čugunnoj dorogi (iron railway) učreždennoj v" Grafstvŧ Surrej, v" Anglii v" 1802 godu'. St Petersburg: at the Medical Printing House.

Veenendaal, A.J., 1995. 'Railways in the Netherlands, 1830–1914', *Railroad History* 173, 5–57.

Viau, R., 1847. *Chemin de Fer du Havre á Rouen: album itinéraire.* Ingouville/Havre: Roquencourt.

Vignoles, O.J., 1889. *Life of Charles Blacker Vignoles.* London: Longmans, Green and Co.

Villefosse, H. de, 1810–19. *Traité de la richesse minerale.* Paris.

Virginskiĭ, V.S., 1971. *Efim Alexeevich i Miron Efimovich Cherepanovy.* Sverdlovsk: Sredne-Ural'skoe knizhnoe izd-vo.

Walker, J. and J.U. Rastrick, 1829. *Liverpool and Manchester Railway: Report to the Directors on the Comparative Merits of Locomotive and Fixed Engines as a Moving Power.* Liverpool: John and Arthur Arch.

Walker, J., R. Stephenson, J. Lock and H. Booth, 1831. *Report to the Directors of the Liverpool and Manchester Railway, on the Comparative Merits of Locomotive and Fixed Engines as a Moving Power/Observations on the Comparative Merit of Locomotive And Fixed Engines as Applied to Railways/An Account of the Liverpool and Manchester Railway.* Philadelphia: Carey & Lea.

Walker, J.S., 1830. *An Accurate Description of the Liverpool and Manchester Railway.* Liverpool: J.F. Cannell.

Wallach, A., 2002. 'Thomas Cole's "River in the Catskills" as antipastoral', *The Art Bulletin* 84, 2, 334–50.

Wang, H.C., 2010. 'Discovering steam power in China, 1840s–1860s', *Technology and Culture* 51, 1, 31–54.

Warren, J.G.H., 1923. *A Century of Locomotive Building by Robert Stephenson & Co.* Newcastle-on-Tyne: Andrew Reid.

Waterhouse, R., 2014. 'The Tavistock Canal plateways and edge railways, West Devon', in *Early Railways 5.* Clare: Six Martlets, 270–94.

Waterhouse, R., 2017. *The Tavistock Canal: Its History and Archaeology.* Camborne: Trevithick Society.

Watkins, J.E., 1891. *The Development of the American Rail and Track, as Illustrated by the Collection in the U.S. National Museum* (from *The Report of the National Museum, 1888–1889*, 651–708). Washington: National Museum.

Watkins, J.E., 1892. *John Stevens and his Sons, Early American Engineers.* Washington: W.F. Roberts.

BIBLIOGRAPHY

Wayt, H., 2016. 'Railroad tracks belonging to the South Carolina Canal and Railroad Company, c.1839–1852', *IA, The Journal of the Society for Industrial Archeology* 42, 1, 19–36.
Webster, C., 1975. *The Great Instauration.* London: Duckworth
Webster, N.W., 1972. *Britain's First Trunk Line.* Bath: Adams and Dart.
Weidmann, F.C., 1842. *Die Budweis-Linz-Gmundner Eisenbahn.* Vienna: Sollinger.
Weidmann, F.C., G. Fobbe and O. Prechtler, 1962. 'Oberösterreichs erste Eisenbahn in zeitgenössischen Schilderungen', *Oberösterreichs Heimatblätter* 16, 2, 107–16 and plates.
Weill, G., 1896. *L'École Saint-Simonienne son histoire son influence jusqu'à nos jours.* Paris: Germer Baillière.
Welch, S., 1835. 'Allegheny Portage Rail-Way', *Hazard's Register of Pennsylvania* 15, 348–51, 353–7.
Whishaw, F., 1842. *The Railways of Great Britain and Ireland Practically Described and Illustrated.* London: Robson, Levey and Franklyn.
White, J.H., 1967. 'James Millholland and early railroad engineering', *Contributions from the Museum of History and Technology Bulletin* 252, 3–36.
White, J.H., 1976. 'Tracks and timber', *IA: The Journal of the Society of Industrial Archaeology*, 1, 35–46.
White, J.H., 1978. *The American Railroad Passenger Car.* Baltimore: Johns Hopkins Studies in the History of Technology.
White, J.H., 1984a. 'Once the greatest of builders: the Norris Locomotive Works', *Railroad History* 150, 17–86.
White, J.H., 1984b. 'Who was Ezra Miller?' *Railroad History* 150, 115–17.
White, J.H., 1993. *The American Railroad Freight Car.* Baltimore and London: Johns Hopkins University Press.
White, J.H., 1994. 'Old debts and new visions: the interchange of ideas in railway engineering', in *Common Roots – Separate Branches: Railway History and Preservation*, ed. R. Shorland-Ball. York and London: Science Museum, 65–87.
White, J.H., 1997. *American Locomotives: An Engineering History.* Baltimore and London: Johns Hopkins University Press.
Whiting, W., 1853. *Argument of William Whiting, Esq. in the case of Ross Winans v. Orsamus Eaton et al. for an Alleged Infringement of his Patent for the Eight Wheel Railroad Car.* Boston: Hewes.
Whitney, A., 1848. *Memorial of Asa Whitney for a grant of land to enable him to construct a railroad from Lake Michigan to the Pacific Ocean.* 30th Congress, 1st session, miscellaneous 28.
Wild, J.C., 1840, *Views of Philadelphia, and its Vicinity.* Philadelphia: Wild and Chevalier.
Wilkins, C., 1888. *The South Wales Coal Trade and its Allied Industries.* Cardiff: Daniel Owen.
Williams, W.S., 1908. *Hanes a Hynafiaethau Llansamlet.* Dolgellau: EW Evans.
Williams-Ellis, E., 2004. *The Carrier's Tale: The Hargreaves Family of Bolton and Westhoughton.* Chester: CC Publishing.
Wilmott, M., 2001. 'Early railways in Dorset: the industrial railways of Purbeck', in *Early Railways*. London: Newcomen Society, 103–13.
Wilson, H.M., 1883. 'The Morris Canal and its inclined planes', *Scientific American Supplement* 24, 5943–4.
Wilson, T., 1845. *The Railway System and its Author, Thomas Gray, Now of Exeter: A Letter to the Right Hon. Sir Robert Peel.* London: E. Wilson.
Wilson, W.H. and S. Roberts, 1879. *Notes on the Internal Improvements of Pennsylvania, and Reminiscences of the First Railroad over the Allegheny Mountain.* Philadelphia: Railway World.
Winans, R., 1854. *Ross Winans vs. the Eastern Railroad Company: Evidence for Complainant. October term, 1853.* Boston: printed for the clerk of the court.

Winstanley, D., 2014. 'The evolution of early railways in Winstanley, Orrell and Pemberston, Lancashire, England 1770s-1870s', in *Early Railways 5*. Clare: Six Martlets, 112–31

Withuhn, W.L., 2006. 'Abandoning the *Stourbridge Lion* – business decision-making, 1829: a new interpretation', in *Early Railways 3*. Sudbury: Six Martlets, 153–75.

Wolmar, C., 2010. *Engines of War: How Wars were Won and Lost on the Railways*. London: Atlantic.

Wood, N., 1825. *A Practical Treatise on Rail-Roads, and Interior Communication in General*. London: Knight and Lacey.

Wood, N., 1831. *A Practical Treatise on Rail-Roads, and Interior Communication in General*. London. Longman, Hurst, Chance, and Co.

Wood, N., 1834a. *Traité pratique des chemins de fer*. Paris: Carilian-Goeury.

Wood, N., 1834b. *Traité pratique des chemins de fer*. Brussels: Lejeune.

Wood, N., 1838. *A Practical Treatise on Rail-roads and Interior Communications in General; 3rd edition with additions*. London: Longman, Orme, Brown, Green, and Longmans

Wood, O., 1954. 'A Cumberland colliery during the Napoleonic War', *Economica*, New Series 21, 81, 54–63.

Wood, P.N., 2010. 'Excavation on the Brunton and Shields Railway at Weetslade, North Tyneside', *Industrial Archaeology Review*, 32 2, 77–90.

Woodcroft, B., 1854. *Titles of Patents of Invention, Chronologically Arranged: From March 2, 1617 (14 James I.) to October 1, 1852 (16 Victoriae)*. London: Patent Office.

Wordsworth, W., 1845. *Kendal and Windermere Railway: Two Letters Re-printed from the Morning Post*. Kendal.

Wosk, J., 2013. *Breaking Frame: Technology, Art and Design in the Nineteenth Century*. Bloomington, IN: Universe Inc.

Wyatt, B., 1803. 'Account of the Penrhyn Iron Railway: communicated by the inventor, Mr Benjamin Wyatt', *Repertory of Arts and Manufactures*, 3, second series, 285–7.

Young, R., 1923. *Timothy Hackworth and the Locomotive*. London: The Locomotive Publishing Company.

Online resources

Biographisches Lexikon des Kaisertums Österreich https://austria-forum.org/web-books/kategorie/lexika/wurzbach-lexikon

Dictionary of National Biography https://en.wikisource.org/wiki/Dictionary_of_National_Biography,_1885-1900

Oxford Dictionary of National Biography https://www.oxforddnb.com/

Railway & Canal Historical Society Railway Chronology Special Interest Group. A Chronology of the Cromford & High Peak Railway and associated events. Compiled by P J McCarthy https://rchs.org.uk/wp-content/uploads/2018/02/Cromford-High-Peak-Railway-Nov-2001.pdf

INDEX

Abernant Ironworks 64, 90
accidents 92, 102, 104, 249, 259–60, 279–80, 297
Ackermann, Rudolph 253
Adams, John Quincy 235, 257, 279
Adams, William Bridges 135
adhesion 32, 43, 93, 103–7, 164, 178, 181, 193, 336, 351
Aire and Calder Navigation 56
Alexandria and Cheneyville Railroad 286
Ali, Muhammad, *Wāli* of Egypt 311
Alleghany County (Maryland) 37, 77, 331
Alexander I, Tsar of Russia 111, 114, 119, 336, 337
Alexandrovsk cannon-works and railway workshops 37, 75–6, 295, 308
Allegheny Portage Railroad 126, 134, 142, 147, 148, 149, 152, 154, 155, 266, 274, 275, 282–3, 292, 339
Allcard, William 290
Allen, Horatio 48, 114, 116, 124, 136, 166, 171–2, 209, 239, 241–2, 245, 249
American Philosophical Society 122, 170
Anderson, James 61
Angerstein, Reinhold Rucker 14, 36
articulation, in railway vehicles and locomotives 6, 54, 73–4, 103, 176, 178, 182–4, 186, 198, 246, 247–8, 277–9, 298, 345, 354
architecture of railway buildings 70, 197, 241, 251–7, 259, 260, 280–2, 309–10

Ashby-de-la-Zouche collieries and plateway system 61
Astrakhan 111, 112, 114
atmospheric traction 157, 300
Australian Agricultural Company *see* Newcastle Colliery and railroad, New South Wales
Austrian Empire *see* German Confederation

Baader, Joseph von 47, 111, 119, 120, 121, 150, 185, 300
Babbage, Charles 258
Backhouse, Joseph 196
Bad Gastein (Salzburg) inclined plane 75, 87
Baldwin, Matthias 122–3, 170, 181, 183, 189, 246
Baldwin Locomotive Works 189, 246
ballast, as track component 16, 32, 42, 43, 132, 134, 202
Ballochney Railway 203, 265
Baltimore and Ohio Railroad 46, 113, 125, 133, 134, 136, 143, 148, 152, 156, 179, 180, 181, 182, 187, 233–62, 269, 274, 277, 278, 302, 305, 308, 310
Banks, Joseph 8–9, 14, 19, 21, 34, 43, 44, 61
Banque de France 272
Bassaleg viaduct 64
Batthyány de Németújváar, Count Lajos 258

INDEX

Bavaria *see* German Confederation
Bavarian Ludwig Railway 121, 291
Bayerische Ludwigseisenbahn see Bavarian Ludwig Railway
Beaumont, Huntingdon 12
Beaunier, Louis-Antoine 204–5
Bedlington Ironworks 128–30
Belgium, development of railway systems in 192, 264, 269, 272, 289, 291, 309
Bell Rock lighthouse 59
Belmont inclined plane 148, 155
'Benny's bank' *see* Rainton and Seaham Railway
Bentham, Jeremy 61
Benton, Thomas Hart 288
Benwell staithes 23
Bersham Ironworks 75
Betancourt y Molina, Agustin 67
Bewicke Main Colliery, railway and inclined plane 25, 81, 90, 140
Bidder, George Parker 288–9
Birkinshaw, John 129
Birkinshaw rail 129, 135, 162, 193
Birmingham and Fazeley Canal 55
Birmingham and Liverpool Junction Canal 51
Black, Joseph 61
Blackett, Christopher 96, 98, 104–5
Blaenavon 64
Blenkinsop, John 21, 99–102, 106–7, 114, 119, 160, 161, 201, 214–15
Board of Agriculture and Internal Improvement 22, 62
Bodmin and Wadebridge Railway 131, 178, 264, 270, 271
boiler design 88, 93, 98, 99–100, 107, 152, 160, 162–83, 186, 187, 201, 205, 207, 245, 298, 299
Bolívar, Simón 215, 295
Bolton and Leigh Railway 130, 141, 146, 151, 154, 169, 171, 179, 188, 200, 201, 266
Bonomi, Ignatius 197
Booth, Henry 145, 165, 236, 239, 250, 318
Borsig Works 298, 299
Boston and Lowell Railroad 134, 269, 282
Boulton and Watt fixed engines 88, 91
Bourne, John Cooke 303, 310, 314, 316, 317
Božek, Josef 226
Bradley Ironworks 130, 132

Braithwaite, John 122, 168, 169–70, 173, 177, 202, 266
Brandling, Charles John 99
Brassey, Thomas 290
Bratislava to Trnava Railway 276
Brecknock and Abergavenny Canal 38, 52, 63
Brecon Forest plateway 52, 69, 70, 71, 142, 150, 151
bridges carrying railways 26, 64, 197, 198, 199–200, 205, 211, 212, 229, 240–2, 266, 274, 302–3, 314, 317
Bridgewater, Francis Egerton, Duke of 11, 49, 66, 85
Bridgewater, Rev Francis Henry Egerton, Earl of 66
Bridgewater Canal and lift 11, 66, 67, 85, 200, 236, 254
Brindley, James 66
Brinore plateway 63
Brontë, Charlotte 316
Brougham, Henry 258
Brown, Samuel (inventor of gas vacuum engine) 185–6
Brown, Samuel (designer of Tees suspension bridge) 199
Brown family (bankers of Baltimore) 237–8
Bruce, John (early locomotive fatality) 102
Brunel, Isambard Kingdom 6, 124, 288–9, 297, 303–4
Brunton, John 62–3, 150
Brunton, William 26, 69, 104, 145, 150
Brunton and Shields Railway 141, 195, 227
Bryant, Gridley 77
Buddicom, William Barber 290, 299
Buddle, John (senior) 35
Buddle, John (junior) 83, 103, 121, 129, 131, 160, 193
Bude Canal 68
Budweis to Linz (horse) railway 2, 34, 213–32, 254, 262, 291, 292
Bugsworth Basin 70
Bullo Pill plateway 58, 71
Burgundy Canal 143, 207, 264
Burstall, Timothy 117, 168–9, 179
Bury, Thomas Talbot 253
Bury & Kennedy, locomotive builders 172, 173, 174, 183, 188, 298
Bute Ironworks 69
Butterfield, Herbert 4

389

INDEX

Butterley estate, foundry and ironworks 36, 42, 60, 61, 103, 104, 151, 153, 155, 220–1, 222–3

Caldbeck mine 12,
Caldon Low (limestone quarries) railroad 59, 86
Caledonian Canal 59, 73, 84
Callan, Nicholas 301
Cambridge University 124
Camphausen, Ludolf 112
Canada, development of railway systems in 288
canal acts 44
Canal du Creusot *see* Creusot Canal
canals 5, 9, 11, 14, 20, 21, 23, 29, 30, 35, 36, 37, 38, 39, 43, 44, 46, 49–52, 55–79, 83–90, 93, 95–6, 100, 101, 103, 108–9, 111, 120–1, 123, 133, 141–3, 149, 150, 155, 159, 165, 166, 177, 193, 196, 200, 201, 202, 203, 207, 209, 210, 214–16, 218, 220–4, 228, 229, 231, 233, 236–7, 240, 245, 251–2, 254, 257, 260, 263–6, 271, 275, 285, 291, 294, 303, 305, 312, 321
Canterbury and Whitstable Railway 130, 141, 151, 153, 154, 172, 195–6, 217
Cape Diamond citadel (Canada) inclined plane 76–7, 84
capstans 222
Carmarthenshire Rail Road 57, 63
carriages, passenger 73, 98, 201, 227, 246–8, 259, 262, 277–80; names 248, 278–9
Carroll, Charles 257
Carron Ironworks 75
Carver, Hartwell 288
Cartwright, William 60, 64
cast iron, use of as track component 2, 3, 8, 14, 18, 19, 22, 30–48, 49–50, 78, 80, 99, 105, 126–32, 136–7, 162, 175, 179, 204, 211, 214, 217, 218, 222–3, 225, 230
Catherine the Great, Empress of Russia 75–6
Central Railroad of Georgia 307
ceremonies, civic, associated with railways 57, 71, 113, 198, 201, 207, 257–60, 291
Chaillot Foundry (Paris) 153
chains, haulage 85, 88, 99, 103–4, 107, 139–55, 158

chairs, track component 40, 135
chaldron waggon 16, 17, 33, 72, 90, 194
Chapel-en-le-Frith inclined plane 63, 70, 85
Chapman, William 23, 68, 72–3, 103–4, 106, 160, 227
Charleston and Hamburg Railroad 2, 133, 134, 142, 148, 152, 179–80, 183, 184, 185, 209, 233–62, 273, 309, 310, 318–19
Chartists 312
Chat Moss 239–40, 253
Cherepanov, Yefim Alekseyevich 115, 174–5, 258
Cherepanov, Miron Yefimovich 115, 174–5, 258
Chesapeake and Ohio Canal 251, 257
Cheshunt monorail 47
Chevalier, Michel 112, 312
China, interest in railways 295
Chopwell waggonway 15
Chorzow ironworks 75, 102, 192, 365
civil engineering 17, 59–65, 124, 205, 239–42, 292, 302–3
Clapeyron, Benoît Paul Émile 125
Clarence Foundry 172–3
Clarence Railway 131, 200, 265, 271, 274
Clowes Wood inclined plane and fixed engine 153
coal as fuel 5, 164, 166, 170, 176, 185, 215, 217
coal industry 2–3, 5, 8–29, 33, 35, 49, 66–7, 71, 75, 102, 109–10, 116, 117, 143, 161, 167, 192–212, 264–70
Coalbrookdale 8, 14, 19, 34, 43, 44, 61, 75, 92
coalfields 2, 5, 10, 12–15, 14, 17, 20–5, 35, 42, 49, 75, 116, 117, 146, 167, 192, 193, 196, 200, 203, 207, 208, 212, 264–6, 268
coerced labour and skills 111, 115, 174, 242, 258, 264, 307–9
coke as locomotive fuel 169, 170, 175, 179, 299
Cole, Thomas 317
Columbia College (New York) 123, 149, 209, 245
Compagnie du chemin de fer de Saint-Étienne à Lyon 131
compressed air as railway prime mover 184–5, 191, 300
Congleton colliery and railroad 39

390

INDEX

Conwy tubular bridge 303
Cort Ironworks 130
cotton industry *see* textile industry
Coxlodge and Kenton railway 23, 26, 27, 100, 102, 106, 107, 161
Crawshay, Richard 65, 93
Crawshay, William 295
Creusot Canal 68
Crewe 290, 299, 308
Crich limestone quarries railway 36, 43, 61, 104
Crockett, David 'Davy' 184, 243, 280, 318–19
Cromford canal 36, 43, 60, 61, 222, 228
Cromford and High Peak Railway 42, 59, 141, 147, 151, 153, 155, 213–32, 292
Crown Street inclined plane 147, 151, 154–5, 244, 254
Cuba, sugar industry and railways 295, 298, 304, 306
Cugnot, Nicolas-Joseph 91
Cumming, Thomas G 215
Curr, John 35–6, 43
Cwm Clydach railroad 39
Cyfarthfa Ironworks 64, 65, 93, 127, 149, 187, 295

Dadford, John 39, 60
Dadford, Thomas (senior) 38, 60
Dadford, Thomas (junior) 38–9, 60, 68
Daglish, Robert 100, 201
Danube River 67, 110, 121, 218–30, 291
Danube Steamboat Shipping Company 230
Danville and Pottsville Railroad 208
Davenport, Thomas 301
Davidson, Robert 301–2
Davis, Phineas 180, 249
Dechen, Heinrich von 119, 129, 194, 196, 220
Delaware and Hudson Canal Company and railroads 114, 133, 134, 143, 149, 165, 166, 171–2, 209, 241, 245
Demidov, Count Paul Nikolaievich 115, 174, 258
Denby collieries plateway system 61
depots 70, 249, 251–2, 281–2, 312
Derby Canal 36
Dickens, Charles 316
Dinorwic (slate quarry) railroad 54
Dr Griffiths' plateway 72
Dodds, Ralph 161, 162, 193

Donaudampfschiffahrtsgesellschaft *see* Danube Steamboat Shipping Company
Donnington Wood Canal 66, 89
Dove Holes limestone quarries 58
Dowlais Ironworks 70, 130, 134, 178, 298
Drumglass Colliery 65
Dublin and Kingstown Railway 132, 175, 231, 267, 269, 270, 277, 279, 280, 283, 301
Ducart, Davis 65
Dudley, John William Ward, Earl of 165, 190
Dudley Canal 55
Dundee and Newtyle Railway 130, 141, 148, 176
Dunham, locomotive builders 189
Durham University 124
Durnford, Elias Walker 76–7
Dusseldorf and Elberfeld Railway 296
Dutens, Joseph 67, 119

East Winlaton waggonway 15
Eastern Counties Railway 135
Eastwick & Harrison 294
École des Ponts et Chaussées 125
Edge Hill 142, 147, 148, 151, 157, 254
edge railways 19, 22, 29, 31, 32, 37, 38, 40, 42, 47, 48, 52, 68, 83, 106, 130, 137, 162, 178, 197, 364
Edgeworth, Richard Lovell 44, 45, 91, 140
Edinburgh University 47, 124, 289
Egypt, railways in 295, 311
Eichthal, Adolphe d' 290
Eliot, George 316, 321
Ellicott's Mills 238, 251, 253, 257, 305
Ellis, Thomas 69
Emerson, Ralph Waldo 187, 318–20
engines, fixed steam 2, 20, 25, 35, 44–5, 68, 80, 88–91, 94, 95, 107, 134, 138, 139–57, 169, 185, 188, 192, 194, 195, 200, 201, 202, 205, 209, 215, 233, 244, 275, 300
engines, locomotive, *see* locomotives
Enlightenment 9, 21, 61, 112, 122, 312
enslaved workers *see* coerced labour
enslavement *see* coerced labour
Épinac railway 143–4, 154, 207, 264, 270
Ericsson, Johan (John) 122, 124, 168, 169–70, 173, 177, 259, 266
Erie Canal 120, 237, 266, 271

INDEX

Erie Railroad 304
ethnic mixing and conflict on railway construction and operation 7, 241, 250, 279–80, 295, 306–8, 312–13
Euston Station 157, 296, 310
Exchequer funds 57, 239, 271

Fairbairn, William 151, 303
Falkenberg (Sokolec, Silesia) inclined plane 75
Fawdon waggonway 140, 141, 195
Felling waggonway/railroad 24
Fenton, Murray & Wood, engineering company 151, 188
Ffestiniog Railway 126, 272, 276
finance *see* investment in railways
fish-belly rail 39, 128, 129, 130, 131, 133, 197, 204, 205, 211, 222, 225, 235, 238
Forest of Dean 6, 61, 69, 71
Forrester, George, locomotive builder 175, 188
Foster, John 254–5
Foster, Rastrick & Co 165–6
foundries 22, 32, 34, 40, 50, 62–3, 89, 90, 96, 99, 100, 102, 103, 113, 132, 147, 152–3, 156, 172, 177, 179–84, 188–9, 190, 201, 208, 211, 222–3, 299
France, development of railway systems in 29, 42, 61, 75, 109, 110, 112, 117, 119, 120, 125, 131, 143–4, 164–5, 171–2, 174, 203–7, 214, 263, 264–5, 268–9, 271–2, 289–90, 299, 308
Frederick the Great, King of Prussia 110
French Revolution, 1789–99 5, 21, 110
Fulton, Robert 62, 66, 67–8, 87

Gainschnigg, Joseph 87
Garnkirk and Glasgow Railway 130, 203, 265, 280, 281,
Gateshead Park Ironworks 96
gauge of railways 6, 12, 13, 16, 17, 20, 29, 33, 36, 44, 54, 64, 92, 96, 102, 134, 136, 150, 175, 178, 197, 204, 207, 208, 210, 224–5, 235, 242, 253, 261, 273, 291, 292, 303–4
George, Watkin 64
George III, King of Great Britain and of Ireland, King of the United Kingdom and of Hanover 39, 93
German Confederation 110, 111–12, 263, 272, 308, 309; Austrian Empire 2, 34, 74–5, 110, 112, 114, 119, 213–32,

291–2, 294, 297, 299, 301, 302–3, 306; Bavaria 121, 291, 298; Hanover 39; Prussia 61, 75, 101–2, 109, 110, 112, 119, 192–3, 207, 264, 269, 276–7, 298, 305; Saxony 298
Gerstner, Franz Anton von 224–9, 231, 261–2, 268, 273, 276, 282, 286, 292, 304
Gerstner, Franz Joseph von 120, 221, 224, 226
Ghega, Karl von 292, 297, 301
Giddy, Davies 93, 94, 95
Gilbert, John 66
Gladstone, John 239
Gladstone, William Ewart 309, 316
Glamorganshire Canal 38, 52, 56, 93
Gleiwitz (Gliwice) ironworks 67
Gloucester and Cheltenham Railway 57, 61, 178, 190–1
Glynneath plateway 68, 90, 95
Goethe, Johann Wolfgang von 112, 284
Gongchen, Ding 295
Gooch, Daniel 299
Goold, James 277
Goudriaan, Bernard 304
gradients 17, 33, 43, 63, 68, 69, 73–4, 82, 85, 90, 94, 100, 104, 144–5, 146, 149, 153, 156, 172, 178, 203, 205, 208, 364
'Grand Allies' 18, 19, 27, 141, 160–1, 194–5
'Grand Cross' canal system 60
Grand Junction Railway 288–9, 297–8, 299
Grand Western Canal 68
Granite Railway (Quincy, Massachusetts) 46, 77, 133, 134, 142, 238
gravity, as impellent force 16, 80, 85–6, 137, 142, 143–4, 198, 205, 206, 209, 210
Gray, Thomas 214–15, 237–8
Great Orme Tramway 157–8
Great Western Railway 6, 186, 288, 299, 302, 303, 304, 305, 306, 310, 314, 316–17
Green, James 68
Gregory XVI, Pope 311
Grimshaw, John 154
Grosmont Railway 51–2, 57–8, 70
Guest, Josiah 70
Gurney, Goldsworthy 187

Hackworth, Timothy 105, 107, 115–16, 118, 130, 156–7, 163, 167, 169, 173, 176–7, 181, 196, 294
Hadley Canal and lift 87

392

INDEX

Haigh Foundry (Wigan) 100, 188
Hamel, Joseph Christian 114–15, 121, 292
Hanover (Electorate) 75
Hanover (Kingdom) *see* German Confederation
Harford Davis Ironworks (Ebbw Vale) 131, 134
Hargreaves, John 188, 202
Harkort, Friedrich 207
Harpers Ferry 238, 253
Harrison, Winans & Eastwick 295, 298–9
Hartley, Jesse 114, 116, 131
Harz 39, 75
Hawks Ironworks (Gateshead) 151, 160
Hay Railway 52, 57–8
Hay lift (Coalport) 65–6, 89, 119
Hayle Foundry 62–3
Haytor and Holwell quarries and railway 46, 69–70, 74, 83
Hazard, Erskine 210
Hazlitt, William 316
Heath, William 253
Heaton Colliery waggonway 103, 193
Hedley, William 105, 107, 183
Henderson, John 117, 212
Henry, Charles-Joseph 204
Hereford Railway 52, 70, 214
Heslop, Adam 88, 89
Hetton Colliery railroad 40, 141, 145, 160, 161, 166–7, 194–5, 196
Hewitt, James 102–3
Hick, Benjamin 179, 184, 188, 201
Hill, David Octavius 265
Hill, Richard 93
Hirwaun Ironworks 64, 95
Hodgkinson, John 61, 64, 222
Höganäs Colliery (Sweden) 117
Holyhead harbour 59
Holy Roman Empire 110
Homfray, Jeremiah 90
Homfray, Samuel 20–1, 56, 92–3, 95
Homfray, Samuel George 69
Hopkins, Roger 60, 63
Horloz Colliery 101, 194
Horsehay 8
horses, as draught animals 1–2, 9–10, 12, 16–18, 22, 24, 25, 32–3, 47, 48, 50, 58, 63, 65, 74, 78, 80–4, 93, 100, 118–19, 121, 137, 163, 166, 198, 199, 202, 203, 207, 208, 210, 217, 221, 224, 227–9, 238, 244, 275–7, 302
Horsley Coal and Iron Company 151

Howth harbour 59
Hoxie, Joseph C 281
Hudson, George 289
Hugh's Bridge lift 66
Hughes, John, *y gof* ('the blacksmith') 128, 363
Hugo, Victor 318
Hulton, William 201
humans as motive power 1–2, 80–1
Hund 12
Hungary 47, 76, 258–9, 276, 292, 298
Hunslet 100, 299
Huskisson, William 113, 236, 259–60
Hutchinson, George 102

Imlay, Richard 246, 257, 277, 278
'improvement' 21, 25, 55, 57, 71, 252, 314
inclined planes 2, 14, 20, 23, 24–6, 35, 49, 51, 60, 61–2, 63, 65–70, 73, 75–7, 84, 85–9, 94, 95, 99, 103, 107, 119, 123, 140–58, 194, 201, 203, 204, 208–10, 222–3, 230, 244, 252, 254, 258, 275, 292, 297, 301, 310; catenary section 85
l'Indret ordnance works (France) 39–40, 75
investment in railways 4–6, 11, 14, 21, 54, 55–8, 60–1, 109, 110–12, 113, 114, 116, 118, 123, 134, 161–2, 194, 196, 214, 217, 220–1, 231, 232, 237, 239, 250, 252, 259, 263–4, 269–73, 286, 289, 298, 309, 310, 318, 361
Ireland 1, 13, 31, 51, 65, 74, 119, 129, 215–16, 267–8, 301; development of railway systems in 289
iron, evolving use as structural component 3, 5, 30
iron, cast 30–48
iron, wrought 126–38
Ithaca and Owego Railroad 77, 143, 150, 266

Jackson, Andrew 262
James & Charles Carmichael, locomotive builder 176
James, William 130, 186, 216–18, 232, 236
James, William Henry 186
James, William T. 187
Japan, interest in railways 295
Jars, Gabriel 17, 18, 36
Jervis, John Bloomfield 166, 181–4, 209, 278, 298
Jevons Iron merchants 130

393

INDEX

Jessop, Josias 42, 64, 222–3, 224, 230, 231
Jessop, William 42, 44, 45, 59–61, 64, 65, 72, 214, 222, 224
John (Johann), Archduke of Austria 114
John Bradley's Ironworks (Stourbridge) 130
Jones, Rees, fitter at Penydarren Ironworks 92
Joseph II, Holy Roman Emperor 224
journals, technical and scientific 94–5, 118–19

Kaiser-Ferdinands-Nordbahn 232, 291
Kemble, Frances (Fanny) 172, 258
Kenton and Coxlodge Railway 26, 100, 107
Kenyon and Leigh Junction Railway 173, 202, 266
Ketley–Horsehay–Coalbrookdale railway 8, 14
Ketley canal and lift 35, 65, 85, 88–9, 149
Kidwelly and Llanelly Canal 68
Killingworth Colliery, waggonway and railroad 27, 40, 106–7, 145, 160, 161, 166–7, 188, 193, 197, 216
Kilmarnock and Troon Railway 57, 63, 64, 69, 73, 161, 162
Kingstown (Dun Laoghaire) Harbour 59, 74, 268
Klein, Ludwig 286
Klodnitz Canal (Silesia) 66
Knight, Jonathan 182, 238
Konigsgrube (Kopalnia Krol) to Konigshutte (Krolewska Huta, Chorzow) railway 75
Kraft, Nikolai Osipovich 294

labourers 58, 59, 240, 241, 305–8, 311, 314, 321
Lake Lock Railroad 39, 56
Lambton waggonway 19, 42, 103
Lamé, Gabriel 125
Lancaster Canal, plateway and viaduct 60, 64, 68, 86, 89
Landore (Swansea) 34, 84
Lanna, Karl Adalbert von 228
Lardner, Dionysius 124
Latrobe, Benjamin Henry II 241, 261–2
Latrobe, John Hazelhurst Boneval 261
Le Creusot Foundry (Fonderie Royale, Saone-et-Loire) 39–40, 75, 131
Le Havre to Rouen Railway 290, 310
Leckhampton freestone quarries 57

Leeds and Liverpool Canal 100, 200, 201–2
Leeds and Selby Railway 132, 174, 189–90, 265–6, 278–9, 280
legislation 55–6, 225, 268–9, 290
Lehigh Coal & Navigation Company 209–10
Leicester and Swannington Railway 130, 141–2, 151, 152, 264–5, 269, 270, 283
Leipzig to Dresden Railway 298
Lemon House 282–3
Liège 37, 75, 101, 110, 192, 269, 291
lime, limestone 3, 11, 33, 36, 43, 44, 47, 49–50, 52, 57, 58–9, 61, 64, 66, 70, 71, 72, 79, 83, 92, 104, 142, 214, 216, 218, 220–1, 230
Lings Colliery 35
Linz 2, 218, 227, 229–30, 231
List, Georg Friedrich 111
Little Eaton gangway 36, 61, 72
Little Schuylkill Railroad 264, 277
Liverpool 48, 54, 114, 116, 131, 133, 142, 236, 237, 239, 252, 255–6, 270
Liverpool and Manchester Railway 2, 112, 113, 115, 117, 130, 133, 134, 136, 139, 140, 145–8, 151, 154–7, 159, 165, 166–71, 172–7, 179, 184, 186, 187, 188, 189, 190, 195, 196, 201–2, 205, 216, 217, 220, 233–62, 263, 266, 269, 270, 277, 278, 280, 288–9, 302, 303, 304, 309–10, 314, 318
Llanelly Railway 177
Llangollen Canal 64–5, 71
Llansamlet plateway 58, 161, 162
Llanvihangel Railway 57–8
Locke, Joseph 146, 167, 259, 288, 290, 297–8, 303
locomotive repair and maintenance facilities 180, 248–9, 299–300, 308
locomotives, compressed air 184, 185, 191, 300
locomotives, electric 301–2
locomotives, horse-operated; *Cycloped* 168, 185, 191
locomotives, internal combustion 3, 185–6, 191
locomotives, steam: number in service 107, 159; speed attained 94, 97–8, 121–2, 159, 168, 175, 214, 280; naming 91, 97, 98, 100, 106, 189–90; practicality and operation 190–1; sentimental affection for 190; by name or type:

394

INDEX

locomotives, steam, by name or type contd
Best Friend 180, 189, 245–6
Billy 161
Britannia 175
Brunton locomotive 26, 69, 104
Black Billy 105–6
Black Diamond 163
Blücher 27, 107, 161, 162, 190
Brother Jonathan 183–90
'Buddicombe'/'Crewe' type 299
Catch Me Who Can 96–8
Charleston and Hamburg articulated locomotives 184
Cherepanov locomotives 174–5
Chittaprat 163, 171
Coalbrookdale locomotive 92
'Crampton' type 299
Dart 194
Davy Crockett 184
DeWitt Clinton 181–2
Diligence 163
Dublin 175–6
Earl of Airlie 176
Experiment (Stockton and Darlington Railway) 171
Experiment (Liverpool and Manchester Railway) 175
Experiment (Mohawk and Hudson Railroad) *see Brother Jonathan*
'Firefly' class 299, 314, 316–17, 319
Gateshead locomotive 96
Globe 173, 181, 189
'Grasshopper' type 180, 245
Hercules 178, 190
Hendreforgan colliery plateway locomotive 161
Hetton Colliery railroad locomotives 141, 160, 161, 194
Hibernia 175
Hope 163
Invicta 172, 173, 190
James locomotive proposals 186
John Bull (Camden and Amboy Railroad) 182–3
John Bull (Mohawk and Hudson Railroad) 182
Kenton and Coxlodge colliery locomotives 26, 100, 102, 106, 107
'Killingworth' type 106–7, 122, 145, 160–3, 168, 171, 203, 216–17
Kingstown 175
Lady Mary, see *Puffing Billy*

Lancashire Witch 171, 189, 201
Liverpool 172–4
Liverpool coke engine, see *Twin Sisters*
Locomotion 163, 190, 198
Lord Wellington 100
Lord Wharncliffe 176
'Majestic' class 176–7
Manchester 175
Marquis Wellesley 100
Murray–Blenkinsop type 99, 102, 106, 160, 161, 214, 215
Murray proposals 186
Murébure Colliery locomotive 101, 107
Nantyglo Ironworks locomotive 101, 107
Neath Abbey locomotives 69, 177–8
Northumbrian 2–3, 172, 190, 244, 266
Novelty 122, 124, 168, 169–70, 190
Novelty type 122, 170, 177, 183
Old Ironsides 181, 183
'Patentee' type 176, 299
Perseverance (Burstall) 168–9, 179, 190
Perseverance (Neath Abbey) 178, 190
Phoenix 180
'Planet' class 2–3, 139, 170, 173–6, 181–3, 191, 244–5
Pride of Newcastle 172, 189
Prince Regent 100
Puffing Billy 105–106, 107, 161, 183
Puffing Devil road locomotive 91–2
Queen Adelaide 170, 189
'Remorquer' type 164–5, 204
Robert Fulton see *John Bull* (Mohawk and Hudson Railroad)
Rocket 2–3, 112, 117, 139, 163, 165, 168, 170–2, 173, 179, 188, 190, 191, 201, 244, 259, 319
Royal George 163–4, 169, 176
Royal William 178, 190–1
Saarland locomotive 102
Salamanca 100, 102
'Samson' type 2–3, 139, 173–4, 176, 181, 182, 183, 188, 191, 244–5
Sans Pareil 168, 169, 176, 190
Speedwell 177
Springwell Colliery railway locomotives 161
Star 194
Steam Elephant 160
Stephenson 188
Stephenson locomotives for France 164–5
Stourbridge Lion 165–6, 209, 245
Tallyho 194

INDEX

Tayleur 188
The Agenoria 165, 190
The Duke 162, 189
Tom Thumb 179, 189, 245
'Tram Engine' 92, 95
Tredegar Ironworks locomotives 69, 172
Trotter 176
Twin Sisters 179
Union 179, 201
Vauxhall 175–6
Victory 176
'Waggon Engine' 96
Walking Horse 100
'Wilberforce' class 176–7
William IV 170, 189
Wylam Dilly 106, 107, 161, 189
Yn Barod Etto 178, 190
Zabrze locomotive 102
London and Birmingham Railway 63, 157, 174, 217, 288, 296, 299, 303, 310, 316, 320
London and Blackwall Railway 157, 296
London and Bristol Railway *see* Great Western Railway
London Docks 47, 59, 68, 89
London and Greenwich Railway 288
London to Holyhead Road 45–6
London and North Western Railway 288
London University 124
Long, Stephen Harriman 148
Longridge, Michael 128, 129, 161–2
Losh, William 40, 112, 129, 161, 162, 216
Losh–Stephenson patent 40–2, 131, 136, 160, 162, 193, 194, 197, 204, 211, 216
Louis, Archduke of Austria 114
Louis Philippe, King of the French 290, 310
Ludwig I, King of Bavaria 121, 291
Lunar Society 21, 61, 91

McAdam, John Loudon 10
Mackenzie, William 290
McNeill, William Gibbs 238
Madras, railways in 295
Magdeburg to Leipzig Railway 298
Maillard, Sebastian von 67, 74–5
Manchester 11, 62, 66, 201, 215, 220, 228, 235, 236, 239, 253, 254, 260, 299
Mancot 34
Mansfield and Pinxton Railway 64
Martin, Edward 95
Martineau, Harriet 242–3

Marx, Karl 61
'March of Intellect' 252–3, 318
Mather & Dixon engineering company 151–2, 157
Mattawan Foundry 148, 152, 156
Matthew, David 181, 190
Mauch Chunk Railroad 209–10
Maudslay, Henry 88
Mauthausen 225, 227
Mellet, Francois-Noël 204
Melling, John 248–9
Melnikov, Pavel Petrovich 294
Menai suspension bridge 59, 199
Menai tubular bridge 303
Merthyr Tydfil 20, 38, 52, 56, 64, 69, 92–3, 95, 295, 298
Merthyr plateway 37, 52, 56, 69, 92–3
Metternich, Klemens von 5, 220, 232, 291, 310
Meynell, Thomas 196–7, 198
Middlesbrough 173, 198–9, 265
Middleton Colliery 21, 35, 99
Middleton Railway 21, 22–3, 81, 86, 99–100, 102–3, 115, 146, 292
Middleton Top inclined plane and fixed engine 153
Midland Counties Railway 231
Milholland, James 187
military use of railways 55, 206, 304–5
Mill Dam Foundry 189
Miller, Edward 147,
Miller, Ezra L 245, 249
Minard, Charles-Joseph 125
Mine Hill Railroad 264
Mississippi River 237, 266–7, 286, 313
models 91, 112, 113, 114, 121, 122, 170, 245, 247, 283–4, 295
Mohawk and Hudson Railroad 142–3, 150, 152, 181, 183, 184, 266, 267, 269, 276, 277, 278, 281
Moisson-Desroches, Pierre-Michel 214
Molesworth, William 270
Monkland and Kirkintilloch Railway 130, 163, 203, 227, 265, 280
Monmouth Railway 57
Monmouthshire Canal and Railways 37, 38, 52, 60, 177, 178, 296, 312
monorails 46–7, 111, 112, 119, 185, 207; in Hungary and USA 76, 77
Mont Clare workshops 180, 249
Montefiore, Moses 272–3
Montgomery and West Point Railroad 307

INDEX

Montgomeryshire Canal 39
Moorish arch 151, 255
Morgan, Richard Price 278
Morris Canal 123, 143, 149, 155
Mostyn Colliery and railroads, Point of Ayr 39
Mount Carbon Railroad 264
Mowbray, Arthur 194
Murdoch, William 91
Murray, Matthew 99, 102, 106, 160, 161, 186, 214, 215, 299

Nantlle Railway 54, 130
Napoleon I, Emperor of the French 108, 111, 119, 214, 220, 221
navvies *see* labourers
Neath Abbey Ironworks 69, 129–30, 151, 177–8, 188, 190–1
Neath Canal 60, 90
Nesham, John Douthwaite 25–6, 104
New York and Harlem Railroad 187, 268, 269, 270, 276, 277, 278, 281
Newbottle waggonway 25–6, 104
Newcastle type railways 14–17
Newcastle Colliery and railroad, New South Wales 42, 43, 117–18, 192, 210–12
Newcomen, Thomas 10, 88
Newmarket races 97–8
newspapers 56, 94, 95, 118, 121, 133, 156–7, 240, 306, 313
Newton Colliery railway 193
Nicholas I, Tsar of Russia 5, 114, 115, 125, 231, 276, 292, 294, 304, 305, 310–11
Nicholson, John 95, 112, 120
Nikolaevskaya (St Petersburg to Moscow Railway) 276, 294–5, 298–9, 302, 303, 305, 306, 308, 310, 311
Nimmo, Alexander 129
Norris's American Steam Carriage Company/Norris Locomotive Works 189, 298
Nova Scotia colliery railway 177

Odenburg (Sopron, Hungary) monorail 76
Oder River 75, 192
Odesa 112
Oeynhausen, Karl von 119, 194, 196, 220
Ogden, Francis Barber 259
Ohio River 148, 237, 266, 275, 288
Ottoman Empire 110, 292, 311
Ouseburn Foundry 103

Outram, Benjamin 31, 36–7, 43, 60–1, 63, 72, 85
Outwood collieries 56
Overton, George 60, 63, 64, 93, 196–7
Owen, Robert 62
oxen, as draught animals 1–2, 77, 84, 144, 153, 205, 207, 320
Oxford University 124
Oystermouth (Swansea and Mumbles) Railway 71, 73, 95

packet ships 236, 266, 268, 318
'Padorama' of the Manchester and Liverpool Rail Road 283–4
Page, Charles Grafton 302
Palmer, Henry Robinson 47, 77, 112, 207
Paris to Brussels Railway 292, 312
Paris to Rouen Railway 290, 299–300, 306, 308, 312
Paris to St Germain Railway 290, 301, 310
parliament (British) 11, 55–6, 118, 211, 215, 217, 221, 222, 231, 236, 259, 269, 270, 288, 321
parliament (Irish) 11, 65
passenger facilities 5, 6–7, 108, 146, 147, 151–2, 157, 205, 228–9, 251–2, 253–4, 260, 280–3, 292–4, 296, 309–10, 311–12, 320
passenger services 2, 6–7, 31, 50, 54, 59, 71–2, 73, 78, 98, 108, 146, 173, 191, 195, 199, 202, 204, 205, 206, 208, 210, 213–14, 215, 216, 227, 228, 229, 233, 236, 238, 246–7, 248, 250, 251–2, 253–4, 260, 262, 266, 267–8, 277, 279–83, 294, 309–10
patent system, British 109
Peak Forest Canal and Railway 36, 58–9, 61–4, 67–8, 70, 83, 85, 218, 223–4, 228
Pease, Edward 116, 130, 161–2, 178, 196–8, 214, 226
Pease, Joseph 270
Peel, Robert 311
Penrhyn (slate quarry) railroad 39, 44, 54, 60, 63, 64, 68, 73, 83, 84, 118–19, 227
Pentewan Railway 270
Percy Main colliery 83
Pereire, Émile 289–90
Perkins, Jacob 187, 319
permanent way *see* track
Pesth (Hungary) monorail 76
Peter Clute's foundry (Schenectady) 152

397

INDEX

Pferdeeisenbahn Budweis–Linz *see* Budweis–Linz (horse) railway
Philadelphia and Columbia Railroad 134, 142, 148, 152, 154, 155
Philadelphia and Reading Railroad 181, 187
Phillips, Richard 214
'pillars of Hercules' motif 255–6, 310
Pius IX, Pope 311
plate rails, form, design and adoption 19–20, 31, 34–8, 39, 48, 54, 61, 77
plateway, concept and definition 19–20, 363
Plymouth and Dartmoor Railway 32, 58, 59, 60, 63, 71, 72, 73
Plymouth harbour and breakwater 59, 73
Plymouth Ironworks, Merthyr Tydfil 93
pole roads 6
political conservatism 5, 108, 111, 114, 118, 220, 231, 232, 239, 246, 291, 292, 310, 311–12
political progressivism 5, 62, 112, 204, 207, 246, 289, 311
Pont y Cafnau (bridge) 64
Pontcysyllte Aqueduct 59
Pontchartrain Railroad 266–7, 269, 273, 281–2
Portland, William Henry Cavendish-Scott-Bentinck, Duke of 57
Poussin, Guillaume Tell 155
Prague to Dresden Railway 306
Prinz-Wilhelm-Eisenbahn 207
Prussia *see* German Confederation
Public Works Loans Act 57
puddling process 127

Quakers 8, 14, 61, 113, 129–30, 161–2, 177, 178, 196, 197, 217, 226, 238, 250, 270–1, 361
Quincy Railway *see* Granite Railway (Quincy, Massachusetts)

rack working 21, 26, 99, 178, 179, 185
R. & W. Hawthorn engineering company 151, 177
railroad, concept and definition 38; American sense 363–4
railway, concept and definition 1, 6–7, 363; as tributary to other forms of transport 3, 11, 14–15, 20, 22, 44, 208–10, 265, 292, 321; benefits and disadvantages compared to other forms of transport 3, 49–50, 51–2, 57, 74–5, 78, 95–6, 108–9, 111, 120–1, 159, 193, 209, 216, 221, 231, 233, 236, 237, 252, 263, 294; 'land-bridge' systems (sea-to-sea) 215; portage systems (water-body to water-body) 22, 134, 142, 147, 149, 152, 266–7, 275; 'mania' 215–16, 289; as employer 250–1, 308–9
Railway Clearing House, United Kingdom 309
Railway Regulation Act, United Kingdom 1844 309
Rainhill locomotive trials 113, 117, 139, 1556, 159, 160, 165, 166–70, 172, 173, 176, 177, 178, 180, 238, 244, 248, 257
Rainton and Seaham Railway 141, 195
Rastrick, John Urpeth 145–6, 156, 165, 167
Rastrick and Hazledine Ironworks 96–7
Rennie, George 122, 227, 236, 239
Rennie, John, the elder 59, 60, 64, 73
Rennie, John, the younger 205, 236, 239
Renwick, James 123, 124, 149
Reynolds, William 61, 65–6, 67, 68, 85, 88–9, 82, 92
Richardson, Thomas 116, 214
Richmond, Fredericksburg and Potomac Railroad 312–13
Rideau Canal 120
Risca viaduct 63–4, 314
river transport *see under individual river names*
roads 8–10, 14, 21, 30, 44–6, 49, 50–2, 56–9, 71, 78, 91–2, 106, 108–9, 120, 125, 140, 150, 159–60, 169, 179, 186, 187, 202, 206, 213, 214, 223, 226, 228, 233, 237, 246, 247, 248, 251, 253, 260, 275, 277, 281, 283, 291, 292, 297, 312, 316, 321
Robert Stephenson & Co. 69, 113, 128, 151, 153, 162, 166, 169, 173–4, 176, 177, 179, 188, 201, 205, 299
Robertstown (Aberdare) bridge 64
Robinson, Moncure 147
Roebling, John Augustus 155
rolling mills 34, 74, 88, 128, 133, 222–3, 225
Room Run Railroad 155, 264
rope and chain haulage 20, 23, 24, 35, 65, 85, 86, 91, 107, 139–58, 194, 201, 230, 296–7

INDEX

Rothschild family 109, 221, 232, 272–3, 290
Rothwell & Hicks, locomotive builders 184, 188
Rothwell, Peter 201
Royal Victualling Yard monorail 47
Royal Wurttemberg State Railways 306
Ruabon Brook plateway 64–5, 71
Ruhr 39, 75, 112, 203, 207
Rush & Muhlenberg manufactory 152
Ruskin, John 320
Russian Empire 5, 76, 111, 124–5; development of railways in 114–15, 119, 124–5, 174–5, 231, 232, 258, 261, 263–4, 276–7, 292–5, 298, 303, 304, 305, 306, 311, 313
R.W. Hawthorn, locomotive builder 151, 177

Saint-Étienne to Andrezieux Railway 42, 204
Saint-Étienne to Lyon railway 113, 114, 131, 164, 203–7
St Helen's and Runcorn Gap Railway 131, 134, 141–2, 170, 200, 202
St Petersburg to Moscow Railway *see* Nikolaevskaya
St Petersburg to Tsarskoe Selo Railway 231, 292–3, 304
St Rollox 265
Saint-Simon, Henri de, Comte de Rouvroy 112, 204, 207
Saint-Simonian movement 112, 204, 207, 264, 289, 312
San Francisco cable car 157–8
Saratoga and Schenectady Railroad 184, 267
Savery, Thomas 88
Schmidl, Eduard 226, 228
Schönerer, Matthias 226, 227, 228, 230, 232
Schuylkill Haven Railroad 264
Schuylkill Valley Navigation and Railroad Co 208
Scott, Walter 316
Seguin, Marc 131, 163, 164, 165, 170, 204–7
Semmeringbahn 291–2, 297, 299, 301, 302–3, 306
Semmering Pass 74–5, 292
serfs *see* coerced labour
settlements (alongside railways) 52–4, 252, 267, 268; (of railway workers) 308

Seven Years War 10
Severn and Wye Railway 60, 71, 74, 103–4
Severn River 8, 12, 13–14, 37, 57, 58, 60, 88–9
Sharp, Roberts, locomotive builders 175, 188
Sheffield 20, 21, 35, 211, 214–15
Sheriff Hill waggonway 23
Shropshire Canal 65–6, 67, 84
Shropshire type railways 12–14, 16, 22, 24, 39, 72, 84, 87, 140
Shutt End Colliery Railway 165
signalling 297, 313
Silesia 37, 39, 66, 75, 102, 110, 192, 224, 228, 298
sills 32, 42, 43
Sinclair, John 22, 62
Sirhowy plateway 37, 51–2, 61, 63–4, 69, 73, 79, 103–4
slate industry 33, 39, 44, 54–5, 65, 68, 70, 72, 73, 83, 88, 118–19, 128, 129, 137, 272, 276
Society of Arts 21, 44
Society of Friends *see* Quakers
Somerset Coal Canal 39, 72, 86
Sotteville-lès-Rouen 300, 308
South Carolina Railroad *see* Charleston and Hamburg Railroad
South Devon Railway 301
Southern State Railway (of Austria) *see* Südlichen Staatsbahn
Sparrow of Wolverhampton 209
'spirit of association' 21
staithes 14, 23
stables 69–70, 83–4, 228, 276
Stahl, Philip Ritter von 224
Stanhope, Charles Mahon, Earl of 62, 67, 68
Stanhope and Tyne Railway 65, 142, 149, 151, 153, 154–5, 264–5, 268
stations *see* passenger facilities
Steel, John 96
Stephenson, George 3, 27–8, 29, 40, 43, 54, 89, 98, 106–7, 112, 114, 115–17, 118, 121, 124, 125, 126, 128, 129, 130, 136, 141, 145, 147, 156–7, 160, 161–2, 163, 167, 170, 172, 173, 186, 188, 190, 193, 194–5, 197, 198, 200, 201, 202, 216, 217, 224, 226, 227, 235, 236, 239, 247–8, 261, 262, 288–9
Stephenson, James 198
Stephenson, Robert (the elder) 54, 194

399

INDEX

Stephenson, Robert 2–3, 27–8, 29, 98, 117, 118, 124, 139, 141, 146, 149, 153–4, 156, 157, 161–2, 163, 164, 165, 167, 168, 170, 171, 173, 182, 193, 194, 196, 224, 235, 244, 247–8, 288, 289, 296, 303
Stephenson railway system 54, 77, 79, 125, 134, 136, 138, 162, 174, 194, 262, 263, 273, 291
Stevens, John 121–2, 123–4, 179, 182–3, 301–2
Stevens, Robert Livingston 135, 182–3, 273
Stevens, Thaddeus 275
Stevenson, Robert 129, 193, 197
Stockton and Darlington Railway 29, 40, 48, 54, 60, 113, 118, 127, 129–31, 132, 140, 141–2, 151, 154, 159, 161, 163, 172, 173, 186, 194, 195–200, 205, 207, 237, 240, 247, 250, 254, 265, 270, 276, 278, 288, 302
stone, use of as track component 16, 22, 30, 31, 32, 36, 42, 43–6, 54, 83, 128, 132–4, 197, 204, 120, 225, 235, 269, 273–4, 298
stone roads 44
Stourbridge Canal 55, 60
strap rail 6, 43, 133, 134, 137, 181, 207, 208, 210, 224, 238, 269, 298
Stratford and Moreton Railway 216–17, 218
Stuart, William 59
Strickland, William 113–14, 119–20, 122, 274
Südlichen Staatsbahn see Southern State Railway
Summerfield and Atkinson foundry 89
Surrey Iron Railway 56, 59, 71, 222, 276
Swannington inclined plane and fixed engine 151, 152, 153–4
Swansea Canal 63, 69, 71, 142, 150
Sweden 121, 124 127, 212; railways in 37, 76, 87, 117, 292
Swindon 299, 308
Szechenyi, Count István 258–9

Taff Vale Railway 297, 304
Tanfield waggonway 15, 18
Tavistock Canal 68
Taylor, John 87–8
Tees River 197, 199, 200, 265
Telford, Thomas 10, 45–6, 47, 59, 73, 84, 122, 214, 227, 268

Temple, Simon 24
textile industry 2, 10, 109, 152, 189, 207, 235, 237, 239, 246, 254, 260, 282, 285, 286, 307
Thomas, Evan 237
Thomas, Philip E. 237
Thompson, Benjamin 140–1, 145, 146, 148, 151, 154, 193, 195, 211
Thon, Konstantin Andreyevich 310
Thoreau, Henry David 319–20
Tindale Fell railway 128, 363
Tite, William 310
Tomlinson, William Weaver 149
Torrington (Rolle) Canal 68
tourism 71, 72, 73, 75, 210, 294, 312
track systems 30–48, 54, 107, 126–38, 162, 166, 174, 184, 233, 235, 238, 247, 273–4, 278
Traeth Mawr 'cob' 83–4
traffic 16–17, 25, 38, 58, 63, 70–2, 82–3, 144–5, 192–212, 213, 217, 218–20, 228, 229–30, 238, 241, 246, 254, 264–7, 271, 276, 296
Trautwine, John Cresson 274
Tredegar Ironworks (Monmouthshire) 69, 74, 92, 95, 172
Tredgold, Thomas 112, 120, 122
Trench lift 66
Trent and Mersey Canal 228
Trésaguet, Pierre-Marie-Jérôme 9–10
Trevithick, Richard 20–1, 37, 52, 69, 88, 90, 91–9, 100, 104–5
Trinity College Dublin 124, 280
Tsarskoe Selo 231, 292–4, 304
tunnels 13–4, 58, 64, 93, 97, 142, 147, 151, 155, 185, 204, 205, 239–40, 244, 254–5, 275, 292, 296, 297, 303, 306, 314
Turner, Joseph Mallord William 316–17
turnpikes *see* roads
Tuscumbia, Courtland and Decatur Railroad 150
Tyne River 2, 3, 12, 14–25, 35, 68–9, 81–3, 89, 90–1, 96, 100, 103, 104, 118, 128, 142, 160
Tyrone Navigation 11, 65
Tyrwhitt, Thomas 71

'undulating railways' 186
uniforms 229, 250, 308, 313
Union Foundry *see* Rothwell & Hicks, locomotive builders

INDEX

United States of America 107–10; development of railway system in 1, 6, 43, 46, 50–1, 76–7, 112, 114, 117–24, 125, 133–6, 139, 142–3, 148–9, 152, 154–5, 165–6, 170, 179–84, 187–92, 208–10, 231–62, 266, 269–71, 273–6, 277–83, 285–8, 294–6, 298–9, 302–10, 312–13, 317–19
Utica and Schenectady Railroad 271

Verein Deutscher Eisenbahn–Verwaltungen 309
Viau, Jean-Baptiste-René 310
Vienna to Neustadt waterway 67, 74–5
Viertbauer, Leopold 229
viewers, colliery 15, 21, 22, 23, 35, 81, 83, 100, 105, 112, 117, 129, 161
Vignoles, Charles Blacker 134, 170, 175, 200, 202, 236, 280, 298
Vivian, Andrew 90, 91–2
Volga River 111, 305
Vulcan Foundry 184, 188, 190

waggonway, concept and definition 16–18, 363
Walker, James 145, 146, 156
Walker, John Scott 255
Walker pits waggonway 22–3
Wallsend waggonway 40, 160
Wapping tunnel and inclined plane 142, 147, 151, 155, 157, 239–40, 244, 303
Washington waggonway 23
water power as railway prime mover 62, 68, 87–8, 144, 149–50
Waters, Thomas 105
Watt, James 10, 61, 88, 91, 104, 153
Wear River 2, 14, 15, 19, 25, 90, 118, 141, 194
Weatherill inclined plane and fixed engine 153
Wellington, Arthur Wellesley, Duke of 100, 258
Welsh language 63, 117, 189, 190
Wertheimstein, Leopold 232
West India Docks (London) 45, 96, 222
West Point Foundry (New York) 152, 166, 179–80, 181, 183, 184, 189, 246
West Point (United States Military Academy) 124, 241
Western way 15, 22, 82
Westland Row station 280
Wheal Crebor mine 87–8

Wheal Friendship mine 87–8
Wheatcroft, German 58, 228
wheel coning 136
Whig view of history 4–5
Whims, horse 84, 150, 303
Whinfield, John 96
Whistler, George Washington 182, 238, 294, 304
Whitby and Pickering Railway 150
Whitehaven colliery 83, 103
Wilkinson, William 75
William I (Willem Frederik), King of the Netherlands 310
Willington Quay (ballast bank) 68–9, 89–90
Winans, Ross 180, 238, 247–8, 249, 294–5
Winches 81, 84, 99, 149, 150
Windmill Farm lift 65, 89
Wingerworth furnace 35
Wishaw and Coltness Railway 203, 265
Wolverton workshops 299, 308
Wood, Nicholas 81–2, 112, 118, 122, 135, 145, 146–7, 157, 161, 167, 185, 193
wood, use of as track component 1–2, 5, 19, 20, 23–4, 31, 47, 77, 80, 103, 126, 132, 134, 225, 238, 273–4; late survival as rail component 6; as fuel 110, 121, 183, 184, 295; use in bridges and trestle formations 5, 6, 77, 110, 166, 209, 212, 227, 251, 261, 274, 302–3, 308–9
Wordsworth, William 316
Worsley colliery and inclined lift 11, 66–7, 85
Wright, John 60–1
Wrockwardine Wood lift 65–6
wrought iron, use of as track component 6, 19, 31, 34, 39, 40, 43–4, 46, 47, 48, 54, 113, 125, 126–38, 162, 193, 196, 197, 200, 204, 205, 207, 209, 210, 222–3, 225, 230–1, 233, 235, 238, 273
Wyatt, Benjamin 68
Wye River 52, 214
Wylam waggonway/plateway 26–7, 29, 35, 40–2, 96, 98, 104–6, 107, 114, 161, 163, 165, 183, 189

Ynysgedwyn inclined plane 69, 142, 150, 151

Zabrze collieries 66–7, 75, 102, 192
Zmeinogorsk ironworks 76
Zolla, Francesco 229–30